IET TRANSPORTATION SERIES 39

Electric Vehicle Components and Charging Technologies

Other related titles

Volume 1	**Clean Mobility and Intelligent Transport Systems** M. Fiorini and J-C. Lin (Editors)
Volume 2	**Energy Systems for Electric and Hybrid Vehicles** K.T. Chau (Editor)
Volume 5	**Sliding Mode Control of Vehicle Dynamics** A. Ferrara (Editor)
Volume 6	**Low Carbon Mobility for Future Cities: Principles and applications** H. Dia (Editor)
Volume 7	**Evaluation of Intelligent Road Transportation Systems: Methods and results** M. Lu (Editor)
Volume 8	**Road Pricing: Technologies, economics and acceptability** J. Walker (Editor)
Volume 9	**Autonomous Decentralized Systems and Their Applications in Transport and Infrastructure** K. Mori (Editor)
Volume 11	**Navigation and Control of Autonomous Marine Vehicles** S. Sharma and B. Subudhi (Editors)
Volume 12	**EMC and Functional Safety of Automotive Electronics** K. Borgeest
Volume 15	**Cybersecurity in Transport Systems** M. Hawley
Volume 16	**ICT for Electric Vehicle Integration with the Smart Grid** N. Kishor and J. Fraile-Ardanuy (Editors)
Volume 17	**Smart Sensing for Traffic Monitoring** Nobuyuki Ozaki (Editor)
Volume 18	**Collection and Delivery of Traffic and Travel Information** P. Burton and A. Stevens (Editors)
Volume 20	**Shared Mobility and Automated Vehicles: Responding to socio-technical changes and pandemics** Ata Khan and Susan Shaheen
Volume 23	**Behavioural Modelling and Simulation of Bicycle Traffic** L. Huang
Volume 24	**Driving Simulators for the Evaluation of Human–Machine Interfaces in Assisted and Automated Vehicles** T. Ito and T. Hirose (Editors)
Volume 25	**Cooperative Intelligent Transport Systems: Towards high-level automated driving** M. Lu (Editor)
Volume 26	**Traffic Information and Control** Ruimin Li and Zhengbing He (Editors)
Volume 30	**ICT Solutions and Digitalisation in Ports and Shipping** M. Fiorini and N. Gupta
Volume 32	**Cable Based and Wireless Charging Systems for Electric Vehicles: Technology and control, management and grid integration** R. Singh, S. Padmanaban, S. Dwivedi, M. Molinas and F. Blaabjerg (Editors)
Volume 34	**ITS for Freight Logistics** H. Kawashima (Editor)
Volume 36	**Vehicular Ad Hoc Networks and Emerging Technologies for Road Vehicle Automation** A.K. Tyagi and S. Malik
Volume 38	**The Electric Car** M.H. Westbrook
Volume 43	**AI Techniques in EV Motor and Inverter Fault Detection and Diagnosis** Y. Hu, X. Zhang and W. Lang
Volume 45	**Propulsion Systems for Hybrid Vehicles** J. Miller
Volume 79	**Vehicle-to-Grid: Linking electric vehicles to the smart grid** J. Lu and J. Hossain (Editors)

Electric Vehicle Components and Charging Technologies

Design, modeling, simulation and control

Edited by
Sanjeev Singh, Sanjay Gairola and Sanjeet Dwivedi

The Institution of Engineering and Technology

Published by The Institution of Engineering and Technology, London, United Kingdom

The Institution of Engineering and Technology is registered as a Charity in England & Wales (no. 211014) and Scotland (no. SC038698).

First published 2023

The Institution of Engineering and Technology
Futures Place
Kings Way, Stevenage
Hertfordshire, SG1 2UA, United Kingdom

www.theiet.org

British Library Cataloguing in Publication Data
A catalogue record for this product is available from the British Library

ISBN 978-1-83953-671-7 (hardback)
ISBN 978-1-83953-672-4 (PDF)

Typeset in India by MPS Limited

Cover image: Peter Dazeley/The Image Bank via Getty Images

Contents

About the editors

Sanjeev Singh is a professor in the Electrical Engineering Department at Maulana Azad National Institute of Technology, Bhopal, India. He has 23 years' experience in teaching and research. His research interests are focused on energy management, renewable energy integration, electric vehicles, power electronics, and drives. He is a senior member of the IEEE, fellow of Institution of Engineers (IE) India, The Institution of Electronics and Telecommunication Engineers (IETE) India, and System Society of India (SSI) and a life member of the Indian Society for Technical Education (ISTE).

Sanjay Gairola is a professor in the Electrical Engineering Department at Govind Ballabh Pant Institute of Engineering and Technology (GBPIET), Uttarakhand, India. His fields of interest include power quality, power electronics, electric machines, drives, electric vehicles, and renewable energy. He is a senior member of the IEEE and a life member of the Indian Society for Technical Education (ISTE), India.

Sanjeet Dwivedi is a senior technology leader at green hydrogen based Danish MNC Everfuel A/S. Previously, he worked as a technology leader in Danfoss Global R&D in Denmark. His research interests include control methods of PM motors, induction motors, and synchronous reluctance motors, sensor-less control of AC drives, energy efficient control of drive, and power quality. He is a senior member of the IEEE and fellow of IET (UK).

About the editors

Preface

The last decade of the 21st century has witnessed a profound shift in the way we perceive and interact with transportation. Electric Vehicles (EVs), once considered a niche concept, have now firmly established themselves as a transformative force in the automotive industry. As we stand on the cusp of a greener and more sustainable future, the need for a comprehensive understanding of the intricate components and advanced charging technologies that power this electric revolution has never been more critical.

"Electric Vehicles: Components and Charging Technologies" is an exploration into the heart and soul of the EV revolution. This book is a result of an extensive collaboration among experts and researchers who have dedicated their knowledge, skills, and passion to decipher the intricate working of electric vehicles and the infrastructure that supports them.

In this ever-evolving landscape, it is crucial for industry professionals, engineers, students, and enthusiasts to grasp the fundamental building blocks of electric vehicles. This book offers a comprehensive journey through the various components that make up an electric vehicle, including the battery technology, electric motors, power electronics, and control systems. Each chapter delves into the principles and innovations behind these components, offering insights into the technological advancements that have made EVs not only viable but also superior in many aspects to their internal combustion engine counterparts.

Moreover, we recognize the pivotal role that charging infrastructure plays in the widespread adoption of electric vehicles. This book also extensively covers the myriad charging technologies, from home charging solutions to fast chargers and the emerging field of wireless charging. Understanding these technologies is imperative to address the practical challenges of range anxiety and facilitate the seamless integration of electric vehicles into our daily lives.

As we embark on this educational journey, our aim is to provide readers with the knowledge and tools necessary to navigate the rapidly changing landscape of electric vehicles. We hope to inspire innovation, foster sustainable transportation solutions, and contribute to a cleaner, more environmentally responsible world. This book is designed to serve as a valuable resource, providing both a solid foundation for newcomers to the field and a wealth of insights for seasoned professionals.

We would like to express our sincere gratitude to the authors and contributors who have poured their expertise into this work, as well as the readers who are joining us on this enlightening journey. It is our hope that "Electric Vehicles:

Components and Charging Technologies" will become an essential reference in the ongoing electrification of the automotive industry, paving the way for a brighter, more sustainable future.

Let us embark on this enlightening journey into the world of electric vehicles, with the belief that by understanding their components and charging technologies, we can contribute to a cleaner, more sustainable world for generations to come.

Last but not the least, we are grateful to our colleagues around the world for their hard work in writing the chapters and for going through many stages of checking the final versions of their chapters and for their perseverance through these processes. The team at the IET, authors and editors have done a tremendous job in keeping the authors meet certain deadlines and in proofreading the chapters thoroughly. We would also like to thank our respective families (Mrs Pravina Singh, Mrs Poonam Gairola and Mrs Alka Dwivedi) for their patience and support throughout our involvement with the book over the past three years.

Editors

Sanjeev Singh, Bhopal India

Sanjay Gairola, Paudi, India

Sanjeet Dwivedi, Graasten, Denmark

Chapter 1
Electric vehicles and its components

Sanjeet Dwivedi[1], *Sanjay Gairola*[2] *and Sanjeev Singh*[3]

1.1 Introduction

After the revolution in the field of communication through smartphones and newer generation of mobile communication technology in last decade, the research around the globe is focused on the improvement and commercialization of the latest technologies along with a new mode of transportation, i.e., electric vehicles. Although the mobile communication technology has improved as an extension of previous technology, the electric vehicle (EV)-based transportation is entirely a new market. Suddenly everyone has started talking about EVs and the din of EVs is fast reaching its crescendo. With the success of electric cars from TESLA, the initial skepticism has started fading, leading to sure-footed march towards routing of the transportation system that is based on petroleum fuel employing an internal combustion engine (ICE). Hyundai, Chevrolet, Volkswagen, Nissan, Toyota, BMW, Mitsubishi, GM, and Audi along with Tesla are some car manufacturers who are already competing to improve these vehicles.

In fact, EVs are not a recent occurrence as the first EV was developed before the ICE that came into the picture in 1885 [1–5]. However, the last century witnessed the dominance and maturity of ICE-based vehicles due to the availability of crude oil in abundance, lower cost, efficacy, and reliability of the technology. The major electric transport that was surviving is one that used rails or ropes/belts such as electric trains, elevators lifts, and ropeways. With time, the depletion of petroleum fuel started while electrical and electronics for power control algorithms improved drastically and so the ICE-based vehicles started losing attraction. Moreover, the pollution due to fossil fuel has increased to levels dangerous for human life itself and, therefore, the committees dealing with environmental concerns are regularly demanding to check the use.

The need for EVs was worldwide felt in 1970s when the demand for gasoline went up and prices also started soaring high. The fear of fast depletion of crude oil fueled the thought and, since 1990s, serious efforts have been made by researchers, scientists, industrialists, and governments. The protocols for EVs were discussed in

[1]Technology Management, Everfuel A/S, Denmark
[2]Department of Electrical Engineering, Govind Ballabh Pant Institute of Engineering & Technology, India
[3]Electrical Engineering Department, Maulana Azad National Institute of Technology (MANIT) Bhopal, India

Figure 1.1 Problems with vehicles

different platforms and strategies for safety, reliability, and cost were analyzed. The motivation for EVs is such that the industrialists are focusing on grabbing this opportunity in every domain of transportation through the adoption of smart technologies and the manufacturing revolution. The competition to claim the future market of electric one/two-wheelers, three-wheelers, cars, trucks, buses, airplanes (small and big), and ships has reached a level where the companies have already chalked out long-term design, testing, and manufacturing plans.

It is the right time not only for the researchers to look deeper into the technological aspects of pure EVs but also for the engineering students who are having some interest in this field. The knowledge of power electronics, motor drives, batteries, fast charging circuits, microcontrollers, programming, optimization, and control are desired along with mechanical and managerial aspects to prosper by working in this sector. The pollution, accidents, and traffic congestion on roads are the problems with the vehicles as shown in Figure 1.1 which need resolution for the safety and healthy life of any society that can be done through smart EVs and management.

1.2 History of EVs

The first EV was a small toy car that employed a primitive electric motor, built in 1828 by a Hungarian named Anyos Jedlik [1]. After that, some better working models of electric car, driven mainly by primary cells, was built in Scotland, Vermont (USA) and Netherlands by individuals. Gaston Plante invented rechargeable lead–acid battery in 1859, making EVs viable for commercial purposes. The first electric car production was started in London by inventor Thomas Parker in 1884. In Germany, first electric car was made by Andreas Flocken in 1888. In the United States, William Morrison built the first car in 1891 that was powered by lead–acid storage cells, having steering and a top speed of 20 miles/h [5].

EVs powered by lead–acid cells became popular in the late 1890s and early 1900s. Some of the popular names associated with EVs during this period are Karl Benz of Germany, Walter Bersey of Britain, Dr Ferdinand Porsche of Austria, Walter Baker of Ohio, and many more.

The decline in the popularity of EVs started by the 1920s when the discovery of large petroleum reserves all over the world led to the availability of cheap gasoline. This led to the acceptance of ICE for vehicles. In the year 1912, the electric starter was invented by Charles Kettering to eliminate the difficult hand crank needed for cars with ICE. Moreover, the mass production of ICE cars was started by Henry Ford in 1913, making them much cheaper than EVs which was another major blow to EVs. By the end of 1935, almost all EVs had disappeared from roads [1].

1.3 Conventional vehicles and hybrid EVs

The ICE is considered conventional while EVs are modern vehicles as per the new needs of society. The transition from conventional to EV is happening through a middle path chosen by a number of vehicle manufacturers like Hyundai, Honda, etc. The philosophy for hybrid vehicles is to take benefit of both modes, reducing the weaknesses existing in electrical vehicle technology. The main drawback is the capacity of battery and its charging time and the related infrastructure. EVs are much expensive than their gasoline counterparts, and that is the main barrier for its adoption as pure EVs at present. The economics of these vehicles have forced the car manufacturer to adopt the hybrid approach. The day cost of gasoline reaches sufficiently high and the cost of battery and its charging cost reduce sufficiently low that the pure EVs shall dominate the vehicle market. At present, the hybrid EV is the preferred choice.

The four types of hybrid vehicles are possible—series hybrid, parallel hybrid, series–parallel, and complex hybrid.

1.4 EV and major components

A pure EV essentially has a motor, a battery, and power electronic circuitry for efficient control. An electric motor is used for propulsion that receives power from an onboard source of electricity which is a rechargeable battery or fuel cell [2]. The increase of efficiency in some EVs is done by additionally employing ultracapacitors or flywheels. It may or may not have the gear mechanism in it for power transmission. Even differential axles may also be absent in some designs. The layout of a typical EV is shown in Figure 1.2 with its components.

Important components of the EV and their basic functions are discussed below.

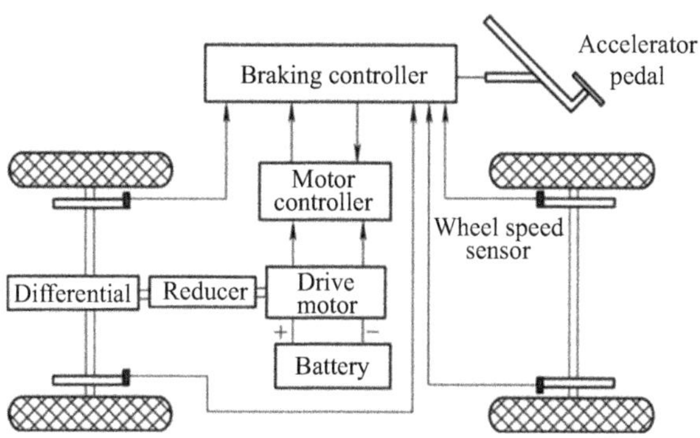

Figure 1.2 Layout of an EV

1.4.1 Drive motor

The drive motor employed for an EV may be any motor among induction motor, permanent magnet brushless (PMBL) motor, switched reluctance motor (SRM), synchronous reluctance motor (SyRM), and other axial flux and magnet less motors. The motor to be employed must have high efficiency and power density; therefore, brushless motors employing rare earth material-based permanent magnets like NdFeB are used.

1.4.2 Battery

The popular types of the batteries in the market are Lead–acid battery, Ni–Cd battery, Li–ion battery, metal–air batteries, etc. The desired features of a battery are low cost, high specific energy, pollution free without carbon footprints, low specific weight, easy constituent chemical availability in abundance, fast charging, and long life. The researchers are trying a number of alternatives and the features determining acceptance of a battery are compared with the corresponding characteristics of ICE fuel for it to be acceptable. It seems at present the Li–ion batteries are the most preferred source for electric cars. It is desired that the battery should be able to power an EV for about 620 km continuously on one charge; however, at present, this is a distant reality.

1.4.3 Battery charging circuit

Charging of vehicle battery should be possible at home during the night; however, charging stations are needed during long-distance drives. Alternatively, the power source for charging the battery may be PV supply or wind power. Some wireless battery charging techniques are also proposed by the researchers through which batteries can be charged on roads at traffic lights.

1.4.4 Power electronic converters

Mainly efficient DC to DC or DC to AC or AC to DC converter topologies are needed. Converters based on vector control, direct torque control, etc. may be needed [3]. Regenerative braking must be employed.

1.4.5 Supercapacitors

The dynamic power supplies and sudden transient disturbance recovery should be possible by using suitable value supercapacitors having capacity values in the range of kilo Farads. These can be charged at much higher rates than a battery. This has led to their suitability in power circuit configurations for EVs.

1.4.6 Flywheels

Flywheels may store kinetic energy in the form of their moment of inertia and return power back for charging a battery or driving wheel through suitable circuitry.

1.4.7 Energy management system

The energy management system (EMS) monitors control of required functions in an EV thereby acting as its brain. It is a computer-based system that dynamically optimizes the charge of batteries to maximize the operating range and improve performance.

1.4.8 Regenerative braking system

If a battery has to have a better powering distance range, it must be efficiently used and, therefore, the regenerative braking is a necessary component of EVs.

1.5 Economics and impact of EVs

It is important to understand the problems of EVs for its realization. There are two major concerns of EVs, namely, cost and range. The cost of an EV depends on the motor and its controller; battery, its charging, and battery management system; in addition to the safety and luxury aspects of the vehicle. To compete with the present ICE-based vehicles, EV manufacturers have the following concerns as challenges for researchers.

- Initial cost.
- Vehicle range on full-battery charge.
- Battery charging infrastructure.
- Battery charging time.
- Performance to accelerate, cruise, and climb.
- Fear of running out of power in the middle of drive.
- Safety and reliability.
- Recurring battery cost for EV.

With an increase in EVs, the load on the conventional grid shall also increase because the primary source of power is still the conventional grid. The use of renewable energy sources specifically solar and wind are being employed either in islanded mode or for supporting the grid. The technologies employed to overcome these difficulties shall decide the economic viability of the EVs based on EVs individual performance and cost; and, hence, their acceptance by customers.

1.6 Important aspects of EV technologies

The important aspects of EV technologies that must be discussed are as follows.

1.6.1 Motor drive technology

The motor drive technology depends on the type of electric motor employed for the EV. At present, induction motor drive and PMBL motor drive technologies are being preferred for most of the EVs [3].

1.6.2 Energy source technology

The researchers around the globe are rigorously searching energy source technologies that are feasible for EVs in terms of high energy storage with less volume and weight. As of date, batteries are the most preferred energy source for EVs. Other sources may be fuel cells, ultracapacitors, and renewable energy if found suitable and economical.

1.6.3 Battery charging technology

The batteries can store a fixed amount of charge, which will deplete after running the EVs; therefore, the charging of batteries is an essential part. The battery charging takes longer time in conventional mode and needs infrastructure for faster charging, if wants to compete with ICE-based vehicles. Wireless charging technology can be used while the vehicle is moving or standing. The conventional source for charging the EV battery shall be a grid only and this charging method is called as grid-to-vehicle (G2V) mode.

1.6.4 Vehicle-to-grid (V2G) technology

The conventional charging of the EV battery from the grid imposes additional load on the grid, if it happens during the peak load hours. Therefore, the EVs can be used as a storage of power if they can be used to feed the grid using inverters during peak load hours. This mode of operation is called V2G mode and the technology used for this is known as V2G technology.

1.7 Main configurations of EVs

The EVs run solely on electric propulsion or it may also have ICE working alongside it. The EVs can be classified based on source of power employed in it and the charging method employed.

1.7.1 All EV (AEV)

The EV that gets it power from battery only is known as all EV (AEV) and sometimes battery EV (BEV). It can also be called as pure EV. Typically, these vehicles can cover a range of 100–250 km on one charge. The recharging time for vehicle batteries may be as high as 8–11 hours to completely replenish the battery.

1.7.2 Hybrid EV (HEV)

To overcome the fear of running out of power in the middle of the drive, some of the manufacturers have used an ICE additionally with the EV. Such vehicles are known as hybrid EVs (HEVs) and lead to huge initial costs [4]. HEVs are further classified based on their arrangement configuration, as (1) series hybrid, (2) parallel hybrid, (3) series–parallel hybrid, and (4) complex hybrid. In HEVs, the electric propulsion is used to reduce the greenhouse gas (GHG) emissions and fuel consumption, especially at low speeds.

1.7.3 Plug-in hybrid EV (PHEV)

PHEVs mainly employ electric propulsion and, therefore, have bigger capacity battery compared to HEVs [4]. It has the ability to run solely on electricity for most of the time thereby reducing the carbon footprint drastically.

1.7.4 Gridable EV

The AEVs and PHEVs, having vehicle-to-grid capability, are known as gridable EVs. These vehicles must have bidirectional converters and can be used as energy storage devices during the off-peak hours of the grid.

1.7.5 Fuel cell EV

Fuel cell has been used as one of the energy sources for providing electric power. In the place of batteries, fuel cell can be used and it avoids long charging hours, but the safety and voluminous space requirements are constraints of such vehicles. These vehicles may be a game changer in the future, as they have high-energy density and no pollution at all.

1.8 Summary

The EVs were around us even before the turn of the twentieth century and it has taken sufficient time to come up to the present stage. The government policies to reduce pollution and dependence on the petroleum products coupled with the enhanced use of renewable energy sources for preserving the nature are the boosting factors for the EVs acceptance. The modern EV is a completely new machine with special features, even then, its acceptance in the society shall be based on many factors including its cost, ease of operation, and other convenience features. Based on the discussions presented in the chapter, the following can be summarized.

- EV is an electric propulsion vehicle that consists of the electric motor, power converter, and energy source, with its own distinct characteristics.
- It is not just a vehicle but a new system for our society, realizing clean and efficient road transportation.
- It must have an intelligent system that can be easily integrated with modern transportation and electrical networks.
- Future designs must involve the integration of art and engineering to fulfill EV user's expectations.

Symbols

Ni	Nickel
Cd	Cadmium
Li	Lithium
NdFeB	Neodymium iron boron

Glossary

Bayerische Motoren Werke GmBH (BMW) A popular German motor manufacture, producing elegant cars since 1917 from Munich.

Permanent Magnet Brushless (PMBL) motors These are special motors having permanent magnets on rotor arranged to produce trapezoidal or sinusoidal flux. The motor is brushless and requires a power electronics controller along with information of rotor position/speed.

Direct Torque Control (DTC) A type of control technique employed in efficient motors like induction motors, controlled through power electronic controllers.

Photo Voltaic (PV) Photo voltaic cells convert solar irradiation into electrical energy employing photoelectric effect.

Grid to Vehicle (G2V) The fast charging strategies have been developed for supplying power to battery from grid. The protocols and standards have been decided internationally.

Vehicle to Grid (V2G) The parked electric vehicles can supply power from their battery back to power grid to meet demand during peak loads.

References

[1] *History of Electric Vehicles* – en.wikipidea.org
[2] *How Electric Cars Work* – auto.howstuffworks.com
[3] C. C. Chan and K. T. Chau, An overview of power electronics in electric vehicles, *IEEE Trans. Ind. Electron.*, vol. 44, no. 1, pp. 3–13, 1997.
[4] M. Ehsani, Y. Gao, S. E. Gay, and A. Emadi, *Modern Electric, Hybrid Electric, and Fuel Cell Vehicles: Fundamentals, Theory, and Design*, CRC Press, Boca Raton, FL, 2005.
[5] A. C. Madrigal, All the promises automakers have made about the future of cars, *The Atlantic*, July 7, 2017.

Chapter 2

Electric vehicle fundamentals

Sanjay Gairola[1], *Sanjeev Singh*[2], *Sanjeet Dwivedi*[3] *and Rahul Dubey*[4]

Conventional vehicles use an internal combustion engine (ICE) which burns one of the petroleum-based fuels, petrol, or diesel mainly, inside a cylinder with a moving piston in it. The burning of fuel thereby produced motion of piston (and, hence, tires connected to the piston through some transmission arrangement rotate) and burnt fuel was exhausted in the form of gases causing pollution. Because of the fast depletion of petroleum and rising concern for pollution, conventional vehicles are in the process of evolution by slowly discarding the ICE and adopting its electric counterpart. This has led to newer vehicle configurations marching towards full electric versions which have better efficiency, controllability with challenges of battery energy storage, and associated difficulties.

In the current scenario, a number of opportunities are available for the transportation due to drastic improvements in the technology in the field of electrical machines, control through power electronics, high-speed computers, Internet connectivity for remote access/poling/driverless trips, high-speed and wireless communication systems, etc. It is important to glance through the history of electric vehicles (EVs) to understand their initial fall and now rise again due to advancements and social needs.

2.1 Evolution of Electric Vehicles

An EV essentially comprises of a battery as a source of electric power for the motor that generates the propulsion torque. At times, ICE may also be working alongside it, aptly called a hybrid EV (HEV). Moreover, fuel cells can also be employed as a power source.

In actual practice, the EVs came into the picture before the ICE vehicles (ICEVs) became popular, as shown by the timeline in Figure 2.1. It is said that the first crude EV was developed in the 1830s while an ICEV was in 1885. The early days of EVs employing lead–acid batteries started mainly from the 1890s [1–5]. By the year 1920, its popularity started declining due to ICEV that was mass produced by Henry Ford. At

[1]Department of Electrical Engineering, Govind Ballabh Pant Institute of Engineering & Technology, India
[2]Electrical Engineering Department, Maulana Azad National Institute of Technology (MANIT) Bhopal, India
[3]Technology Management, Everfuel A/S, Denmark
[4]Edge-AI, Doulos, USA

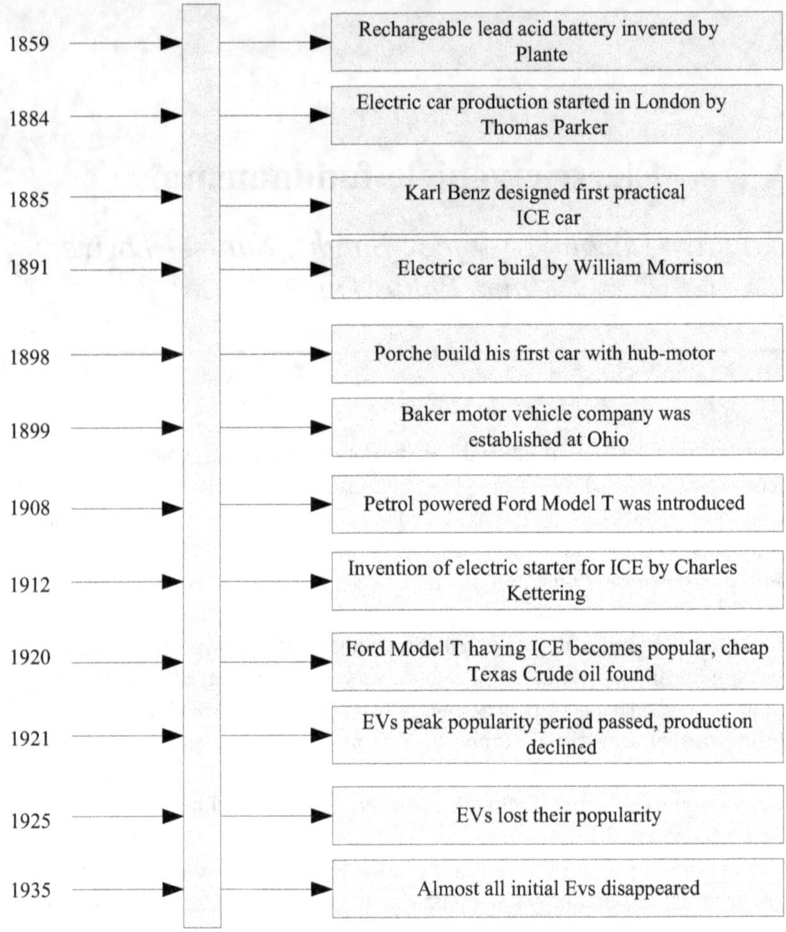

Figure 2.1 Timeline showing the rise and fall of early EV development between 1859 and 1935

the same time, huge reserves of petroleum were found all over the world. The ICEV had better mileage, speed, low capital, and running cost. These advantages led to the initial death of EVs and, by the year 1935, they almost disappeared from the market.

Once again, the oil crisis upsurge of the 1970s generated fluctuations in gasoline in the market, and this led to the renewed interest in EVs [5–8]. By the year 1995, the digital systems, microprocessors, power electronic circuits, motors, control, and battery technology had also improved which created a pathway for high-performing EVs. The international oil shortage and pollution due to ICEV (in the form of air pollutants and greenhouse effect) also dictated the demand for EVs once again. Tesla, Toyota, Honda, Nissan, and GM were among the first few manufacturers who make HEV or pure battery EV (BEV). Some important developments are benchmarked in Figure 2.2. It is predicted that the gasoline vehicles will be completely wiped out of the market by the year 2042.

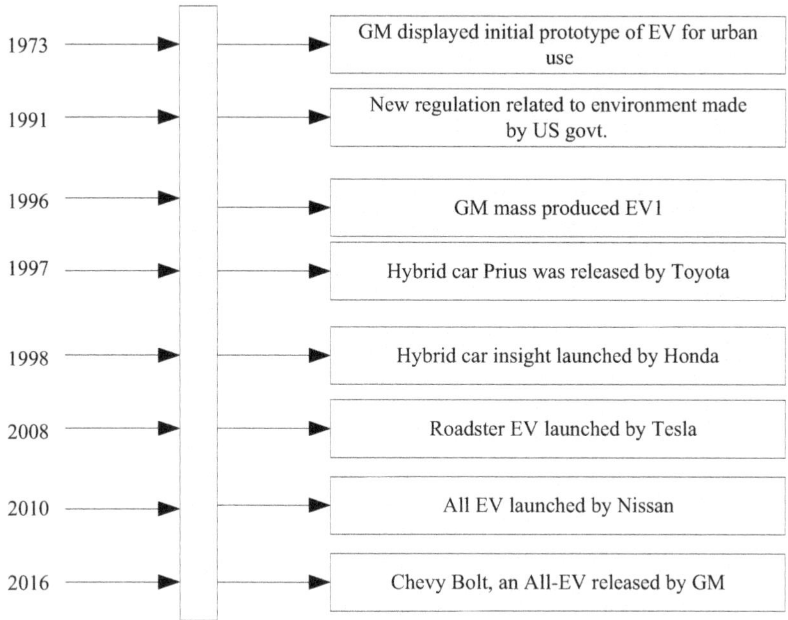

Figure 2.2 Some important events of modern EV development

The vehicle dynamics and various types of arrangements and components are discussed in the following text.

2.2 General vehicle dynamics

It is well known that the smooth movement of any vehicle requires a smooth road or track so that the machinery developing tractive effort in it (an ICE or electric motor) can be accommodated easily at a cheaper cost. The road may take vehicles uphill or downhill with some gradient, generally limited to 15°. To analyze general vehicle dynamics, a vehicle going uphill with some gradient is shown in Figure 2.3, with various forces marked on it [9,10].

For analysis, let us assume the following:

v is the climbing vehicle speed going, in m/s
M is the vehicle mass, in kg
g is the acceleration due to gravity in m/s^2
L is the distance between the front and rear wheels
L_A and L_B are distance of front and rear tires from point C, in m
H_{cg} is the height of center of gravity of vehicle mass above the ground level, in m
α is the gradient of the road, in degrees
v_w is the wind speed in the direction opposite to motion
A_F is the vehicle's frontal area

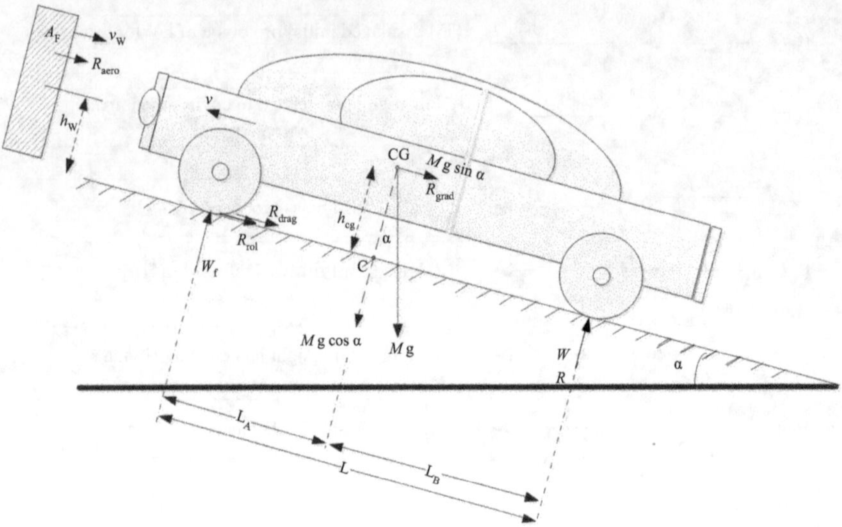

Figure 2.3 Various forces on a climbing vehicle

The dynamics can be analyzed by employing the force balance of various forces marked in the figure. The developed engine tractive force which is in access to the forces opposing it shall cause the acceleration of vehicles.

The forces that are opposing the motion can be put in four parts: rolling resistance, aerodynamic resistance, gradient resistance, and acceleration resistance. The value of these four resistances can be determined from the following equations:

Rolling resistance, R_{rol}: The component of force in the vertical direction due to the mass of vehicle is the cause for frictional opposition at the point of contact with the ground, given as

$$\mathbf{R}_{rol} = \boldsymbol{C}_{rol}\, \mathrm{M\ g} \cos \alpha \tag{2.1}$$

where C_{rol} is the coefficient of rolling.

Aerodynamic resistance, R_{aero}: The air in front of the vehicle opposes the motion of vehicle and it increases with air density, frontal area, and effective velocity of vehicles:

$$\mathbf{R}_{aero} = \frac{\mathbf{1}}{\mathbf{2}} \rho\, \boldsymbol{C}_{drag}\quad \boldsymbol{A}_F\, (\boldsymbol{v} + \boldsymbol{v}_w)^2 \tag{2.2}$$

where C_{drag} is the coefficient of drag by wind and is the density of air, in kg/m_3.

Gradient resistance, R_{grad}: The component of force due to the mass of vehicles in the inclined direction of the road always acts in the downward direction, which is given as

$$\mathbf{R}_{grad} = \mathrm{M\ g} \sin \alpha \tag{2.3}$$

Acceleration resistance R_{acc}: When the vehicle is accelerating, there is opposition to movement which is proportional to the rotor inertia constant and total force in the forward direction

$$\mathbf{R}_{acc} = \boldsymbol{C}_{rotInt}\,\mathrm{M}\,\frac{\boldsymbol{dv}}{\boldsymbol{dt}} \tag{2.4}$$

where C_{rotInt} is the coefficient rotor inertia.

Road resistance R_{road} is the addition of rolling resistance and gradient resistance:

$$\mathrm{R_{road}} = \mathrm{R_{rol}} + \mathrm{R_{grad}} \tag{2.5}$$

Therefore, the total resistive force offered to the vehicle going upward is expressed as the sum of four resistances, given as:

$$\mathbf{R}_{driving} = \mathbf{R}_{rol} + \mathbf{R}_{aero} + \mathbf{R}_{grad} + \mathbf{R}_{acc} \tag{2.6}$$

$$\mathbf{R}_{driving} = \boldsymbol{C}_{rol}\,\mathrm{M}\,\mathrm{g}\cos\alpha + \frac{1}{2}\rho\,\boldsymbol{C}_{drag}\,\boldsymbol{A}_F\,(\boldsymbol{v} + \boldsymbol{v}_w)^2 + \mathrm{M}\,\mathrm{g}\sin\alpha + \boldsymbol{C}_{rotInt}\,\mathrm{M}\,\frac{\boldsymbol{dv}}{\boldsymbol{dt}}$$

Vehicle performance parameters

The capacity of a vehicle to speed up or down on a road with varying gradients characterizes its performance. These characteristics are governed by its type, size, mass, aerodynamics, and the engine driving it. The performance parameters of a vehicle are the following: (i) maximum speed, (ii) gradeability, and (iii) accelerating performance.

2.3 Economic and environmental impact of conventional vehicle leading to selection of EV

- The modern world is employing technology in every aspect of life, and the use of chemicals, composites, and plastics is helping in newer developments. This is cause for newer challenges, global warming, and irreversible climate change draws [11,12]. It has only been recently acknowledged that some regulations have been decided internationally for stopping the global warming.
- The petroleum fuel-based cars and trucks are responsible for almost 25% of CO_2 emissions and other such vehicles account for about 12%.
- A hybrid vehicle essentially combines petrol/diesel fuel along with one different power (energy) sources. The possible second source is generally one of the following: battery, flywheel, and fuel cell (FC) (sometimes CNG is also considered by the people).
- The major economic parameters for a customer to purchase any vehicle are the following: (i) vehicle price, (ii) fuel cost, (iii) maintenance costs and service station availability, (iv) driving range, (v) availability of the fueling station, (vi) time required for the fueling, and (vii) battery replacement cost.

2.3.1 Environmental impact that is handled by EVs

During the manufacturing process of a vehicle, the involved harmful gases contribute to the environmental impact through the pollution associated with mainly three stages: (i) the extraction and processing of material resources, (ii) industrial manufacturing of vehicles, and (iii) the disposal of vehicles.

Two elements are accounted for deciding the impact on the environment due to conventional petroleum-based vehicles: (i) air pollution (AP) and (ii) greenhouse gas (GHG) emissions.

- CO, NO_x, SO_x, and VOCs fall under the category of APs. The weighting coefficients for these APs are 0.017, 1, 1.3, and 0.64, respectively.
- CO_2, SF_6 (sulfur hexafluoride), N_2O, and CH_4 fall under the category of GHGs which have different GHGs impact weighting coefficients. These coefficients relative to CO_2 are 1, 24,900, 310, and 21 in sequence.

Additional sources of environmental impact are also associated with the fossil–fuel production and consumption. These stages of various environmental impacts are known to cause numerous health hazards such as birth defects and infections in the lungs, kidneys, nervous system, etc. Additionally, the PM10 emission from exhaust pipes and non-exhaust emissions (from tire, brake, and road surface wear) also led to premature death due to respiratory problems, breathing diseases, cancer, and lung-associated issues.

Global warming due to GHG emissions has led to record high temperatures, severe floods, droughts, and rising sea levels. The seashore cities are among the prime endangered human habitats.

2.4 Types of EVs

An EV essentially comprises of a battery as a source of electric power for the motor that generates the propulsion torque. At times, ICE may also be working alongside the motor to make it a HEV. An HEV may have better efficiency, depending on the prices of battery, battery charging, and petroleum fuel. Therefore, the EVs are broadly classified as HEVs and pure EVs as shown in Figure 2.4.

The categorization of EVs based on the power source is:

- BEV
- HEV
- Plug-in HEV (PHEV)
- Fuel-cell electric vehicle (FCEV)

The salient features of these different types of EVs are given in the following text.

2.4.1 BEVs

- BEVs rely solely on power from rechargeable battery packs like Li–ion battery or metal–air battery.

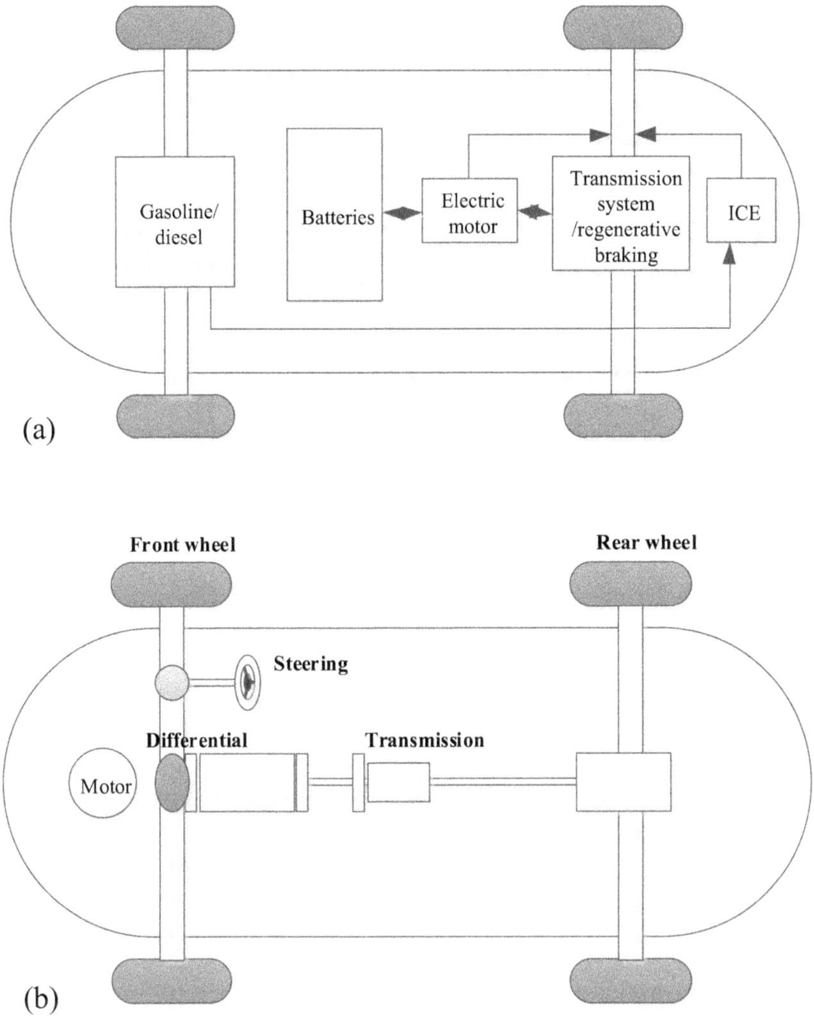

Figure 2.4 Two configurations of EVs: (a) HEV and (b) pure EV

- The presence of regenerative braking in BEVs improves its performance significantly.
- Battery charging takes a significant amount of time.
- Fast battery-charging stations are required, as a substitute or in parallel to existing gasoline fuel stations.
- BEVs technology is yet to mature and at present range anxiety is associated with it.
- The BEVs are fast replacing the existing ICE-based vehicles due to the fast depletion of gasoline and the steep rise in its prices.

2.4.2 HEVs

– HEVs employ ICE along with the electric power train.
– The main power for vehicle propulsion is from ICE and the battery is employed to improve the vehicle performance by suitable switching of power source.
– Electric propulsion is used when power demand is low.
– HEVs have reduced fuel consumption and GHG emission compared to a pure ICEV.

2.4.3 PHEV

– In a PHEV, electric propulsion is the main driving force.
– It has bigger batteries compared to an HEV.
– The batteries are required to be charged at regular intervals.
– Batteries can be charged directly from grids.

2.4.4 FCEV

– FCEVs employ fuel cells to generate electric power for the propulsion of vehicles.
– Fuel cells mainly employ hydrogen as the fuel for the chemical reaction to produce electricity.
– High-pressure tanks are required in FCEVs to store hydrogen which is inflammable.
– The hydrogen tank is a substitute for the gasoline tank of ICEV.
– The by-product from a fuel cell is water and, therefore, it has no pollution.
– A FCEV has the advantages of BEV as well as ICEV; however, the technology is yet to mature.
– Moreover, the cost of hydrogen production is high and the most efficient and viable process for it is being researched.

2.5 Configurations of EVs

The main components involved in a power train inside an EV besides the battery are the clutch, differential, gearbox or fixed gearing, and electric motor. The clutch, gearbox, and differential may be absent depending on the configuration employed in the architecture of an EV. These vehicles may be front-wheel driven or rear-wheel driven and, generally, the front engine or rear engine is accordingly installed. In some designs, in-wheel motors with fixed gearing may be employed thereby eliminating the need for differential. Based on these choices, the configuration of an EV may be one of the following:

1. Front (or rear) engine-front (or rear) wheel drive with a single motor drive: Motor–clutch–gearbox–differential

2. Front (or rear) engine-front (or rear) wheel with a single motor drive:
   Motor–gearbox–differential (clutch absent)
3. Front (or rear) engine-front (or rear) wheel with single motor drive:
   Motor–fixed gearing–differential (clutch and gearbox are absent)
4. Front (or rear) engine-front (or rear) wheel drive with two motors:
   Motor1–fixed gearing–left wheel
   Motor 2–fixed gearing–right wheel
   (clutch, gearbox, and differential are absent)
5. Front (or rear) engine-front (or rear) wheel drive EV with two motors (the same as 4 without fixed gearing):
   (Clutch, gearbox/fixed gearing, and differential are absent)
6. All-wheel drive (AWD) configuration having in-wheel motors.
7. Wireless in-wheel drive configuration having two/four in-wheel motors.

2.6 General EV setup

As seen from Section 2.3, an EV essentially has a battery, a motor or motors, transmission system having the clutch/gearbox/fixed gearing/differential carrying propulsion power to the wheels. Additionally, control and management of various components are done with the help of an electronic controller, a driver circuit, and a suitable power electronic converter. The electronic controller and driver shall also require power that is generated by an auxiliary supply fed from the batteries. Moreover, the controller requires feedback through sensors from the motor, transmission system and wheel about speed/torque, temperature, pressure, and battery status. Accordingly, the general setup is shown in Figure 2.5.

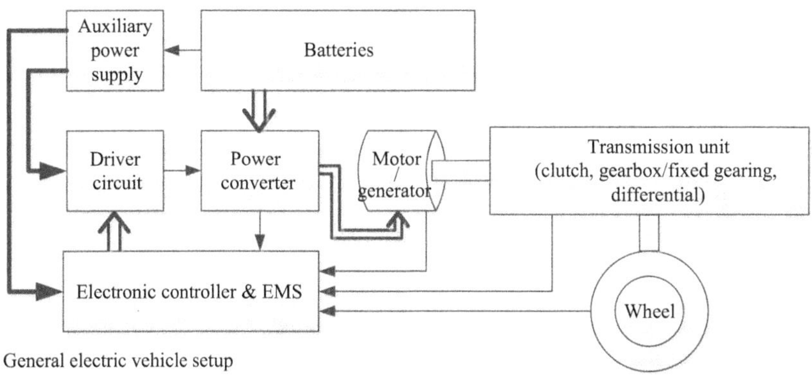

Figure 2.5 Important components of an EV

2.7 Energy sources

The energy required to power the drive motor in an EV [13] can be obtained from one or a combination of the following:

1. Battery
2. Ultra-capacitors
3. Flywheels
4. Fuel cells

The important characteristics of any energy source for an EV are as follows:

- It must have high specific energy and high energy density.
- Moreover, high values of specific power and power density are essential.
- Time required to recharge the battery must be small.
- The battery must have the capability to overcome the deep discharge.
- It must have a long life and a high value of the charging/discharging cycle.
- It must be safe for humans, less susceptible to temperature variations, free from fire, gas emissions, and radiation hazards. The material used in it should be environmentally sound and recyclable.
- The capital cost, maintenance cost, recharging cost, and replacement cost must be small.
- The raw material used in its production must be abundant and locally available.

The historic EVs of the nineteenth century (1880–1924) employed batteries like lead–acid battery, Ni–Cd battery, Ni–iron battery, etc. But these had poor energy density and power density and therefore could not compete with the advantages offered by ICEV at that time. Therefore, the modern EVs are employing newer batteries such as Li–ion battery, metal–air batteries, etc., which have higher life cycle and energy density.

The battery-operated vehicles are expected to provide the propulsion power and range similar to the well-developed ICEV counterpart. This implies that the batteries in an EV on single charge should be able to run continuously for more than 500 km. Moreover, the maximum speed and acceleration should also match. The EV like Rivian R1T will do 0–60 mph in 3 s.

2.8 Electric motors

The requirements of the EV motor are similar to that of the energy source. These are listed as follows:

- The motor must have high instant power and a high value of power density. This means that compact motors with smaller weights shall be preferred. Generally, brushless permanent magnet (PM) motors have high-power density.
- It should be capable of high torque and wide speed range. The conventional AC motors with conventional controls are not suitable.
- High efficiency of the motor is very important for long-distance driving on a single charge.

- High efficiency of regenerative braking is also desired.
- The motor must have a long life and high reliability. This implies that the rugged motors in which the commutator, slip rings, and carbon brushes are absent shall be preferred. Synchronous reluctance motor and squirrel cage induction motor have these features.
- Low capital cost and maintenance-free operation are very important. The conventional DC motors do not suit this requirement.
- The raw material used in it must be abundantly available and recyclable.
- The motor component must not be affected by temperature variation. The motors having magnets in them may suffer from this problem.

The main motors that are found suitable for EVs are the following: squirrel cage induction motor (SCIM), permanent magnet brushless motor (PMBLDC or PMSM), synchronous reluctance motor (SyRM), and switched reluctance motor (SRM). All these motors are having a rugged structure and employ some suitable power electronic controlled converter configuration [14], digital processors generating control signals dynamically based on feedback from position/speed sensors. However, motors combining the features of these motors are being investigated and a lot of optimizations are being carried out by researchers in terms of cost, size, efficiency, and power density.

These motors that have tailored characteristics in general operate in constant torque mode below the base value of its velocity and in constant power mode above the base value, as shown in Figure 2.6.

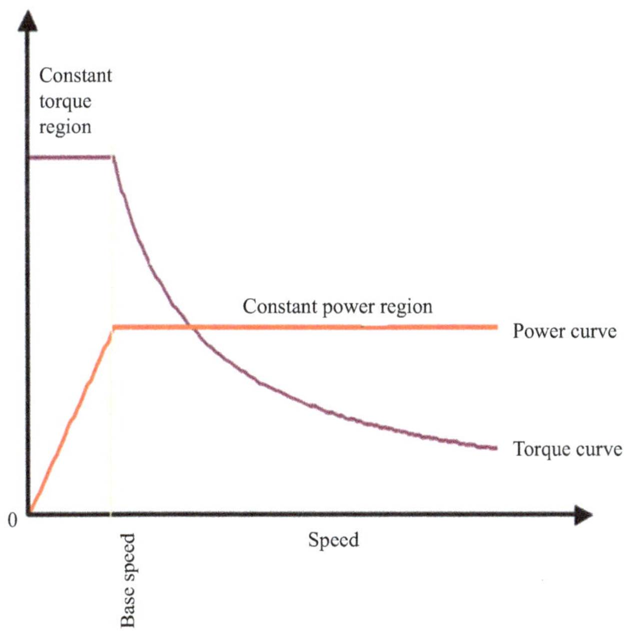

Figure 2.6 Operation ranges of EV motor

2.9 High-voltage system in EV

The EV batteries are charged at level 1 or 2 or 3 depending on the manufacturer's specification or charging available. For fast charging, level 3 is employed. Three-phase AC voltage is converted by the state-of-the-art fast EV chargers to about 480 V DC for high-voltage battery charging. The conversion stages are made up of two power electronic circuits. The rectification stage having power factor correction (PFC) provision changes the input voltage to an intermediate DC, and the second DC/DC stage changes the intermediate DC voltage into regulated output for charging a battery.

The grid and EV battery generally have a galvanic isolation to protect it from spurious transients in power systems leading to instability and unwanted disturbances near the point of common coupling. The first option for isolation is to employ a line-frequency transformer before the rectification and another option for isolation is to put a high-frequency transformer suitably in the DC/DC converter. A DC fast charger system of higher power rating can be met by employing multiple modules in parallel.

2.10 EV automakers legacy

The governments of the developed countries have declared restrictions on the use of petroleum-based vehicles and this has led to the awakening of global auto manufacturers. The United Kingdom plans to have 100% ban on petrol and diesel car sales by year 2040 [5]. These manufacturers work for profit, and therefore, the innovation is focused not only on technology but also on the economic factors, market trends, customer behavior, and geo-politics. The threat to ICE from oil price rise, government restrictions, and the rise of Tesla has initiated a new struggle for existence among automakers. The legendary automakers have now brainstormed to overcome these new challenges to retain their market share.

A number of carmakers have announced to release all-EV by certain target date, if not already done. Volkswagen Group started Roadmap E drive in 2017 and already spent more than $40 billion (by the end of 2022) on electric cars. It envisages to roll out 80 new electric cars by the end of the year 2025.

The iconic brand Tesla has already crossed 1 million annual EV sales, with its Model 3 production to about 10,000 per week. The other projects of Tesla include the solar roof, "Model Y," and driverless cars which are setting a benchmark for other automakers. T has also planned to develop more giant Gigafactory for batteries in Nevada with a capacity of 35 GWh production.

Toyota started with the Hybrid Prius and wants its cars to be emission free by the year 2050. It has planned big for fuel cell-based technology.

Other car manufacturers who have fallen in line are the following: Renault-Nissan-Mitsubishi, General Motors, Ford, BMW, Audi, Hyundai & Kia motors, Volvo, Daimer, etc. All the car manufacturers have planned their projects with challenges to meet battery supply, high-performance motors with their controllers, charging stations, maintenance teams and strategies, government regulations, soaring high prices, market trends, and customer behavior.

2.11 Market and technology trends for EVs

Mature technology and large-scale production of EVs are one of the early achievements of the twenty-first century. The number of EVs is poised to grow in the coming years. This section discusses the market potential of EVs along with the technology trends of different subsystems inside an EV. Our reference to EVs in this chapter is for BEVs.

2.11.1 Market and energy usage

The energy consumption by different modes of transport gives us an indirect estimate of the market size for EVs. In the USA, about 53% of energy consumed by different modes of transport is by light trucks, cars, and motorcycles. This is seen by the bar graph in Figure 2.7. The number of EVs sold and used across the world is growing rapidly and has the potential to capture the existing market of cars, motorcycles, light trucks, other trucks, and buses. Once we add up the vehicle categories mentioned above, the EV market size in the USA consists of vehicles that can use about 80% of transportation energy.

2.11.1.1 The case for electric vehicles

Before the arrival of EVs, improved fuel economy of light vehicles has kept in check the consumption of gasoline [15]. Daily gasoline consumption in the USA is close to 400 million gallons per day (Figure 2.8). Due to gasoline price volatility and environmental concerns from gasoline vehicle emissions, there is a move towards EVs.

Three categories of EVs are becoming popular across the world – electric cars, buses, and two wheelers.

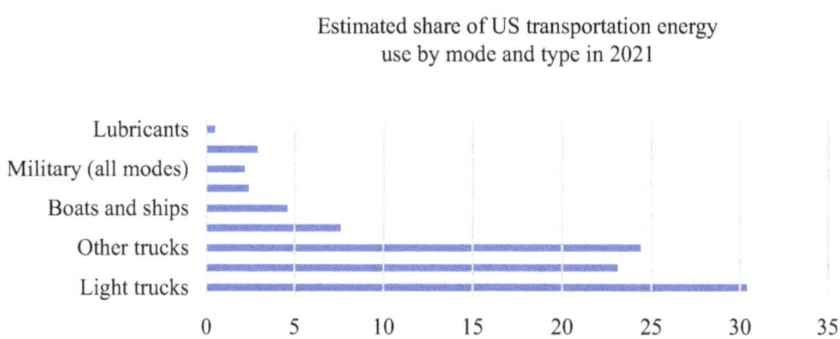

Figure 2.7 Light trucks, cars, and motorcycles account for more than 50% energy usage. Viable electric options are now available for this group of vehicles [16].

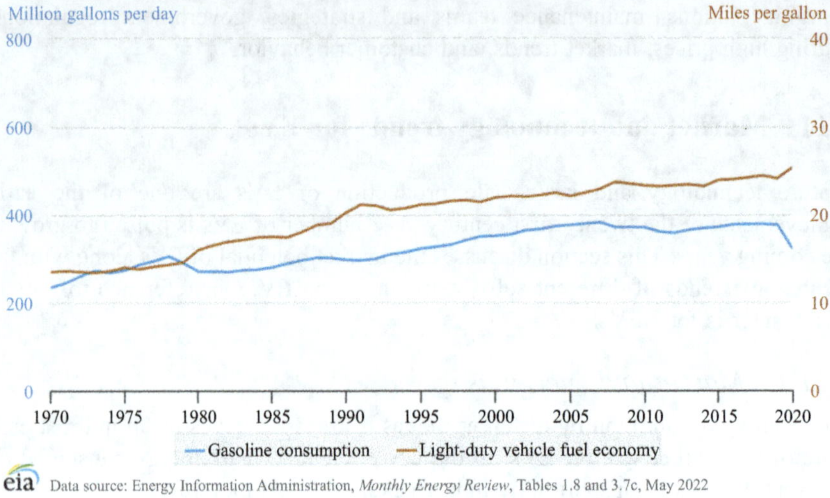

Figure 2.8 Improved fuel economy of light-duty vehicles has limited the usage of gasoline to 400 million gallons per day [16]

Cars: Electric cars are becoming popular in China, Europe, and the USA. In the USA, the number of electric cars on the road increased with the introduction of Tesla Model 3 in 2017. In the broad category of EVs, electric car is the most popular EV. There are several reasons for this. Contemporary battery storage capacity provides a similar driving range for a car as a full tank of petrol/diesel. The location and weight of the battery provide a low center of gravity for the car and increases its stability.

Buses: In most parts of the world, buses are used for public transport and school. These buses have a diesel engine, which emits pollutants. Battery-powered electric buses use similar approach of an electric car. China has taken the lead in adopting electric buses.

Two wheelers: Electric two wheelers are growing at a good pace (upwards of 100%) in developing countries [15]. Even though the charging infrastructure and robustness of electric two wheelers are not as mature as electric cars, it is likely to be in place within a few years.

EVs such as cars, light trucks, and buses are seeing the maximum growth. The increased energy density offered by battery, higher efficiency drive train, and lightweight aluminum body have made EVs at par with their conventional counterparts. Let us evaluate the efficiency of electric cars.

2.11.1.2 Efficiency of EVs

Fuel efficiency of an EV is much higher than gasoline engines [15–17]. In a conventional gasoline-powered vehicle, only about 12%–30% energy is available for motion, while in an EV up to 87%–91% energy is available for movement. Table 2.1 tabulates energy losses for a conventional gasoline car, hybrid, and electric car as shown in Figure 2.9.

Table 2.1 Energy usage of different types of cars [19]

Car type	Engine losses	Parasitic losses	Drivetrain losses	Auxiliary electrical losses	Idle losses	Energy to wheels
Gasoline	68%–72%	4%–6%	3%–5%	0%–2%	3%	16%–25%
Hybrid*	65%–69%	4%–6%	3%–5%	0%–3%	~0%	24%–38%
Electric**	18%	0%–4%	~0%	87%–91%		

*Hybrid vehicles recover 5%–9% energy by regenerative braking.
**EVs recover up to 22% energy using regenerative braking.

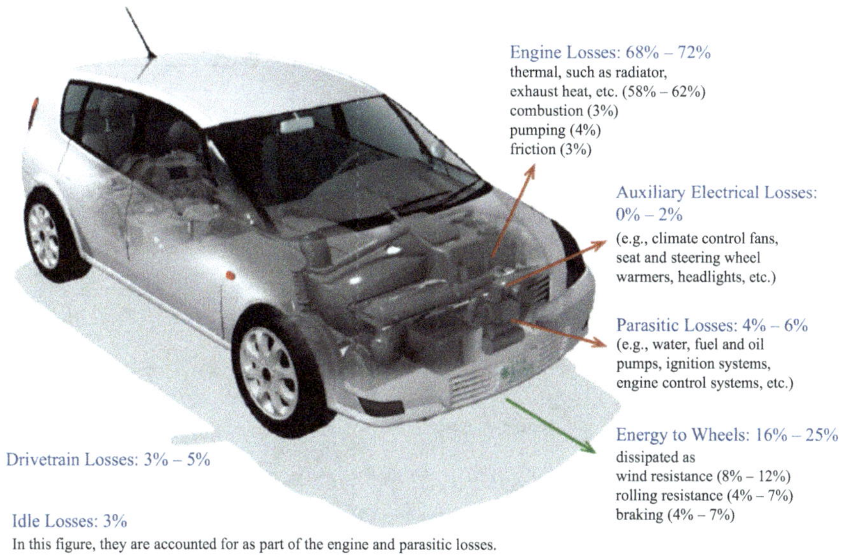

Figure 2.9 Energy consumption by different parts of gasoline car. The bulk of the energy goes to meet engine losses. Only a quarter of the energy is used for locomotion [19].

The data in Table 2.2 clearly show how hybrid and electric cars provide larger percentage of energy to wheels. Improved efficiency and less pollution on streets make EVs, the future of road transportation.

2.11.2 Technology trends

Before we get into details of technology trends in the EV space, let us understand the building blocks of the drive system in an EV. As seen in Figure 2.10, the left side consists of a circuit that is used to charge the battery. The charging circuit is

Table 2.2 Technology status for EV components

Subsystem	Existing technology	Existing applications	Improvements made for EV
Three-phase electric motor	Stable motor technology for synchronous and asynchronous AC motor	Industry workhorse for pumps, servo control, and variety of applications	Light weight, more efficient
Power electronics converter/ inverter	Inverter, convertible used for rectification as well as inversion	Uninterruptible power supply (UPS), variable frequency AC motor drives (VFD)	Robust packaging – compact, vibration, and temperature tolerant design
Battery	Lead-acid battery	Used for UPS, car battery for starting	Modified lithium-ion with larger energy storage and longer battery life retention

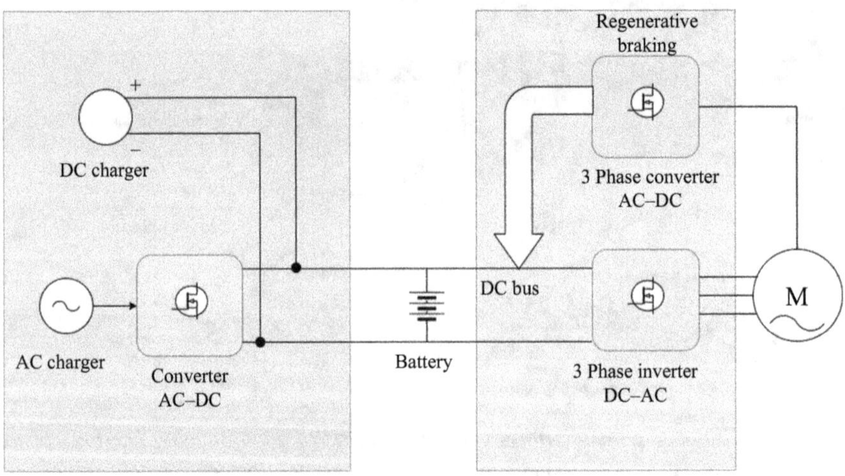

Figure 2.10 Block diagram of electric drive system

active when the vehicle is at a standstill. The right side of the circuit is used to power the EV's motor and to recollect energy when the car brakes. Recovered energy in a conventional vehicle is lost as heat in the wheel's brake pads. In an electric or hybrid vehicle, braking energy is fed back to the battery. This is done by the regenerative braking module.

Location of all subsystems of an electric drive system can be seen in a transparent picture of an electric car (Figure 2.11). The traction battery pack is the battery that runs the EV. The power electronics controller consists of the inverter and the regenerative braking modules.

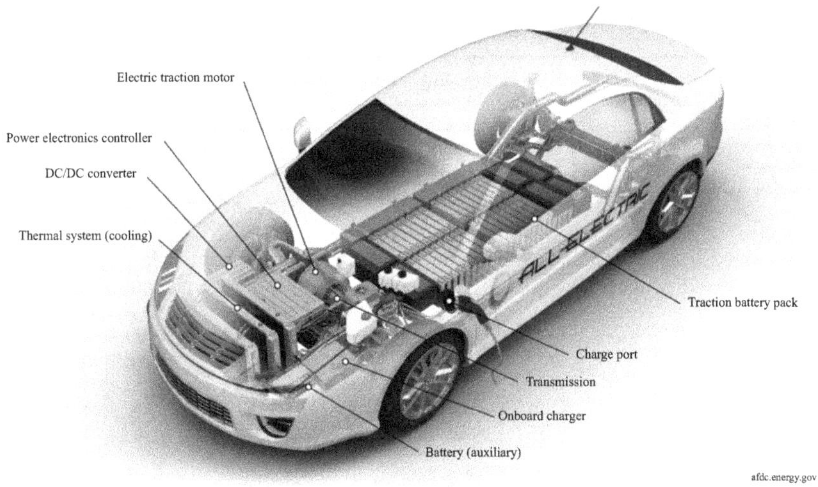

Figure 2.11 Location of electric drive system parts inside an electric car [18]

2.11.2.1 Electric drive technology

Even though EVs have become popular since the start of the twenty-first century, most of the technology powering has been around for quite some time. The electric motor and the power electronics to control the speed of the electric motor have been battle tested in industries throughout the world. Table 2.2 mentions the improvements carried out to make existing technology suitable for EVs. The most prominent on-going improvement is the battery technology and its charging.

Recent advances in real-time telemetry have made it possible to analyze battery performance. Data from laboratory and fleet of EVs has shown that heat and rate of charge play an important role in life of battery. Higher heat and fast charging accelerate Li-ion battery degradation. Contemporary EV battery retains 90% of charge storage capacity after 7 years of service. Research for low-cost, high-density battery storage technology is ongoing. Lithium-ion is a stable but expensive battery technology [17–19].

Standardization of battery charging connector is converging (see Table 2.3). High voltage direct current chargers provide fast charging for larger distances. In the USA, electric car companies have set up large numbers of fast charging stations in parking lots. It remains to be seen how this is replicated in developing countries.

2.11.2.2 Auxiliary technology trends in telemetry and autonomous driving

Electric cars have been bought in new auxiliary technology such as vehicle telemetry (remote connectivity) and basic level of freeway autonomous driving. Vehicle telemetry provides over-the-air update of software along with real-time monitoring of vehicle parameters. New technology such as battery, power electronics, and motor are remotely monitored and tuned for better performance.

Table 2.3 Electric car battery charging technology [17]

Charging level	Connector type	Charging voltage and charging rate
1	J1772	AC charger (120 V). Approximately 5 miles of range per 1 h of charging
2	J1772, Tesla connector	AC charger (240 V). Approximately 25 miles of range per 1 h of charging
3	CCS, CHAdeMO, Tesla connectors	High voltage (480 V) direct current fast charging. Approximately 100–200+ miles of range per 30 min of charging

Multiple camera sensors mounted at different locations in the car provide it with the ability to follow lane markings and have the car maintain its path on a freeway.

Even though auxiliary technology trends such as autonomous driving are not unique to EVs, they provide an incentive for many customers to switch from gasoline run vehicles to electric ones.

2.12 Summary

The electrical vehicle market is the focus of the researchers who are working to overcome the difficulties associated with the aging of ICE-based vehicles. The challenge is manifold, struggling not only with the prices of petroleum fuels that are depleting fast but also the environmental and economic impact to meet the social demand and safety of life on earth. The history of EVs and their modern development has been addressed, which clearly highlights the focus needed on vehicle dynamics, energy sources, charging technology, EV motors, controllers, engine management, optimal configuration, battery management systems, connectivity with utility, and power quality. These aspects shall be discussed in the following chapters in greater depth.

Symbols

M	Vehicle mass, (kg)
g	Acceleration due to gravity, (m/s^2)
L	Distance between the front and rear wheels
LA, LB	Distance of front and rear tyres from point C, (m)
v	Climbing vehicle speed, (m/s)
Hcg	Height of centre of gravity of vehicle mass above the ground level, (m)
a	Gradient of the road, (degrees)
vw	Wind speed in the direction opposite to motion
AF	Vehicle's frontal area
ρ	Density of air, (kg/m^3)

R*rol*	Rolling resistance
R*aero*	Aerodynamic resistance
R*grad*	Gradient resistance
R*acc*	Acceleration resistance
R*road*	Road resistance
C*drag*	Coefficient of drag by wind
C*rol*	Coefficient of rolling
GWh	Giga Watt hour

Glossary

Internal Combustion Engine (ICE)	The conventional vehicles used everywhere have internal combustion engine in which mainly petrol or diesel is used as fuel.
Compressed natural gas (CNG)	This gas is being used in many vehicles as a cheaper substitute of petrol and diesel for about last two decades.
All-Wheel Drive (AWD)	The vehicles in which propulsion power is provided to all its wheels is an all-wheel drive. Such vehicles may have separate electric motor independently controlled for each wheel.
Squirrel Cage Induction Motor (SCIM)	This type of Induction motor has a copper cage on its rotor rather than windings. This enables it to become free of brushes, slip-rings and therefore making it robust.
Permanent magnet brushless DC (PMBLDC)	These are special motors having permanent magnets on rotor arranged to produce trapezoidal flux. The motor is brushless and require power electronics controller along with information of rotor position/speed
Permanent Magnet Synchronous Motor (PMSM)	These are high power density motors having permanent magnets on rotor arranged to produce sinusoidal flux and emf. The motor is brushless and require power electronics controller along with information of rotor position/speed at all speeds.
Synchronous reluctance motor (SyRM)	A motor in which winding is absent and rotor construction is such that reluctance variation provides the fixed high and low reluctance paths to flux thereby causing field to vary sinusoidal.
Switched reluctance motor (SRM)	A motor having salient rotor in which windings or magnets are absent and capable to run at much higher speeds. It works on principle of reluctance torque.

References

[1] *The History of Electric Vehicles* – www.electricvehicles.com
[2] *History of Electric Vehicles* – en.wikipidea.org
[3] *Basics of Electric Vehicles – Service Training, Self-Study Program*, Volkswagen
[4] *How Electric Cars Work* – auto.howstuffworks.com
[5] A. Singh, *Electric Vehicles and the End of ICE Age*, Adhyyan Books, New Delhi, 2019.
[6] M. Ehsani, Y. Gao, S. E. Gay, and A. Emadi, *Modern Electric, Hybrid Electric and Fuel Cell Vehicles – Fundamentals, Theory, and Design*, 2nd ed., CRC Press, Boca Raton, FL, 2010.
[7] A. J. Hawkins, *How Tesla Changed the Auto Industry Forever*, www.theverge.com, July 28, 2017.
[8] A. C. Madrigal, All the promises automakers have made about the future of cars, *The Atlantic*, July 7, 2017
[9] G. Lechner and H. Naunheimer, *Automotive Transmissions: Fundamentals, Selection, Design and Application*, Springer, Berlin, 1999.
[10] NPTEL – Electrical Engineering, "Dynamics of electric and hybrid vehicles," *Introduction to Hybrid and Electric Vehicles*, Joint Initiative of IITs and IISc – Funded by MHRD, India. Available online https://archive.nptel.ac.in/courses/108/103/108103009/.
[11] A. Xue, M. Strabala, D. Lee, P. Lou, and A. Qiu, *The Economics and Future of Electric Powered Automobiles*, University of Chicago, Chicago, BPRO 29000 – Energy and Energy Policy, 2015.
[12] A. Nordelof, M. Messagie, A.-M. Tillman, M. L. Söderman, and J. Van Mierlo, *Environmental Impacts of Hybrid, Plug-in Hybrid, and Battery Electric Vehicles – What Can We Learn from Life Cycle Assessment*, Springerlink.com, August 2014.
[13] S. F. Tie and C. W. Tan, A review of energy sources and energy management system in electric vehicles, *Renew. Sustain. Energy Rev.*, vol. 20, pp. 82–102, 2013.
[14] C. C. Chan and K. T. Chau, An overview of power electronics in electric vehicles, *IEEE Trans. Ind. Electron.*, vol. 44, no. 1, pp. 3–13, 1997.
[15] Auto Economic Times. (2022 January 6). Electric two wheelers. Retrieved from https://auto.economictimes.indiatimes.com/news/two-wheelers/scooters-mopeds/electric-two-wheelers-register-a-staggering-132-growth-in-2021-but-2022-promises-to-be-even-better/88734671
[16] EIA, US Energy Information Administration. (2022 June 28). Energy use for transportation. Retrieved from https://www.eia.gov/energyexplained/use-of-energy/transportation-in-depth.php

[17] US Department of Energy – Energy Efficiency and Renewable Energy. (2022 August 4). Alternative Fuels Data Center. Retrieved from https://afdc.energy.gov/fuels/electricity_infrastructure.html
[18] US. Department of Energy – Vehicle Technologies Office. (2022 August 4). How do all electric cars work. Retrieved from https://afdc.energy.gov/vehicles/how-do-all-electric-cars-work
[19] US. Department of Energy. (2022 July 5). Fuel economy. Retrieved from https://www.fueleconomy.gov/feg/atv.shtml.

Chapter 3
Electric energy sources and storage device

R. Kalpana[1], Kenguru Manjunath[1], Vinod R. Chiliveri[1] and R. Kiran[2]

3.1 Introduction

Electric energy sources are the various methods and technologies that generate electrical energy to power our modern society. These sources play a crucial role in meeting the increasing demand for electricity across residential, commercial, and industrial sectors. They provide the necessary power for lighting, heating, cooling, transportation, communication, and powering electronic devices.

3.1.1 Electric energy sources

Electric energy sources can be categorized into two main types: conventional and renewable. Figure 3.1 shows the different types of energy sources to generate electricity.

Conventional energy sources rely on the extraction and combustion of fossil fuels or the use of nuclear reactions to generate electricity. These sources include the following: *fossil fuels*: coal, oil, and natural gas are burned in power plants to produce steam, which drives turbines connected to generators. This process converts the chemical energy stored in fossil fuels into electrical energy [1]. In nuclear power plants, it utilizes nuclear reactions, particularly nuclear fission, to generate heat. This heat is used to produce steam, which drives turbines to generate electricity.

Renewable energy sources derive power from natural resources that are constantly replenished. They have gained significant attention and popularity due to their environmental benefits and potential for long-term sustainability. Solar panels capture sunlight and convert it into electricity through the photovoltaic effect. This technology has seen significant advancements and cost reductions, making it one of the fastest growing renewable energy sources. In a wind energy system, wind turbines harness the kinetic energy of the wind and convert it into electrical energy

[1]Department of Electrical and Electronics Engineering, National Institute of Technology Karnataka, India
[2]Buymyev Technology Pvt Ltd, Bengaluru, India

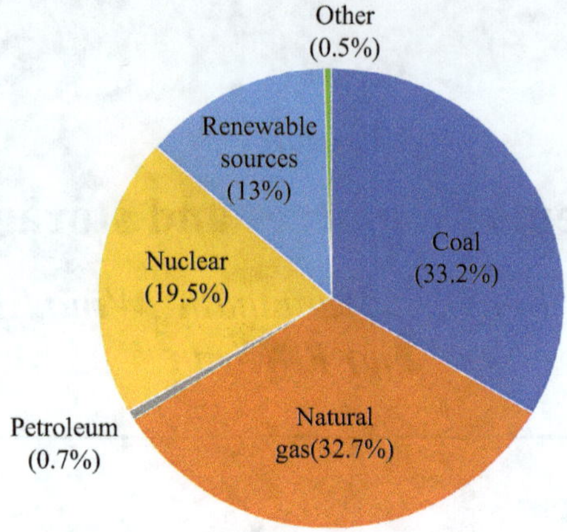

Figure 3.1 Energy sources used to generate electricity

through the rotation of turbine blades. Wind farms are often installed in areas with high wind speeds to maximize power generation. Whereas hydropower plants generate electricity by utilizing the gravitational force of falling or flowing water. Dams or other water storage systems are used to control the flow and release of water, which drives turbines to generate power [2].

Geothermal power plants extract heat from the Earth's internal sources, such as hot water or steam, to generate electricity. This energy source is particularly viable in regions with geothermal activity, and biomass energy involves using organic materials such as wood, agricultural residues, and dedicated energy crops to generate heat or produce biogas through processes like combustion, gasification, or anaerobic digestion. The heat or biogas is then used to generate electricity.

3.1.2 Storage devices

Energy storage is an essential process to make the most of energy resources. It involves the conversion of different forms of energy, such as electricity, chemical, and thermal energy, since some forms are more suitable for efficient storage than others. There are several mechanisms that are employed to carry out this process. Figure 3.2 presents the different types of energy storage technologies.

Electric energy storage devices play a crucial role in modern energy systems by storing electricity generated during times of low demand or from intermittent renewable sources and releasing it when needed. These devices help balance energy supply and demand, enhance grid stability, improve renewable energy integration, and provide backup power in case of outages.

3.1.2.1 Electrochemical storage

The most common form of energy storage is through electrical means. Batteries are the most widely used example, as they can be used to store electricity for a wide range of applications.

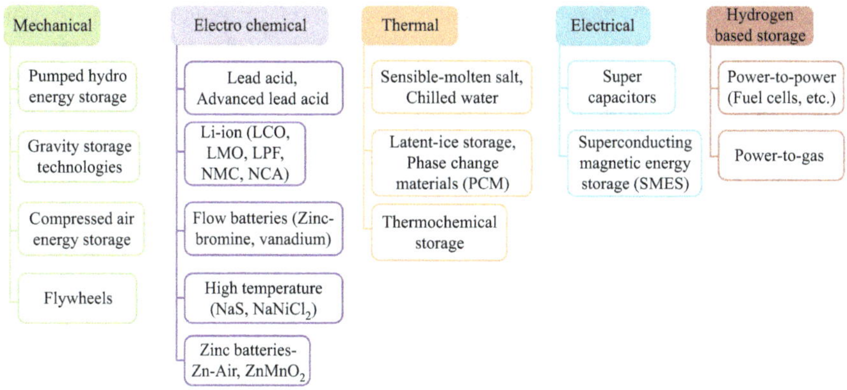

Figure 3.2 Classification of energy storage technologies

Electrochemical storage is a form of energy storage that uses batteries to convert chemical energy into electricity and vice versa. These batteries are available in a wide range of power capacities, from a few watts to hundreds of kilowatts, depending on the specific requirements of the application. Rechargeable batteries are a great way to store energy in a cost-effective manner. They are not only reusable, meaning they can be used multiple times, but they come in a variety of different types. Lead–acid, lithium-ion, nickel–cadmium (Ni–Cd), and nickel–metal hydride are just a few examples of the many types of rechargeable batteries available.

1. Lead–acid batteries have been used for decades and are known for their low cost, wide availability, and robustness. They are commonly used in automotive starting batteries, uninterruptible power supply (UPS) systems, and off-grid applications.
2. Lithium-ion batteries are the most widely used rechargeable batteries due to their high energy density, long cycle life, and relatively low self-discharge rate. They are commonly found in portable electronics, electric vehicles, and grid-scale energy storage systems.
3. Ni–Cd batteries: Ni–Cd batteries have a long cycle life, high discharge rate capability, and good performance in extreme temperatures. However, they are less commonly used today due to environmental concerns related to cadmium content.
4. Flow batteries: Flow batteries store energy in liquid electrolytes contained in external tanks. They offer scalability and long cycle life, making them suitable for grid-scale energy storage applications.

3.1.2.2 Pumped hydro storage

Pumped hydro storage is currently the most extensive energy storage method available. It operates by using excess electricity to transfer water from a lower-level reservoir to a higher-level reservoir, effectively converting the electricity into potential energy [3]. When electricity is in demand, the water is released from the higher reservoir to the lower one, thereby releasing the stored potential energy as electricity. This method is highly efficient and cost-effective, making it the most popular option for storing energy. It is also capable of providing energy in a short span of time, making it an ideal choice for emergency situations. With the increasing demand for energy, the use of pumped hydro storage is likely to increase in the near future.

3.1.2.3 Compressed air energy storage

In power plants, excess energy can be stored using a technique called compressed air energy storage. This method involves storing the surplus energy in an underground chamber. When electricity is required, the stored air is heated, and the resulting hot air is directed through turbines. The rotation of the turbines converts the heat into electricity, providing a reliable source of energy. This method of converting surplus energy into electricity is efficient and cost-effective for power plants.

3.1.2.4 Flywheel energy storage

Flywheel technology is increasingly being used for quick backup power in the event of a power outage. It is a system that involves a rotor in a vacuum enclosure. During periods of excess electricity, the flywheel is accelerated to high speeds, thus storing the energy in the form of kinetic energy. The rotor is connected to a motor/generator setup, which allows for electricity to be extracted from the flywheel when needed. This is done by slowing the wheel down, which causes a discharge of the stored energy. Flywheel technology has become a reliable source of backup power and is becoming a more popular option for those looking to ensure continuity of energy supply [4].

3.1.2.5 Supercapacitors

Supercapacitors serve as a link between traditional capacitors and rechargeable batteries, providing a higher energy density and faster charging and discharging cycles. While supercapacitors are a notable form of energy storage, they are not the sole option available. Other storage methods encompass fossil fuel storage, thermal storage, biological storage, and chemical storage. Each has its own advantages and disadvantages, depending on the application. For example, thermal storage is an efficient way to store energy over long periods of time, while fossil fuel storage is more suitable to generate short-term power. Biological and chemical storage are more suitable for smaller-scale applications, such as those in devices or portable energy sources.

3.2 Electric batteries

Electric batteries are a crucial energy source and storage device that has revolutionized the way our modern world is powered. Batteries have emerged as a crucial technology that enables the transition to a greener and more efficient energy ecosystem in response to the growing demand for clean and sustainable energy solutions. Batteries use electrochemistry to convert chemical energy into electrical energy, making them a portable and dependable power source for a wide range of applications.

The first practical battery, the voltaic pile, developed by Alessandro Volta in the early nineteenth century, is the origin of the concept of electric batteries. Since then, battery technology has made significant progress, and a variety of battery types have been developed to meet a variety of energy requirements. Batteries have evolved in terms of efficiency, energy density, and safety from the lead–acid batteries that powered the first electric vehicles to the lithium-ion batteries that are driving the electric mobility revolution today.

In this section, we will investigate the background of electric battery and its structure. We will also explore the working of battery cell and its various types, advantages and limitations. By understanding the role of electric batteries, we can pave the way for progress in a cleaner and sustainable energy future.

3.2.1 Preliminaries

Cells are the smallest electrochemical unit that acquires, stores, and transmits the energy that depends on the mixture of chemicals and substances used for cell development. Battery or battery packs are made up combination of a number of cells connected in series or parallel. In terms of IEEE Std. 446, battery is defined as "Two or more cells electrically connected for producing electric energy." Most people define "cell" and "battery" interchangeably; however, this is not always clear. The schematic diagram of the cell and the battery is shown in Figure 3.3. It will be always confusing whether the correct term is cell or battery because batteries are sometimes available in a single unit. For instance, a 12-V lead–acid battery comprises of six "2V" cells connected in series and also for high-capacity batteries consist of cells connected in series–parallel connection [5].

Primary cells and secondary cells are the two types of cells that are accessible. Primary cells are those that are only accessible for one-time use, whereas secondary cells are those that can be recharged. Cell (nominal) voltage depends on the combination of active chemicals used in the cell (e.g. Ni–Cd-1.2 V, lithium-based cell >3 V). The cell voltage also depends on various factors such as temperature and the state of charge. The list of cell voltages for various electrochemistry is listed in

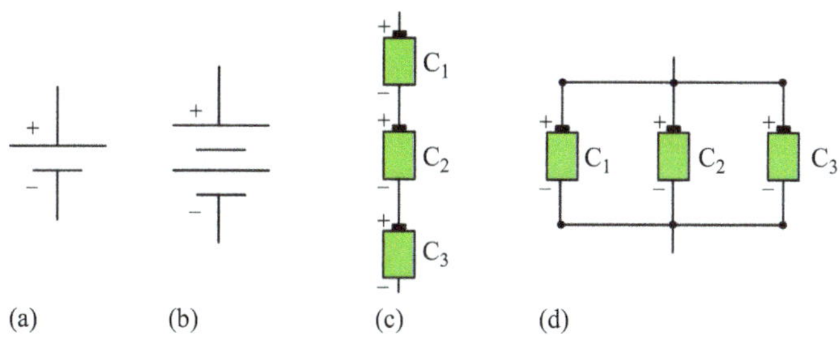

Figure 3.3 Symbol of a (a) cell, (b) battery; (c) and (d) series and parallel connection of three cells

Table 3.1 List of cell voltage with different electrochemistry

Electro-chemistry	−ve Electrode	+ve Electrode	Electrolyte	Nominal voltage (v)
Lead–acid	Pb	PbO_2	H_2SO_4	2.1
Ni–Cd	Cd	NiOOH	KOH	1.35
Dry cell	Zn	MnO_2	$ZnCl_2$	1.6
Alkaline	Zn	MnO_2	KOH	1.5
Zinc air	Zn	O_2	KOH	1.65

Table 3.1. The cell's nominal charge capacity is usually rated in C-rate. The C-rate is mostly related to the cell current. The current required to charge/discharge a cell in an hour is known as the C-rate (1C). It is usually rated in ampere hour rating. For illustration, a cell with 10 Ah rating will be able to deliver 10 A for 1 h ("1C"-rate) or 2 A for 5 h ("C/5"-rate) before the cell is completely discharged. If the cell is drained at a rate of 10C, it should discharge completely in 6 min. On a note, the discharge time and C-rate are not obvious a linear relation, it varies based on operating in different C-rate. On instance, if cell is operated at 1A ("C/10" rate) then to reaching minimum voltage cell may take slightly more than 10 h and if it is discharged at 10C-rate, then it may come to a minimum voltage slightly prior to 6 min.

The energy is stored inside the cell in electrochemical form and then transmit electrical energy whenever its required. This energy is rated in watt-hour (Wh) or kilowatt-hour (kWh) and termed as cell's nominal energy capacity. It can also be defined as the cell nominal ampere-hour multiplied by cell nominal voltage. For example, a 2-V lead–acid cell with 20 Ah will have an energy storage capacity of roughly as 40 Wh. It can also be noted that the energy is usually measured in joules (J) (SI unit), which is equal to watt-sec (Ws). The joule is a very small unit of energy, compared to the energy that the battery holds. Thus, it may be the reason that SI units are not usually used for batteries.

When cells are connected in series, as shown in Figure 3.3(c), then the sum of all the cell voltages is equal to the battery voltage. The battery capacity is equal to the individual cell capacity because the same current will flow through all cells. For example, three Li-ion cells with 3.6 V and 20 Ah are connected in series as shown in Figure 3.2. The battery voltage is equal to 10.8 V and the battery charge capacity is 20 Ah. The energy capacity of the battery is given as 216 Wh.

When cells are connected in parallel, as shown in Figure 3.3(d), then the sum of all cell charges is equal to the battery charge capacity since the battery current is equal to the sum of all cell currents. The battery voltage is equal to the individual cell voltage. For example, three Li-ion cells with 3.6 V and 20 Ah are connected in parallel as shown in Figure 3.5. The battery voltage is equal to 3.6 V and the battery charge capacity is 60 Ah. The energy capacity of the battery is given as 216 Wh.

The other terms of the cell to measure the approximate mass and volume for a given energy capacity are specific energy and energy density. The specific energy of a cell is measured in Wh/kg and is defined as the amount of energy stored per kilogram of battery mass. The energy density is defined as the amount of energy stored per cubic meter of battery volume and is measured in Wh/m^3. Additionally, a cell design with a higher specific energy exhibits the ability to store a greater amount of energy while maintaining a fixed weight, resulting in a reduced weight for a given storage capacity. Similarly, a cell with a higher energy density possesses the capability to store a larger amount of energy within a given volume, leading to a reduced size for a given storage capacity [6].

3.2.2 Working of a cell

Cells are made up of several major components. A negative electrode, a positive electrode, an electrolyte, and a separator are among them. Some types of cells additionally contain current collectors that are separate from the electrodes. In Figure 3.4, a Li-ion cell schematic is shown, although the essential concept is applicable in general.

In an electrochemical cell, the negative (−ve) electrode is typically an alloy, a pure metal, or even hydrogen. In the discharge process, the −ve electrode releases electrons to the load circuit; this process is called oxidation: the loss of electrons from the electrode or the rise in the oxidation state of the electrode, i.e., it is more positively charged. In the charging process, the −ve electrode receives electrons from the load circuit; this process is called reduction: the gain of electrons or the decrease of the oxidation state, i.e. it is more negatively charged. Therefore, the process during charging and discharging is called redox reactions or reduction–oxidation. Similarly, the positive (+ve) electrode is typically a sulfide, metallic-oxide, or oxygen. In the discharge process, the +ve electrode receives electrons from the load circuit, the process in which the electrode is reduced. In the charge process, the +ve electrode releases the electrons to the load circuit and this process is referred as oxidization [7]. Moreover, technically, the oxidized electrode is termed as an anode and the reduced electrode is termed as a cathode. And the materials used for positive and negative electrodes for various cells are listed in Table 3.2.

The electrolyte in a cell acts as an ionic conductor, facilitating the internal transfer of charges between the electrodes. Typically, the electrolyte consists — of

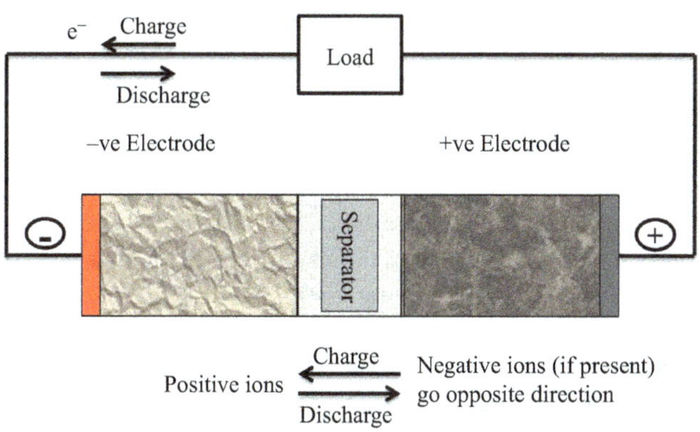

Figure 3.4 Schematic of Li-ion cell

Table 3.2 Comparison of different batteries technology characteristics

Application	Lead–acid	Ni–Cd	NiMH	Li-ion
Specific energy (Wh/kg)	30–40	35–80	55–110	100–300
Cell voltage	2 V	1.2 V	1.2 V	2.4–3.8 V
Energy density (Wh/l)	50–90	50–70	160–420	125–600+
Maximum discharge (rate)	6–10 C	20 C	15 C	80 C
Power density (W/kg)	100–200	100–150	100–500	500–5,000
Charge efficiency (%)	60–80	60–80	70–90	>95
Useful capacity (DOD %)	50	50	50–80	>80
Temperature range (°C)	−40 to 60	−20 to 70	−20 to 65	−30 to 70
Self-discharge (months)	3–4	15–20	15–30	2–3
Cycle life (no. of cycles)	200–400	300–1,000	500–1,000	>2,000
Memory effect	No	Yes	Yes	No
Robustness (over/under voltage)	Yes	Yes	Yes	Need BMS

a liquid solvent containing dissolved chemicals that enable ionic conductivity. However, solid polymer electrolytes can also serve this purpose. In Table 3.2, the listed chemistries utilize aqueous electrolytes, where water serves as the solvent, and the ionic charge transfer occurs through the presence of an acid (H_2SO_4), a base (KOH), or a salt ($ZnCl_2$). Aqueous electrolyte cells have a voltage limitation of approximately 2 V due to the dissociation of oxygen and hydrogen in water under higher voltages [8]. On the other hand, Li-ion cells necessitate nonaqueous electrolytes due to their overall voltages exceeding 2 V.

During the discharge process of a cell, cations (positively charged ions) traverse the electrolyte, moving towards the +ve electrode. Simultaneously, anions (negatively charged ions) travel through the electrolyte towards the ve electrode. Conversely, during the charging process, the opposite occurs: cations migrate towards the ve electrode while anions move towards the +ve electrode.

To ensure the separation and isolation of the +ve and −ve electrodes, a separator is employed. The separator acts as an ionic conductor but functions as an electronic insulator. Its primary purpose is to prevent internal short circuits between the electrodes, which would lead to rapid self-discharge and render the cell unusable.

If current collectors are present, their role is that of electronic conductors. The electrode materials either adhere to or mix with these current collectors. While current collectors do not participate in the chemical reactions within the cell, they facilitate electronic connections to materials that might otherwise pose challenges in establishing a connection to a cell terminal [9]. Alternatively, current collectors may be included to reduce the electronic resistance of an electrode. For instance, in a Li-ion cell, the ve electrode current collector is typically composed of copper foil, while the +ve electrode current collector is usually made from aluminum foil. In the +ve electrode of a dry cell, carbon serves as the current collector.

3.2.3 Different types of batteries

3.2.3.1 Lead–acid batteries

The lead–acid battery is one of the most widely used and the oldest rechargeable battery technologies. It has been a reliable source of energy for various applications, including automotive, backup power systems, and renewable energy storage. Lead–acid batteries employ a chemical reaction between lead, lead oxide, and sulfuric acid to store and release electrical energy. They consist of a series of lead plates submerged in a sulfuric acid electrolyte solution. The plates are made of a lead-based alloy, with the positive plate coated in lead dioxide (PbO_2) and the negative plate composed of spongy lead (Pb). These plates are separated by porous separators to prevent short circuits while allowing the flow of ions:

$$Pb + PbO_2 + 2H_2SO_4 \leftrightarrow 2PbSO_4 + 2H_2O \tag{3.1}$$

During discharge, the lead–acid battery undergoes a chemical reaction that converts the lead and lead di-oxide into lead sulfate ($PbSO_4$) and releases electrons. The discharge process is illustrated in Figure 3.5(a). Similarly, when the battery is charged, the lead sulfate is converted back into lead and lead dioxide, reversing the chemical process. Figure 3.5(b) demonstrates the charging process of lead–acid cells.

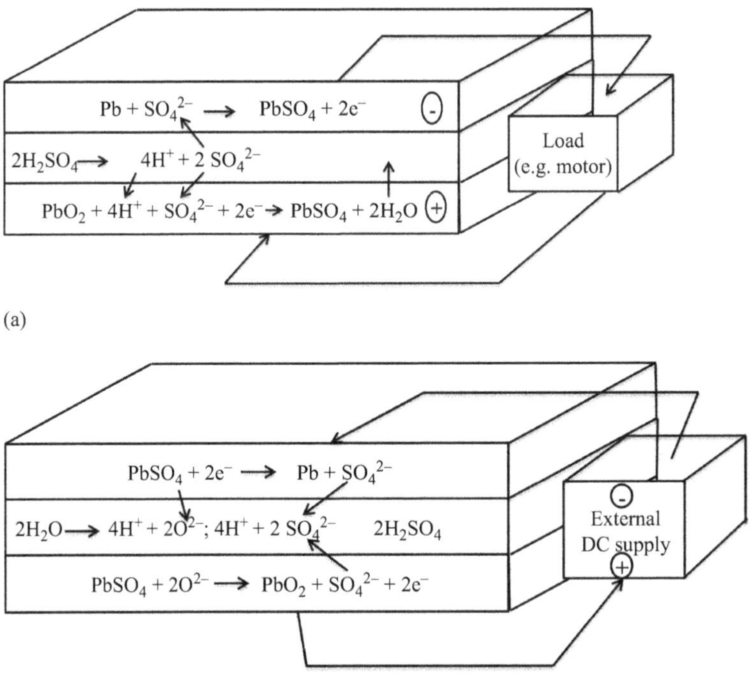

(a)

(b)

Figure 3.5 Lead–acid cell reaction during (a) discharging and (b) charging

Lead–acid batteries are known for their robustness, ability to deliver high currents, and cost-effectiveness. They have a relatively low-energy density compared to some other battery types, but they compensate with their high-power density. Lead–acid batteries can provide a large amount of power for short durations, making them suitable for applications that require bursts of energy [10]. One of the notable characteristics of lead–acid batteries is their ability to withstand deep discharges without significant damage to their overall performance. They can recover well from deep discharge cycles, making them suitable for applications that demand occasional deep cycling, such as automotive starting batteries. The typical lead–acid battery parameters are listed in Table 3.2.

However, lead–acid batteries do have some limitations. They are relatively heavy and bulky compared to other battery technologies. They also require regular maintenance, including checking the electrolyte level, adding distilled water, and periodic equalization charging. Additionally, lead–acid batteries contain toxic materials such as lead and sulfuric acid, necessitating proper recycling and disposal practices to mitigate environmental impact. Despite these limitations, lead–acid batteries continue to be widely used due to their low cost, reliability, and well-established infrastructure for manufacturing, recycling, and maintenance. They are especially prevalent in automotive applications, where they are used for starting, lighting, and ignition (SLI) systems, as well as in backup power systems and off-grid renewable energy storage.

In summary, lead–acid batteries have a long history of reliable service in various applications. They are known for their high-power density, robustness, and cost-effectiveness. While they have some limitations, lead–acid batteries continue to be a popular choice for applications that require a reliable and economical energy storage solution.

3.2.3.2 Nickel–cadmium

The Ni–Cd battery is a type of rechargeable battery that has been widely used for several decades. It operates based on the electrochemical reactions between nickel oxide hydroxide (NiOOH) in the +ve electrode, metallic-cadmium (Cd) in the +ve electrode, and an alkaline electrolyte, typically potassium hydroxide (KOH). The overall reaction of Ni–Cd battery in the alkaline electrolyte is given below. The discharge process is elaborated in Figure 3.6:

$$Cd + 2NiOOH + 2H_2O \leftrightarrow Cd(OH)_2 + 2Ni(OH)_2 4 \tag{3.2}$$

Ni–Cd batteries are known for their high energy density, good cycle life, and ability to deliver consistent power output. They have a relatively stable voltage output throughout discharge, making them suitable for applications where a steady power supply is required.

One of the significant advantages of Ni–Cd batteries is their robustness and durability. They can withstand a wide range of temperatures and environmental conditions, making them suitable for use in harsh environments. Additionally, Ni–Cd batteries can endure a large number of charge–discharge cycles without

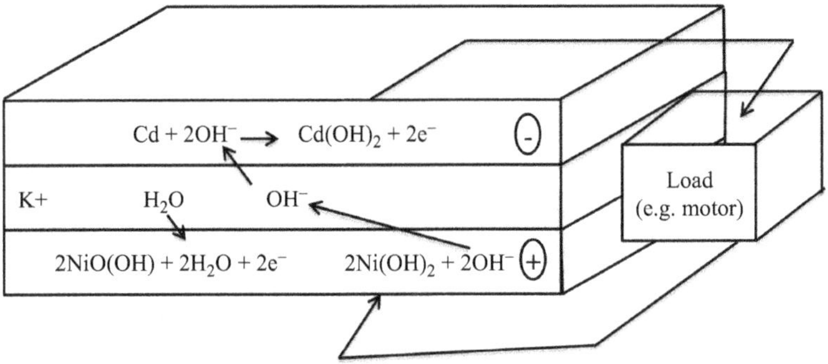

Figure 3.6 Ni–Cd battery reaction during the discharging process

significant capacity loss, making them reliable for long-term use. Ni–Cd batteries also have a rapid charge and discharge capability, allowing for quick recharge times and high discharge rates when needed. This characteristic makes them suitable for applications that require high-power demands, such as electric vehicles and power tools.

However, Ni–Cd batteries do have some limitations. They suffer from the "memory effect," which can reduce their usable capacity if they are not fully discharged before recharging. To mitigate this effect, periodic deep discharge and full recharge cycles are recommended. Another drawback of Ni–Cd batteries is their cadmium content. Cadmium is a toxic heavy metal, and its presence in batteries raises environmental concerns, particularly during disposal. Proper recycling and disposal practices are essential to minimize the environmental impact associated with cadmium. Over time, the use of Ni–Cd batteries has declined due to the emergence of newer battery technologies, such as nickel–metal-hydride (Ni–MH) and lithium-ion (Li-ion) batteries, which offer higher energy densities and improved environmental characteristics. However, Ni–Cd batteries still find application in specific industries and niches, such as aviation, telecommunications, and emergency backup power systems. The nominal parameters of Ni–Cd battery are listed in Table 3.2.

In summary, Ni–Cd batteries have been widely used for their high energy density, good cycle life, and robustness. They are known for their rapid charge and discharge capabilities and can withstand harsh environments. However, their environmental impact and the emergence of newer battery technologies have led to a decline in their usage. Nonetheless, Ni–Cd batteries still serve specific applications where their unique characteristics are valued [11,12].

3.2.3.3 Nickel–metal-hydride battery

The Ni–MH battery is a type of rechargeable battery that has gained popularity as a replacement for older battery technologies, such as Ni–Cd batteries. Ni–MH batteries operate based on the electrochemical reactions between a nickel oxyhydroxide

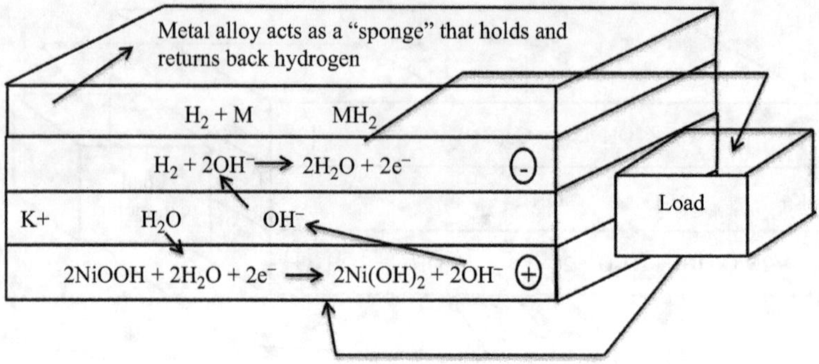

Figure 3.7 Reaction during discharge of Ni–MH battery

(NiOOH) positive electrode, a hydrogen-absorbing alloy negative electrode, and an alkaline electrolyte, typically KOH. The overall reaction is given as follows:

$$MH + NiOOH \leftrightarrow M + Ni(OH)_2 \tag{3.3}$$

The reaction at the +ve electrode is the same as for the Ni–Cd cell; the NiOOH becomes nickel hydroxide during discharge. At the ve electrode, hydrogen is released from the metal to which it was temporarily attached, and reacts, producing water and electrons. Figure 3.7 shows the discharge process of Ni–MH battery.

Ni–MH batteries offer several advantages over Ni–Cd batteries. They have a higher energy density, meaning they can store more energy in a given volume or weight. This increased energy density allows Ni–MH batteries to provide longer runtimes and higher capacities, making them suitable for applications that require extended periods of use [13]. One of the key benefits of Ni–MH batteries is their reduced environmental impact compared to Ni–Cd batteries. Ni–MH batteries do not contain toxic cadmium, eliminating concerns associated with cadmium disposal. This makes Ni–MH batteries a more environmentally friendly choice. Ni–MH batteries also exhibit a lower susceptibility to the memory effect, a phenomenon in which the battery's capacity is reduced if it is not fully discharged before recharging. While Ni–MH batteries can still experience a memory effect to some extent, it is less pronounced compared to Ni–Cd batteries. This allows for more flexible charging patterns and improved overall performance. The following nominal parameters for the Ni–MH battery are listed in Table 3.2.

However, Ni–MH batteries do have some limitations. They have a slightly lower energy density compared to Li-ion batteries, which are now the dominant rechargeable battery technology. Ni–MH batteries are also more prone to self-discharge, meaning they lose their charge over time even when not in use. This can limit their suitability for applications that require long periods of battery storage.

Overall, Ni–MH batteries have found extensive use in various applications, including hybrid electric vehicles (HEVs), cordless power tools, portable electronics, and renewable energy systems. While they have been largely overshadowed by Li-ion batteries in recent years, Ni–MH batteries continue to be a reliable and widely available option for those seeking a rechargeable battery with a good balance of energy density, cycle life, and environmental friendliness.

3.2.3.4 Li-ion battery

The Li-ion battery is a popular and widely used rechargeable battery technology known for its high energy density and excellent performance. Li-ion batteries operate based on the movement of lithium ions between the +ve and ve electrodes during charge and discharge cycles. The chemical reactions that occur in a Li-ion battery are responsible for the storage and release of electrical energy. These reactions take place at the positive and negative electrodes, involving the movement of lithium ions (Li^+) and electrons (e^-) during the charge and discharge cycles of the battery.

Li-ion batteries avoid the use of metallic lithium by employing lithium-intercalated carbons (Li_xC), such as graphite or coke, as the ve electrode. These materials have the ability to absorb lithium ions. As for the +ve electrode, lithium metallic-oxides are utilized. The most commonly used +ve electrode material in Li-ion batteries is cobalt oxide, despite its high cost. It has been proven to provide satisfactory performance. An alternative +ve electrode option is nickel oxide, specifically $LiNiO_2$, which is structurally more complex but more cost-effective. The performance of nickel oxide electrodes is comparable to that of cobalt oxide electrodes. Another area of research involves +ve electrodes based on manganese oxides, such as $LiMn_2O_4$ or $LiMnO_2$. Manganese is a cheaper and more readily available material compared to cobalt or nickel, and it is also less toxic. These manganese-oxide-based electrodes show promise and are being actively studied.

During the charging process, an external power source applies a higher voltage to the battery, causing the lithium ions to move from the +ve electrode (cathode) to the ve electrode (anode) through the electrolyte. At the cathode, a transition metal-oxide material, such as lithium cobalt-oxide ($LiCoO_2$), is commonly used. The Li-ions intercalate into the crystal structure of the cathode material, which is accompanied by a reduction reaction. This reduction involves the transition metal ions in the cathode material accepting electrons and undergoing a change in oxidation state. At the anode, typically made of graphite, the Li-ions are deintercalated from the graphite structure and enter the electrolyte. This process is accompanied by an oxidation reaction, where the graphite releases electrons to the anode.

During the discharging process, when the battery is being used to power a device, the reverse reactions occur. The Li-ions move back from the ve electrode to the +ve electrode, while the electrons flow through an external circuit to provide the desired electrical current. The overall chemical reaction of the Li-ion battery is obtained as:

At negative electrode:

$$Li_xC_6 \leftrightarrow 6C + xLi^+ + xe^- \quad \text{where } 0 < x < 1 \tag{3.4}$$

At positive electrode:

$$xLi^+ + xe^- + Li_{(1-x)}CoO_2 \leftrightarrow LiCoO_2 \tag{3.5}$$

Overall, the movement of lithium ions and the associated reduction and oxidation reactions at the electrodes are the fundamental chemical processes that allow a Li-ion battery to store and release electrical energy. The reversible nature of these reactions enables the repeated charge and discharge cycles of the battery. A Li-ion battery's nominal cell voltage is 3.6 V, which is the same as the voltage of three Ni–MH or Ni–Cd battery cells. Figure 3.8 shows the Li-ion cell structure.

Li-ion batteries offer several advantages that have made them the preferred choice for numerous applications. They have a significantly higher energy density compared to other rechargeable battery types, allowing them to store more energy in a smaller and lighter package [14]. This high-energy density makes Li-ion batteries well-suited for electric vehicles, portable electronics, and energy storage systems where maximizing energy capacity is crucial. Another key advantage of Li-ion batteries is their low self-discharge rate. They can retain their charge for longer periods when not in use, which is particularly beneficial for devices that require occasional use or backup power. Li-ion batteries are known for their high efficiency and excellent voltage stability throughout the discharge cycle. They deliver a consistent and stable power output, ensuring reliable performance for a wide range of applications. Additionally, Li-ion batteries have a long cycle life, allowing them to endure hundreds to thousands of charge–discharge cycles before experiencing a noticeable capacity loss. This durability contributes to their overall cost-effectiveness and longevity. In Table 3.2, the nominal parameters of Li-ion batteries are listed.

Despite their many advantages, Li-ion batteries have some limitations. They are sensitive to high temperatures, and exposure to extreme heat can degrade their performance and reduce their lifespan. Overcharging or deep discharging Li-ion batteries can also lead to performance degradation or irreversible damage. Proper charging and discharging practices are essential to ensure their optimal performance and longevity. Modern Li-ion battery designs incorporate various safety mechanisms to prevent overcharging, overheating, and short circuits. These safety

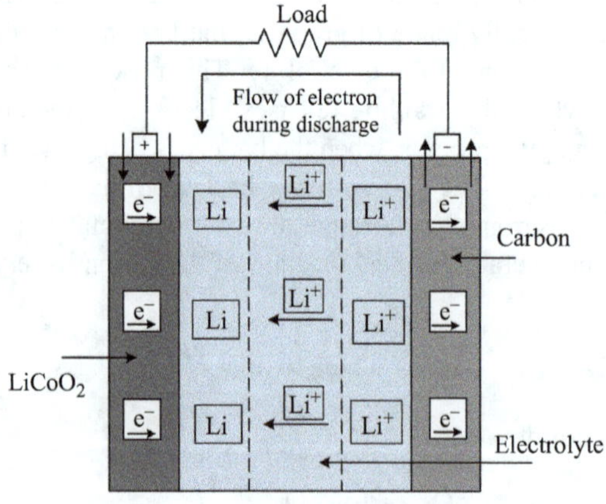

Figure 3.8 Structure of a Li-ion cell

features help mitigate the risk of thermal runaway or fire hazards, making Li-ion batteries relatively safe for everyday use.

3.3 Fuel cell

3.3.1 Introduction

A fuel cell is a device that utilizes a chemical reaction between a fuel and an oxidizing agent to convert their energy into electrical energy. Unlike energy storage devices that store energy for later use, a fuel cell continuously converts energy from a fuel source into electricity. Fuel cells are highly efficient, typically achieving an energy efficiency of 40%–60%. Additionally, the fuel itself holds the energy, as it is chemically bound to it. As such, fuel cells are a great option for a sustainable energy source.

Fuel cells are renowned for their ability to generate electricity through the use of a variety of different fuels, including the commonly referred hydrogen and oxygen (H_2O) fuel cell. In addition to hydrogen, fuel cells can utilize a diverse range of fuels, including methanol, butane, and natural gas. This wide array of fuel options offers a variety of technological possibilities, which can be further categorized based on the type of electrolyte used. These different types of fuel cells offer different operating temperatures and pressures, meaning that there are a range of options available to those looking to create energy through fuel cells.

Fuel cells have been used in many industries, ranging from residential and commercial applications to automotive and aerospace applications. They are also used in remote locations where grid power is not available. With the increasing availability of renewable energy sources, fuel cells are becoming an increasingly attractive option due to their ability to store energy and provide backup power. With the right type of fuel cell, they can provide a reliable and cost-effective source of energy. Furthermore, the use of renewable fuels has the potential to reduce their environmental impact and make them even more attractive.

Fuel cells possess a distinct advantage over thermal engines in that they directly convert chemical energy into electrical energy, whereas thermal engines necessitate the combustion of fuel to convert chemical energy into thermal and mechanical energy. Subsequently, the mechanical energy is utilized for electricity generation, while the thermal energy is often wasted in numerous applications. This means that fuel cells can be more efficient in terms of energy conversion, as all of the chemical energy is converted into electrical energy. In addition, fuel cells produce no emissions of combustion-related pollutants, whereas thermal engines can produce significant amounts of emissions. This makes fuel cells a much cleaner and more sustainable energy source. Fuel cells offer significantly higher theoretical efficiency compared to thermal engines, as the useful work they can generate is not constrained by Carnot efficiency but rather by the free enthalpy of the chemical reaction, which can exceed the potential efficiency of a thermal engine. This is because a thermal engine needs to be coupled with a generator to generate power, which reduces its efficiency. This improvement in efficiency has a huge impact on the environment, making fuel cells a more desirable option for power generation.

The use of thermal engines powered by fossil fuels has many advantages, one of which is the avoidance of undesirable emissions. Moreover, the chemical fuel used by thermal engines has a much higher storage density than electric energy storage solutions such as batteries and allows for faster refueling times and reduced self-discharge. This makes thermal engines ideal for applications where fast refueling and higher energy densities are needed. In applications that demand extended operation durations with minimal downtime between intervals, fuel cells are the preferred option [15]. Not only are they more reliable and efficient but they also require less maintenance and have a longer lifespan. The primary drawbacks of using hydrogen as a fuel source are its difficulty in storage, the costly production from renewables such as electrolysis, and the current investment costs in fuel cells which are significantly higher compared to that of batteries and thermal engines.

3.3.2 Working of fuel cell

A fuel cell is an electrochemical energy conversion device that utilizes the reaction between hydrogen and oxygen or air to produce electrical energy. This is done by two electrodes the cathode and anode being placed in an electrolyte. During the process, an electrochemical reaction occurs when hydrogen (H_2) is broken down into its components: hydrogen ions (H^+) and electrons (e^-). These ions then travel through the electrolyte, and the electrons travel through an external circuit from the anode to the cathode, creating an electric current. The electrons then combine with oxygen at the cathode, and the hydrogen ions and oxygen combine to form water or steam. This process occurs in a continuous loop, making fuel cells an effective and efficient way of producing electrical energy. The following equations show the electrochemical reactions:

$$H_2 \rightleftharpoons 2H^+ + 2e^- \tag{3.6}$$

$$\frac{1}{2}O_2 + 2H^+ + 2e^- \rightleftharpoons H_2O \tag{3.7}$$

Electrolytes are substances that allow ions to pass through them but not electrons. This means that electron transfer between two electrodes is possible only by way of an electric connection. As electrons transfer from one electrode to another in a fuel cell, it creates an electric current and generates heat as a by-product. This electric current and heat can be harnessed for various applications (Figure 3.9).

3.3.2.1 Types of fuel cells

There are three main types of fuel cells that are currently being developed. The basic scheme of fuel cell is shown in Figure 3.9.

(a) Proton-exchange membrane fuel cell

A proton exchange membrane fuel cell (PEMFC) is a type of fuel cell that uses a semi-permeable membrane as an electrolyte. Within the fuel cell, there exists a membrane situated between two electrodes. This membrane is constructed

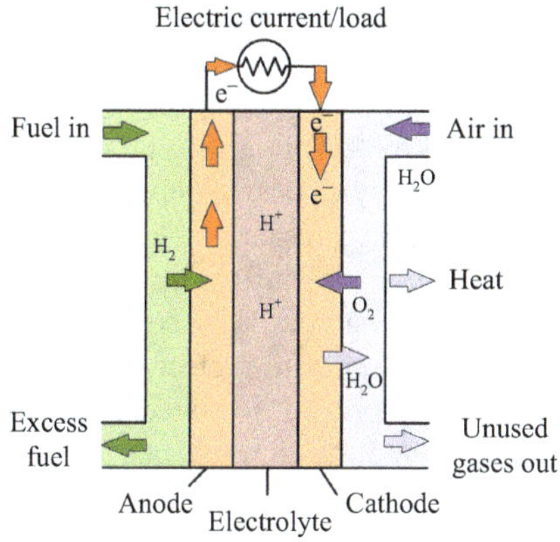

Figure 3.9 The basic scheme of fuel cell

from carbon paper coated with a hydrophobic polymer and embedded with platinum particles.

The platinum particles act as catalysts, enabling the electrochemical reaction to take place between the electrodes. The reaction produces electricity and is typically carried out at temperatures below 80 °C:

$$\text{Anode}: 2H_2 \rightarrow 4H^+ + 4e^- \tag{3.8}$$

$$\text{Cathode}: 4H^+ + O_2 + 4e^- \rightarrow 2H_2O \tag{3.9}$$

The electrodes within a fuel cell can be categorized into two separate zones. The gas diffusion layer is specifically designed with porosity, enabling the diffusion of oxygen and hydrogen feed streams towards the catalytically active material found in the second zone. The gas diffusion layer is an important component of the fuel cell and plays a key role in the electrochemical reaction. It should provide good electrical conductivity to allow efficient electron transport to the bipolar plates. Furthermore, it is also crucial to maximize the catalytic active surface in the catalytic active zone, which is where the electrochemical reaction takes place.

The proton exchange membrane (PEM) plays a crucial role in the operation of fuel cells. It acts as a separator between the two electrodes, providing an electrical isolation while still allowing for ionic conduction. Additionally, the membrane also acts as an electrolyte, allowing for the transfer of protons from one side of the cell to the other. To ensure efficient operation, the membrane must be highly conductive to protons. Nafion is a membrane material commonly used in PEM cells, and it is well known for its desirable properties. Its conductivity is better when the membrane is kept thin, thus making a thickness of less than 200 μm preferable [16].

Bipolar plates have a crucial role in fuel cell stacks, serving as a structural support and separating the feed streams. In addition, they must exhibit excellent electrical and thermal conductivity to facilitate electron transfer between electrodes and assist in system cooling.

(b) Phosphoric acid fuel cell

Phosphoric acid fuel cells (PAFCs) are a specific type of fuel cell that employs phosphoric acid as its electrolyte. These cells typically operate at approximately 200 °C and atmospheric pressure, exhibiting an electric efficiency of approximately 40%. The electrodes in PAFCs commonly consist of a combination of Pt, Fe, or Co catalyst materials supported on carbon paper. It is essential for the gas chamber separating electrodes to be hydrophobic to prevent corrosion. Ongoing research aims to enhance system stability and lifespan by utilizing materials that demonstrate resistance to corrosion when exposed to phosphoric acid electrolytes.

The equation for the electrochemical reactions inside a PAFC is given as follows:

$$\text{Anode}: H_2 \rightarrow 2H^+ + 2e^- \tag{3.10}$$

$$\text{Cathode}: \frac{1}{2}O_2 + 2H^+ + 2e^- \rightarrow H_2O \tag{3.11}$$

(c) Solid oxide fuel cell

Solid oxide fuel cells (SOFCs) consist of three primary components arranged in layers: two electrodes and a ceramic electrolyte positioned between them. What sets SOFCs apart from other fuel cell types is their high operating temperature, which can reach up to 1,000 °C. This characteristic makes them particularly well-suited for stationary applications. Within the SOFC, oxygen ions are transported through the electrolyte from the cathode to the anode, where they react with hydrogen.

The equation for the electrochemical reactions inside a (SOFC) is given as follows:

$$\text{Anode}: 2H_2 + 2O^{2-} \rightarrow 2H_2O + 4e^- \tag{3.12}$$

$$\text{Cathode}: O_2 + 4e^- \rightarrow O^{2-} \tag{3.13}$$

Compared to PEMFCs, solid oxide fuel cell (SOFC) stacks offer several advantages, including the aforementioned qualities and high electric efficiency. Another advantage is their compatibility with biogases. Unlike PEMFCs, which primarily rely on hydrogen, SOFCs have the flexibility to utilize methane, biogas, or carbon monoxide-rich synthesis gas derived from solid biomass gasification as renewable alternatives to natural gas as fuel.

In addition to the primary fuel cell technologies, there are various other fuel cell technologies available that operate at low, moderate, and high temperatures. Among these, two notable ones are the molten carbonate fuel cell (MCFC) and the alkaline fuel cell (AFC). The MCFC is classified as a high-temperature fuel cell that utilizes natural gas and hydrogen as fuel, while the AFC is categorized as a low-temperature fuel cell that relies on hydrogen as its fuel source.

3.3.3 Applications of fuel cell

The two main applications of the fuel cell are the following.

3.3.3.1 Power technologies

1. Fuel cells are increasingly being used to generate power and to store energy in various applications. The choice of fuel cell technology and fuel, as well as the heating and cooling of the stacks, are the main differences between fuel cell systems. In addition, stationary fuel cells can be used as a primary source of power. They are often used to provide electricity for remote locations that are not connected to the grid, as well as to provide additional energy when the main power source is insufficient.
2. The SOFC has demonstrated a high level of efficiency and reliability, making it an attractive choice for power generation. Additionally, its high-temperature waste heat production capability allows for further integration into existing power systems, providing even greater efficiency benefits. The SOFC's potential for high efficiency, combined with its compatibility for mobility, makes it an ideal choice for power systems.
3. An alternative application of fuel cells involves integrating an SOFC with a gas turbine. This integration holds the promise of enhancing the overall efficiency of the gas turbine process by capturing energy losses caused by irreversibilities. It is estimated that the gas turbine–SOFC system could achieve an efficiency of up to 60%.
4. Unlike batteries, fuel cells do not store energy in the same manner. Instead, fuel cells store energy in the form of hydrogen. This characteristic makes fuel cells particularly appealing for applications involving energy storage, such as standalone power plants that rely on intermittent sources like solar or wind power. In such scenarios, fuel cells can be integrated with electrolyzers and storage systems to create a more efficient and comprehensive energy storage system [17].

3.3.3.2 Fuel cell vehicles

1. Hydrogen fuel cells have made significant progress in the transportation sector, and there are already several vehicles, including cars and buses, that utilize this technology for operation. Fuel cells offer several advantages for transportation applications, such as zero-emission operation, longer range compared to batteries, and quick refueling times.
2. Fuel cells are particularly well-suited for certain sectors of mobility where long-range capabilities and refueling challenges are significant factors. Transit vehicles such as ships, trucks, and non-electric trains, as well as cars requiring extended ranges without frequent refueling, can benefit from fuel cell technology.
3. Fuel cell vehicles predominantly rely on hydrogen as their fuel source, which powers an electric engine either by charging a battery or directly. PEMFCs offer numerous benefits, including quick start-up times, high power density, and efficient operation at low temperatures. The hydrogen is stored in a pressurized tank within the vehicle, with standard pressures of 70 megapascals (MPa) or 700 bar for cars,

and 35 MPa or 350 bar for buses. The choice of pressure tank significantly impacts the vehicle's range. The fuel cell system in a motor vehicle typically operates within a temperature range of 60 °C–80 °C, as the use of a polymer membrane in the fuel cell stack restricts the operating temperature to below 100 °C.

3.4 Ultracapacitors

Ultracapacitors, also known as supercapacitors or electrochemical capacitors, are energy storage devices that fill the void between normal capacitors and batteries. Unlike traditional capacitors that store energy electrostatically, ultracapacitors store energy through a combination of electrostatic and electrochemical processes, allowing them to deliver high power density and exhibit remarkable charge–discharge characteristics.

The structure of the ultracapacitor is similar to Li-ion cell structure. It also consists of +ve electrode, ve electrode, separator, and current collector. Figure 3.10 shows the structure of the ultracapacitor. The core component of an ultracapacitor is the electrode material, which is typically composed of porous carbon with a large surface area. This high surface area enables the adsorption and desorption of ions, facilitating rapid charge and discharge cycles. The electrodes are separated by an electrolyte solution, which acts as the ion conductor. Commonly used electrolytes include organic solvents with dissolved salts or ionic liquids.

Ultracapacitors excel in providing bursts of power for short durations, making them suitable for applications that require rapid energy delivery and high-power density. They exhibit an excellent power-to-weight ratio, allowing for quick charging and discharging cycles. However, their energy density, or the amount of energy stored per unit mass or volume, is lower compared to batteries. This limitation restricts their ability to store large amounts of energy for extended periods. Ultracapacitors exhibit power density and energy density in the range of 10^6 W/m^3 and 10^4 Wh/m^3, respectively. While their energy density is lower compared to

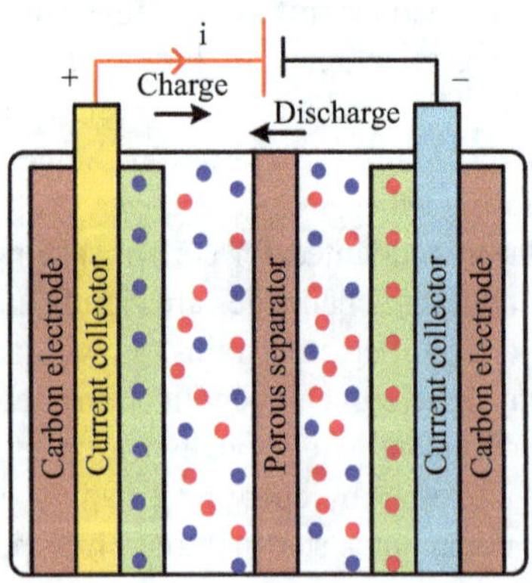

Figure 3.10 Structure of ultracapacitor

batteries, which typically range from around 5 to 25 × 10^4 Wh/m^3, ultracapacitors have significantly faster discharge times (around 110 s) compared to batteries (approximately 5 × 10^3 s). Moreover, ultracapacitors boast a much longer cycle life, with a durability of approximately 10^5 cycles, whereas batteries typically have a cycle life of 100–1,000 cycles.

Ultracapacitors find numerous applications across various industries. They are commonly used in hybrid and electric vehicles to provide quick bursts of power during acceleration and absorb energy during braking, improving overall energy efficiency. They also find applications in renewable energy systems, grid stabilization, and power quality enhancement by compensating for rapid load changes and providing short-term power backup.

Ongoing research and development efforts focus on enhancing the energy density of ultracapacitors, exploring new electrode materials, and improving their overall performance. As ultracapacitors continue to evolve, they hold the potential to play a vital role in shaping the future of energy storage, complementing and even replacing certain applications currently dominated by batteries.

In summary, ultracapacitors offer unique advantages such as high-power density, rapid charge–discharge capabilities, and exceptional cycle life. While their energy density is lower compared to batteries, their ability to deliver bursts of power quickly makes them valuable in various applications where high-power demands are critical. Continued advancements in ultracapacitor technology will unlock their full potential and expand their applications in a wide range of industries [18].

3.5 Fundamentals of electric battery

Batteries are an essential component of many electronic devices, providing the stored energy necessary to power them. They are composed of one or more cells, each of which is an electrochemical unit that stores the chemical energy that can be converted to electrical energy upon demand. Cells are often connected in series to form a battery module, with the entire module being enclosed in a protective casing. A battery pack is an assembly of multiple individual battery modules connected in various series and parallel combinations to provide the necessary voltage and energy to a power electronic drive system.

3.5.1 Battery cell structure

Battery structure refers to the physical components and arrangement of a battery. The cell symbol and cross section of a cell are shown in Figure 3.11. While the exact structure can vary depending on the type and design of the battery.

1. Electrochemical cells: The fundamental building blocks of a battery are the electrochemical cells. Each cell consists of two electrodes: an anode (negative electrode) and a cathode (positive electrode) immersed in an electrolyte solution. These electrodes and the electrolytes facilitate the chemical reactions that generate electrical energy.
2. Separator: The electrodes within an electrochemical cell are separated by a permeable barrier called a separator. The separator prevents direct contact

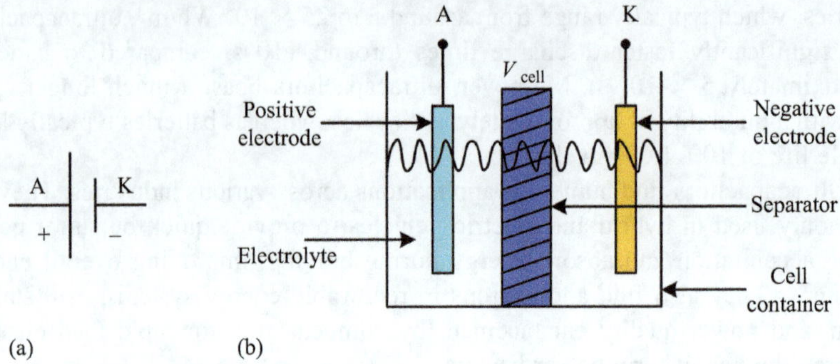

Figure 3.11 (a) Cell symbol and (b) cross-section of a cell

between the electrodes while allowing the movement of ions between them. It helps maintain the integrity of the cell and prevents short circuits.

3. Electrode materials: The anode and the cathode are typically made from different materials to enable the desired electrochemical reactions. Common anode materials include graphite, lithium metal, or various metal alloys, while cathode materials can range from lithium cobalt oxide to lithium iron phosphate, depending on the battery type.
4. Electrolyte: The electrolyte serves as a medium for ion transport between the electrodes. It is usually a liquid or gel substance containing ions that facilitate the flow of charge during the battery's operation. The electrolyte can be aqueous (water-based) or non-aqueous (organic solvent-based) depending on the battery chemistry.
5. Current collectors: To extract electrical energy from the battery, current collectors are connected to each electrode. These collectors act as conductive pathways, allowing the flow of electrons between the electrodes and the external circuit.

3.5.2 Battery parameters

Battery parameters refer to the characteristics or properties of a battery that are used to describe its performance, capacity, and behavior. These parameters provide important information for selecting, using, and evaluating batteries.

Battery capacity:

Capacity refers to the amount of electrical energy a battery can store and deliver. It is usually measured in Ah or mAh and represents the total charge a battery can provide over a specified period. Higher capacity values indicate that a battery can store more energy and provide longer runtime.

Open circuit voltage (OCV):

1. OCV refers to the voltage across the terminals of a battery when no load or external circuit is connected to it. In other words, it is the voltage output of a battery when there is no current flowing through it. The OCV versus SoC characteristics for new versus aged batteries are shown in Figure 3.12.

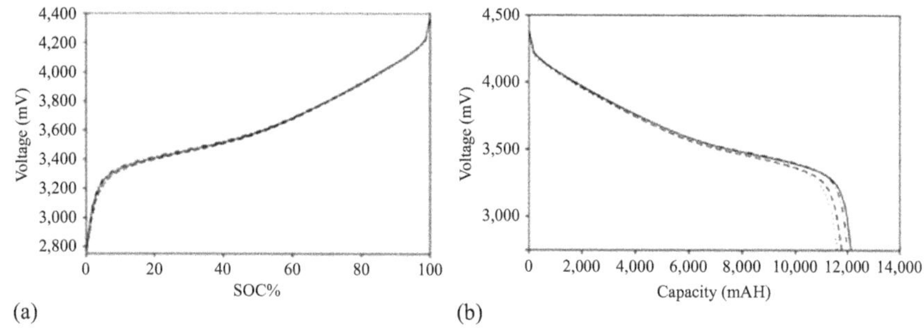

Figure 3.12 *(a) OCV versus SoC characterization curves of new versus aged batteries and (b) battery terminal voltage characteristics in relation to capacity*

2. OCV of a battery is influenced by various factors, including the chemistry and state of charge (SoC) of the battery. Different battery chemistries have different voltage characteristics. For example, a Li-ion battery typically has an OCV range of around 3.0–4.2 V per cell, depending on the specific chemistry and SoC.
3. The OCV is often used to estimate the state of charge of a battery. By comparing the OCV to a reference voltage or using voltage measurements over time, it is possible to estimate the remaining capacity or charge level of a battery. However, it is important to note that the relationship between OCV and SoC is not linear, and other factors like temperature and battery history can also affect the accuracy of this estimation.

Terminal voltage:

1. The terminal voltage of a battery refers to the voltage measured at the battery's terminals when a load or external circuit is connected to it. It represents the electrical potential difference between the positive and negative terminals of the battery.
2. The terminal voltage of a battery depends on various factors, including the battery's chemistry, SoC, and the current being drawn from it. Different battery chemistries have different voltage characteristics. For example, a fully charged Li-ion battery typically has a terminal voltage of around 3.6–4.2 V per cell, while a lead–acid battery may have a terminal voltage of around 2 V per cell when fully charged.
3. As the battery discharges and its stored energy is depleted, the terminal voltage gradually decreases. The rate at which the voltage drops can vary depending on the battery chemistry and the load being applied.

Energy density: Energy density describes the amount of energy stored per unit volume or mass of a battery. It is typically measured in Watt–hours per liter (Wh/L) or Watt–hours per kilogram (Wh/kg). Higher energy density means that a battery can store more energy in a given volume or weight, which is crucial for applications that require high energy storage in a compact space.

Power density: Power density relates to the amount of power a battery can deliver per unit volume or mass. It is measured in Watts per liter (W/L) or Watts per kilogram (W/kg). Higher power density indicates that a battery can supply more power quickly, which is important for applications that demand high power output or rapid charging.

Internal resistance: Internal resistance refers to the opposition to the flow of current within a battery. It arises from the resistance of the battery's components, including the electrolyte, electrodes, and current collectors. Higher internal resistance results in energy losses and voltage drops during discharge or charging processes.

Cycle life: Cycle life represents the number of charge and discharge cycles a battery can undergo before its capacity significantly degrades. It is a measure of a battery's durability and longevity. Batteries with higher cycle life can sustain more charge–discharge cycles, making them suitable for long-term and frequent use.

Self-discharge rate: Self-discharge rate refers to the loss of battery capacity over time when not in use. It is typically expressed as a percentage per month or year. Batteries with low self-discharge rates retain their charge for longer periods, making them suitable for applications where infrequent use or long storage periods are common.

C rate:

1. The C rate of a battery refers to the rate at which it is charged or discharged relative to its rated capacity. It is a measure of the current flow in relation to the capacity of the battery.
2. The C rate is defined as the ratio of the current (in amperes) to the battery's rated capacity (in ampere-hours).
3. For example, if a battery has a rated capacity of 1,000 mAh (1 Ah) and a discharge current of 500 mA (0.5 A), the discharge rate would be expressed as 0.5C (500 mA/1 Ah). Similarly, if the battery is discharged at 2 A, the discharge rate would be 2C (2A/1 Ah).

SoC:
The SoC of a battery refers to the amount of electrical energy that remains in the battery compared to its maximum capacity. It is expressed as a percentage, indicating the level of charge remaining in the battery.

Methods used to estimate SoC for certain battery chemistries:

1. Coulomb counting:
$\text{SoC} = (Q_r - \int i\,dt)/Q_r$, where SoC is the state of charge in percentage, Q is the rated capacity of the battery in ampere-hours, $\int i\,dt$ is the integral of the current over time, representing the cumulative charge or discharge.
2. Voltage based:
$\text{SoC} = (V - V_{\text{min}})/(V_{\text{max}} - V_{\text{min}}) \times 100$, where SoC is the state of charge in percentage, V is the measured battery voltage, V_{min} is the minimum voltage observed at a known low SoC, and V_{max} is the maximum voltage observed at a known high SoC.

3. OCV method:
   SoC = *f*(OCV), where SoC is the state of charge in percentage, and *f*(OCV) represents a lookup table or mathematical model that relates the OCV of the battery to the corresponding SoC.

Remaining useful life:
The remaining useful life (RUL) of a battery refers to the estimated time or capacity that remains before the battery reaches the end of its useful lifespan. There are multiple approaches and algorithms used to estimate the RUL of a battery, including empirical models, data-driven models, and physics-based models.

RUL = (capacity new − capacity present)/rate of capacity loss capacity new is the original capacity of the battery when new, capacity present is the current capacity of the battery, and rate of capacity loss is the rate at which the battery capacity is decreasing over time.

Depth of discharge (DoD):
DoD refers to the amount of a battery's capacity that has been discharged relative to its total capacity. It is expressed as a percentage and indicates the extent to which the battery's energy has been used:

$$\mathrm{DoD(t)} = \frac{Q_T - \mathrm{SoC}_T(t)}{Q_T} \times 100 \tag{3.14}$$

$$\mathrm{DoD} = \frac{\int_{t_0}^{t_1} i(t)}{Q_T} \times 100 \tag{3.15}$$

For example, if a battery with a total capacity of 100 Ah has discharged 50 Ah of energy, the DoD would be: DoD = (50 Ah/100 Ah) 100 = 50%.

3.6 Modeling of electric battery

Battery modeling refers to the process of creating mathematical or computational models that represent the behavior and characteristics of batteries. These models are used to simulate and predict the performance, dynamics, and interactions of batteries in various applications. Battery modeling serves several purposes. It allows researchers, engineers, and manufacturers to understand and optimize battery performance, design more efficient battery management systems (BMSs), and assess the impact of different operating conditions on battery behavior. It also aids in the development of battery control strategies, state estimation algorithms, and optimal battery usage. Battery models typically capture the electrical, thermal, and electrochemical processes that occur within a battery. They consider factors such as voltage, current, SoC, state of health (SoH), temperature, and other relevant parameters. Various types of battery models exist, ranging from simple empirical models to more complex physics-based models [19,20].

Empirical models are based on experimental data and mathematical equations that describe the voltage and capacity characteristics of batteries under different conditions. These models are relatively simple and provide a practical representation of battery behavior but may lack accuracy for specific applications. Physics-based models, on the other hand, take into account the fundamental electrochemical processes occurring inside the battery. They incorporate equations that describe the chemical reactions, ion diffusion, and heat transfer within the battery cells. Physics-based models can provide more detailed and accurate representations of battery behavior but require more computational resources and detailed knowledge of battery chemistry.

In recent years, advanced modeling techniques, such as electrochemical impedance spectroscopy (EIS) and equivalent circuit models (ECMs), have gained popularity. EIS involves measuring the response of a battery to small alternating current signals at different frequencies to extract information about its internal impedance. ECMs use electrical components, such as resistors, capacitors, and inductors, to represent the equivalent behavior of a battery. These techniques provide a balance between accuracy and computational complexity. Battery modeling plays a crucial role in the development and optimization of battery technologies for applications such as electric vehicles, renewable energy systems, portable electronics, and grid-scale energy storage. By accurately simulating battery behavior and performance, modeling enables better battery design, management, and control, ultimately leading to improved efficiency, safety, and overall battery performance.

3.6.1 Equivalent circuit model

Equivalent circuit-based models are commonly used for battery modeling as they offer the advantage of using lumped parameters, which makes them easily integrated into larger system simulation models. These models utilize a combination of circuit elements such as capacitors, inductors, resistors, and dependent sources to represent the functionality and behavior of the electrochemical cell. The model parameters are determined from output data obtained from the battery, eliminating the need for detailed knowledge of the chemical processes and design specifics. ECMs can range from simple linear-resistive models to more complex ones that capture the chemical processes using lumped parameters. While the precision of these models falls between empirical models and theoretical models, they are highly useful for both system simulation and design purposes. They strike a balance between accuracy and practicality, making them valuable tools for analyzing battery behavior and designing battery-dependent systems.

To effectively evaluate the application aspects of batteries, it is essential to gain insights into both the device's operation and its interaction within the system. Circuit models that are complex will be utilized to investigate dynamic responses, and also the characteristic features including pulse discharges in EVs. The electrochemical activities within the battery are observed by two important relationships: The Butler–Volmer relationship, which describes the exchange of electrons at the electrolyte–electrode interface, and Faraday's electrolysis law, which establishes that the current governs the reaction. By establishing connections between these principles and the diffusing charge and stored charge within the electrochemical cell, it becomes possible to develop an electric-circuit model. The parameters of this model can be determined through validated data. When constructing battery models, it is more suitable to consider the charge stored and diffused near the electrode surface, rather than focusing on surface activities or species concentration. The instantaneous stored charge is represented as $q_s(t)$ and the instantaneous diffusion charge as $q_d(t)$, both near the electrode surface. Suppose Q denotes the total capacity of the cell, the charge in the non-energized species can be expressed as $Q - q_s(t)$. One challenge lies in finding the Butler–Volmer equation's inverse to represent the terminal voltage in terms of electrode current accurately. The Tafel and Nernst equations are approximations that provide an estimation of the terminal current; however, they have limitations. By understanding the underlying principles and relationships involved in the electrochemical processes, battery models can be developed, allowing for analysis of the device's behavior and its impact on the larger system. These models provide valuable insights into battery performance, aiding in the design, optimization, and evaluation of battery applications in various fields. An approximate representation of battery behavior is provided by the simplified Unnewehr universal model, expressed as

$$E(t) = E_0 + R_\Omega i(t) + k_1 q_s(t) \tag{3.16}$$

where E_0 represents the cell's initial voltage, R_W denotes the resistance, and k_1 is the constant.

To address the solution of the Butler–Volmer equation in a generalized form, Hartley and Jannette introduced an equation as follows:

$$\begin{aligned} E(t) = {} & E_0 + R_\Omega i(t) + k_1 \ln(1 + |i|)\text{sign}(i) + k_2 \ln(1 + |q_d|)\text{sign}(i) \\ & + k_3 \ln(1 - q_s) \end{aligned} \tag{3.17}$$

The constants k_1, E_0, R_Ω, k_2, and k_3 are determined based on the properties of the particular electrochemical cell and can be obtained from validated data. Although the Hartley model provides a terminal voltage in mathematical representation, it is often more practical to find an ECM for simulating and analyzing battery cells. This allows for easier integration into simulation platforms and facilitates the analysis of battery performance. In the subsequent discussion, various ECMs representing electrochemical cells are explored, initially with the Hartley model considered as the basic model. These circuit models provide a practical approach to simulate and analyze battery behavior, offering insights into the dynamics and characteristics of the battery under different operating conditions.

Simple battery model:

We can start by considering a straightforward electrical equivalent circuit model, which captures the fundamental principles of battery operation and is suitable for characterization based on the discharge data of the cell, as depicted in Figure 3.9. Among the critical dynamics that need to be modeled is the process of diffusion. While more intricate representations involving Constant Phase Element or Warburg impedance can be employed, an approximate solution can be obtained by considering the change in diffusing charge in a similar fashion to the voltage across an RC circuit element. As a result, the impact of the diffusion charge on the terminal voltage can be characterized by a first-order differential equation:

$$\frac{dv_d(t)}{dt} = \frac{1}{C_d}i(t) - \frac{1}{C_d R_d}v_d(t) \tag{3.18}$$

where $v_d(t)$ denotes the voltage drop across the R_d and C_d circuit which is proportional to the diffusion charge $q_d(t)$. While for accurate measurement of the diffusion charge, more RC circuits are added, for the sake of simplicity, our model in Figure 3.13 retains a single RC time constant. Another important aspect to consider is the impact of SoC on the terminal voltage of the battery cell. As shown in Figure 3.13, as the cell is discharged, the terminal voltage drops. In the linear region of this discharge characteristic, we can model the voltage decrease using a capacitor C_s connected in series to denote the charge stored in the cell. Hence, the charge stored q_s t is directly proportional to the voltage across the capacitor C_s. When the SoC of the cell decreases or increases during discharging or charging, the voltage across that capacitor will correspondingly decrease or increase. Additionally, an electrochemical cell naturally loses charge when at rest. To account for this self-discharge, a parallel resistor R_{sd} can be added to the storage capacitor C_s. In Figure 3.13, the R_{sd}, C_s circuit elements, representing the self-discharge resistor and storage capacitor, are connected in series with the diffusion parameters. In terms of the terminal current, this component of the circuit model may be mathematically represented as follows:

$$\frac{dq_s(t)}{dt} = i(t) - \frac{1}{R_{sd}}q_s(t) \tag{3.19}$$

To complete the equivalent circuit model of the electrochemical cell, two additional parameters need to be included: a resistor and voltage source in series to represent the ohmic resistance drop. R_W represents the ohmic resistance and the voltage source is treated as the cell's OCV, E_0. These elements are depicted in Figure 3.13, also in series with the diffusion and storage parameters. This simple ECM provides a representation of an electrochemical cell. The values of the above-defined elements in the circuit can be obtained experimentally by varying a battery

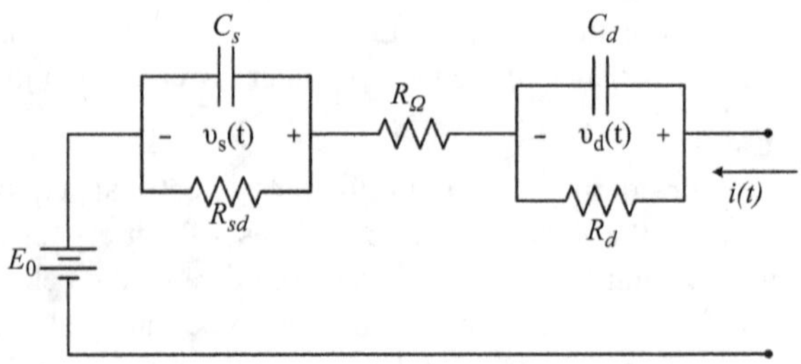

Figure 3.13 Electric equivalent circuit battery model

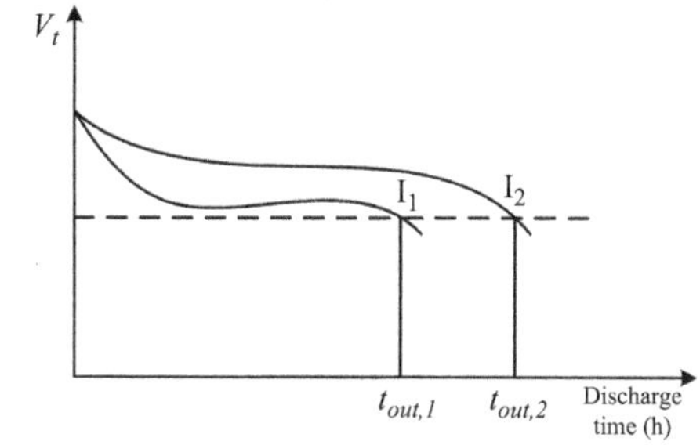

Figure 3.14 Constant current discharge curves

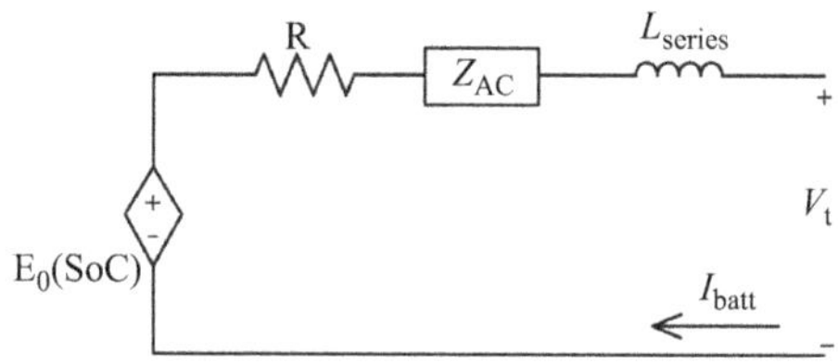

Figure 3.15 Impedance-based equivalent electric circuit battery model

current in a step-change manner, allowing for the characterization and analysis of the battery's behavior (Figure 3.14).

Impedance-based model:
Another approach to battery equivalent circuit modeling is the impedance model, which is based on electrochemical impedance spectroscopy. This technique is utilized to construct an AC impedance-based circuit representation of battery characteristics. Figure 3.15 illustrates a battery model derived from impedance spectroscopy. Impedance-based models provide valuable insights into the electrical behavior of batteries at different frequencies. However, these models are less intuitive compared to other circuit-based models [21]. It is important to note that impedance-based models are applicable only under fixed SoC and temperature conditions. Predicting the direct current (DC) response and runtime of a battery using impedance models can be challenging. Nonetheless, these models are instrumental in analyzing the frequency-dependent behavior of batteries and studying their impedance characteristics.

3.7 Various electric battery technologies

There are several different battery technologies available, each with its own characteristics, advantages, and limitations.

Advanced battery technology encompasses the utilization of cutting-edge techniques and materials to optimize the performance, efficiency, and longevity of batteries. It involves a range of research domains dedicated to creating batteries with improved energy density, extended lifespan, reduced cost, and the exploration of innovative charging methods like rapid and wireless charging. Several examples of advanced battery technologies include silicon and lithium–metal anodes, solid-state electrolytes, advanced designs for lithium-ion batteries, lithium–sulfur (Li–S) batteries, sodium-ion (Na-ion) batteries, redox flow batteries (RFBs), as well as zinc-ion, zinc–bromine, and zinc–air batteries.

Advanced batteries have gained significant importance in modern industries, finding numerous applications across various sectors. These batteries, with their enhanced capabilities and performance, have revolutionized multiple industries, including portable electronics, electric and hybrid vehicles, energy storage systems, medical devices, industrial equipment, and military applications.

Silicon anodes and lithium–metal anodes represent significant material advancements in the field of Li-ion batteries, offering the potential for substantially higher energy density. However, addressing concerns related to rate capability, safety, and cost is crucial alongside these advancements. The commercial utilization of silicon and lithium-metal anodes has been hindered by significant lifespan issues. Similarly, for Li–S batteries, achieving long-term durability remains an even greater challenge.

When considering alternatives to lithium-based chemistries, energy density is often compromised in favor of environmental sustainability, reduced capital or lifetime costs, enhanced rate capability, or longer cycle life. Ultimately, the selection of technology and chemistry depends on the specific requirements of a given application, balancing various performance characteristics [22]. There is a growing demand for longer duration storage solutions, particularly in stationary energy storage applications. This creates opportunities for innovative approaches such as redox flow batteries, which offer the advantage of scalable energy capacity while utilizing cost-effective and widely available active materials.

1. Li–S batteries:
    - Li–S batteries are a type of rechargeable battery that utilizes lithium as the positive electrode (cathode) and sulfur as the negative electrode (anode). They are considered one of the most promising next-generation battery technologies due to their high energy density and potential for significantly higher capacity than traditional Li-ion batteries.
    - The basic principle behind Li–S batteries involves the electrochemical reaction between lithium ions and sulfur. During discharge, lithium ions migrate from the positive electrode (cathode) through an electrolyte to the negative electrode (anode), while sulfur undergoes a series of redox reactions to form lithium sulfide. The reverse reaction occurs during the charging process.
    - However, there are several challenges associated with the practical implementation of Li–S batteries. One significant challenge is the inherent insulating nature of sulfur, which hampers its electrochemical utilization

and causes low electrical conductivity. This leads to issues such as poor cycling stability and limited overall battery performance.

- Overall, Li–S batteries hold great potential for revolutionizing energy storage technologies with their high energy density and low-cost materials. While there are technical challenges to overcome, ongoing research and advancements are bringing us closer to realizing the practical implementation of Li–S batteries in various applications, including electric vehicles, portable electronics, and grid energy storage.

2. Na-ion batteries:
   - Na-ion batteries are a type of rechargeable battery that utilizes sodium ions as the charge carriers instead of lithium ions, which are used in Li-ion batteries. Sodium, being more abundant and less expensive than lithium, makes Na-ion batteries an attractive alternative for large-scale energy storage applications.
   - During discharge, sodium ions move from the positive electrode (cathode) to the negative electrode (anode) through an electrolyte. This process involves the insertion and extraction of sodium ions into and from the anode material, respectively. The reverse process occurs during the charging phase.
   - One of the primary advantages of Na-ion batteries is their cost-effectiveness. The abundance of sodium and its lower cost compared to lithium make Na-ion batteries more economically viable for applications that require large-scale energy storage, such as renewable energy integration and grid-level energy storage.
   - However, Na-ion batteries face several challenges that need to be addressed for their widespread adoption. One of the main challenges is finding suitable electrode materials. Sodium ions are larger than lithium ions, and this size difference presents difficulties in finding materials that can efficiently store and release sodium ions during charge and discharge cycles.

3. Redox flow batteries:
   - Redox flow batteries (RFBs) are a type of rechargeable battery that stores energy in chemical compounds dissolved in liquid electrolytes. Unlike conventional batteries where energy is stored within the electrodes, RFBs store energy in external tanks containing electrolyte solutions.
   - During the charge–discharge cycle, the electrolyte is circulated through the electrochemical cell, and the electroactive species undergo oxidation and reduction reactions at the electrodes, releasing or absorbing electrons. The electricity generated or consumed is proportional to the flow rate and the concentration of the electroactive species [23–25].
   - One of the key advantages of RFBs is their ability to separate power and energy. The power output of an RFB can be easily adjusted by changing the size of the electrochemical cell and the flow rate of the electrolyte. This makes RFBs suitable for applications that require scalable and

flexible power delivery, such as grid energy storage and renewable energy integration.
- There are several types of RFBs, including the vanadium redox flow battery (VRFB), zinc–bromine flow battery (ZBB), and iron–chromium flow battery (ICFB). Each type has its own unique set of electrolyte chemistries and operating characteristics, which determine factors such as energy efficiency, power density, and cost.

4. Zinc–air batteries:
   - Zinc–air batteries are a type of electrochemical energy storage system that utilizes the oxidation and reduction reactions of zinc and oxygen to generate electricity. These batteries are known for their high-energy density and potential for long-duration energy storage.
   - The zinc–air battery consists of two main components: a zinc anode and an air cathode. The anode is made of zinc, which serves as the active material. The cathode is typically a porous material saturated with oxygen from the air, allowing the oxygen to react with the zinc during discharge.
   - During the discharge process, zinc atoms at the anode oxidize and release electrons, forming zinc ions (Zn^{2+}) in the electrolyte. Simultaneously, at the cathode, oxygen from the air combines with water and the electrons to form hydroxyl ions (OH^-). The zinc ions migrate through the electrolyte, while the hydroxyl ions migrate through a separate pathway called the ionic conductor. The migration of ions and the flow of electrons in the external circuit generate electrical energy.
   - One of the main advantages of zinc–air batteries is their high energy density. The energy density is primarily determined by the amount of zinc used as the anode material, making it possible to achieve a high energy-to-weight ratio. This makes zinc–air batteries attractive for applications that require lightweight and long-lasting energy sources, such as electric vehicles and portable electronic devices. However, zinc–air batteries do have some limitations. One significant challenge is the limited cycle life.

3.8 Selection of electric battery

The selection of an electric battery is a critical decision that depends on various factors, including the specific application requirements, performance criteria, safety considerations, and cost-effectiveness [26]. Here are some key aspects to consider when choosing an electric battery:

Energy and power requirements: Determine the energy and power demands of the application. Energy requirements relate to the amount of stored energy needed, while power requirements refer to the rate at which energy must be delivered. Different battery chemistries have varying energy and power densities, so selecting a battery that aligns with the specific energy and power requirements is crucial.

Voltage and capacity: Consider the desired voltage and capacity of the battery. The voltage requirement depends on the system or device being powered, while capacity relates to the amount of energy the battery can store. It is essential to choose a battery with the appropriate voltage and capacity to meet the application's needs.

Cycle life and durability: Evaluate the expected lifespan and durability requirements. Cycle life refers to the number of charge–discharge cycles a battery can undergo before its performance starts degrading. Select a battery with a cycle life that matches or exceeds the application's anticipated usage to ensure longevity and reliability.

Safety considerations: Safety is a critical factor, particularly for applications involving transportation or sensitive environments. Assess the safety features and considerations of the battery, such as thermal stability, overcharge and discharge protection, and robust containment to prevent leakage or thermal runaway [27].

Environmental impact: Consider the environmental impact of the battery chemistry and its disposal. Some battery chemistries may contain hazardous materials or require specific disposal methods. opt for batteries with minimal environmental impact and explore recycling options for end-of-life batteries.

Cost and efficiency: Evaluate the cost-effectiveness of the battery, taking into account the initial purchase cost, operational efficiency, and maintenance requirements. Consider the overall cost of ownership, including factors like energy efficiency, maintenance needs, and any additional system requirements.

Temperature range: Determine the operating temperature range required for the application. Some batteries may perform better at specific temperature ranges, while others may require thermal management systems to maintain optimal performance.

Integration and compatibility: Consider the compatibility and ease of integration of the battery with the application or system. Assess factors such as physical size, weight, and electrical interface to ensure seamless integration and efficient operation.

Available infrastructure: Assess the availability of charging or swapping infrastructure for rechargeable batteries, especially for applications like electric vehicles. Consider the accessibility and compatibility of charging stations or battery-swapping facilities.

Future developments: Stay informed about emerging battery technologies and advancements. Battery technology is evolving rapidly, so consider the potential for future improvements, such as increased energy density, faster charging, or enhanced safety features.

By carefully evaluating these factors and considering the specific requirements of the application, one can select an electric battery that provides optimal performance, reliability, safety, and cost-effectiveness.

3.9 Battery management system

Battery management involves utilizing various techniques, systems, and technologies to effectively monitor, control, and safeguard rechargeable batteries. It encompasses a diverse set of functions that focus on enhancing battery performance, ensuring safety, extending lifespan, and maximizing energy efficiency. Effective battery management is of utmost importance across a range of applications such as electric vehicles, renewable energy storage, portable electronics, and industrial systems [28]. It plays a critical role in ensuring reliable and efficient operation by efficiently managing and optimizing battery systems. An essential aspect of battery management is a well-designed BMS, which assumes a crucial role in monitoring key battery parameters like voltage, current, temperature, and SoC. Additionally, it implements control strategies to ensure cell balancing, prevent overcharging or over-discharging, and deliver vital insights regarding battery health and performance [29].

3.9.1 Cell balancing

Li-ion batteries are widely utilized in diverse applications ranging from electric vehicles to smart grid systems, owing to their favorable characteristics such as high-power density, high energy density, low self-discharge rate, compact structure, enhanced safety, and absence of memory effect. However, a single Li-ion battery cell can only provide a voltage range between 2.5 V and 4.2 V due to its electrochemical limitations. This voltage range falls short of meeting the high-voltage requirements of electric vehicles. As a result, a large number of Li-ion cells must be connected in series and parallel configurations to meet the desired battery voltage and power specifications.

In practice, variations in manufacturing and environmental factors can lead to discrepancies in the electrical properties of individual cells within a battery pack. This can result in voltage mismatches between cells during charging or discharging operations. These mismatches can lead to overcharging or depletion of specific cells, diminishing the usable battery capacity and overall lifespan. To address this issue and improve system performance, a reliable cell balancing system is necessary for the battery pack [30].

3.9.2 Types of cell balancing techniques

There are several different techniques used for cell balancing in BMSs. These techniques aim to equalize the voltages or SoC of individual cells within a battery pack.

(a) Passive cell balancing
Passive balancing is a technique used to equalize the voltage or SoC of cells in a battery pack. It relies on the use of resistors or other passive components to achieve this equalization. When a cell in the pack reaches a higher voltage or SoC than the others, excess energy is dissipated through the resistor. This dissipation allows the voltage or SoC of the cell to decrease, bringing it closer to the levels of the other cells.

(b) Active cell balancing
Active cell balancing is a technique that actively transfers energy between cells to achieve voltage or SoC equalization. It employs additional circuitry, control algorithms, and monitoring systems to facilitate the redistribution of energy. Compared to passive balancing, active cell balancing is more efficient as it involves energy transfer rather than dissipation. This approach offers enhanced flexibility and adaptability, enabling effective balancing across different operating conditions.

(c) Active cell balancing:
Figure 3.16 illustrates the active cell balancing topology, consisting of two stages of balancing circuits. In Figure 3.16(a), the second stage balancing circuit is depicted, which includes three cells, two inductors (L_1 and L_2), and four switches (S_1, S_2, S_3, and S_4) to form a single-cell equalization unit. On the other hand, the structure of the first stage resembles that of a buck-boost converter topology. The complete configuration of this two-stage balancing topology, where six cells are connected in series, comprises five inductors (L_1, L_2, L_3, L_4, and L_m) and ten switches (S_1, S_2, …, S_8, S_{m1}, and S_{m2}) along with the body diodes, as shown in Figure 3.13(b).

(d) Operating principle:
To illustrate the operational principle of this active cell balancing topology, a battery pack consisting of six cells is considered, where each module comprises three cells. This cell balancing topology incorporates two stages of equalization circuits: module balancing and cell balancing. Achieving SoC equalization is a

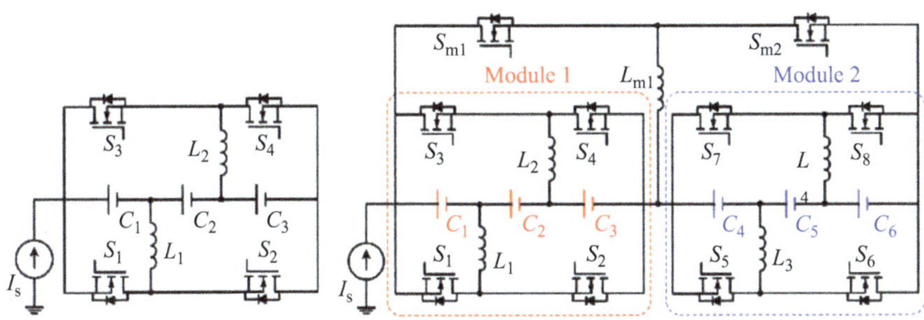

Figure 3.16 The active cell balancing circuit: (a) cell balancing circuit and (b) two-stage cell balancing circuit

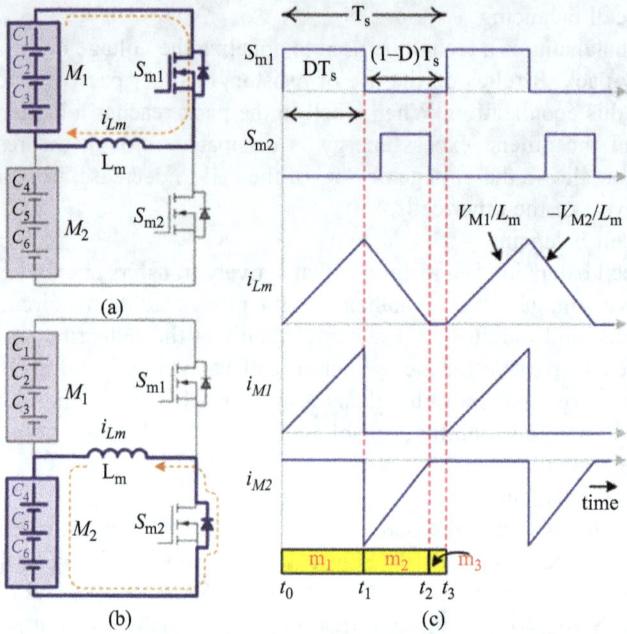

Figure 3.17 The operational principle of module balancing: (a) S_{m1} operating, (b) S_{m2} operating, and (c) key waveforms in the discontinuous current mode of operation

common technique for balancing a battery pack, but accurately determining the exact SoC of each cell presents significant challenges. However, the SoC is directly proportional to the battery's OCV. Therefore, the equalizer in this cell balancing topology determines the operating modes and switching patterns based on the voltage differences among the cells.

1. First stage (module balancing):
   The module balancing stage functions based on the operational principle of the buck-boost converter topology. The decision regarding the source and target modules is made by the module itself, taking into account the voltage levels. In Figure 3.17, it can be observed that the switches S_{m1} and S_{m2} are controlled to enable the transfer of energy from the source module to the target module. This energy transfer is achieved by monitoring the voltage levels of two adjacent battery modules [31].
2. Second stage (cell balancing):
   In contrast to module balancing, the cell balancing process occurs among the three cells within the second stage to achieve balance. The working principle of cell balancing is categorized into four distinct cases. Figure 3.18 illustrates the operational principle of cell balancing for these four cases.

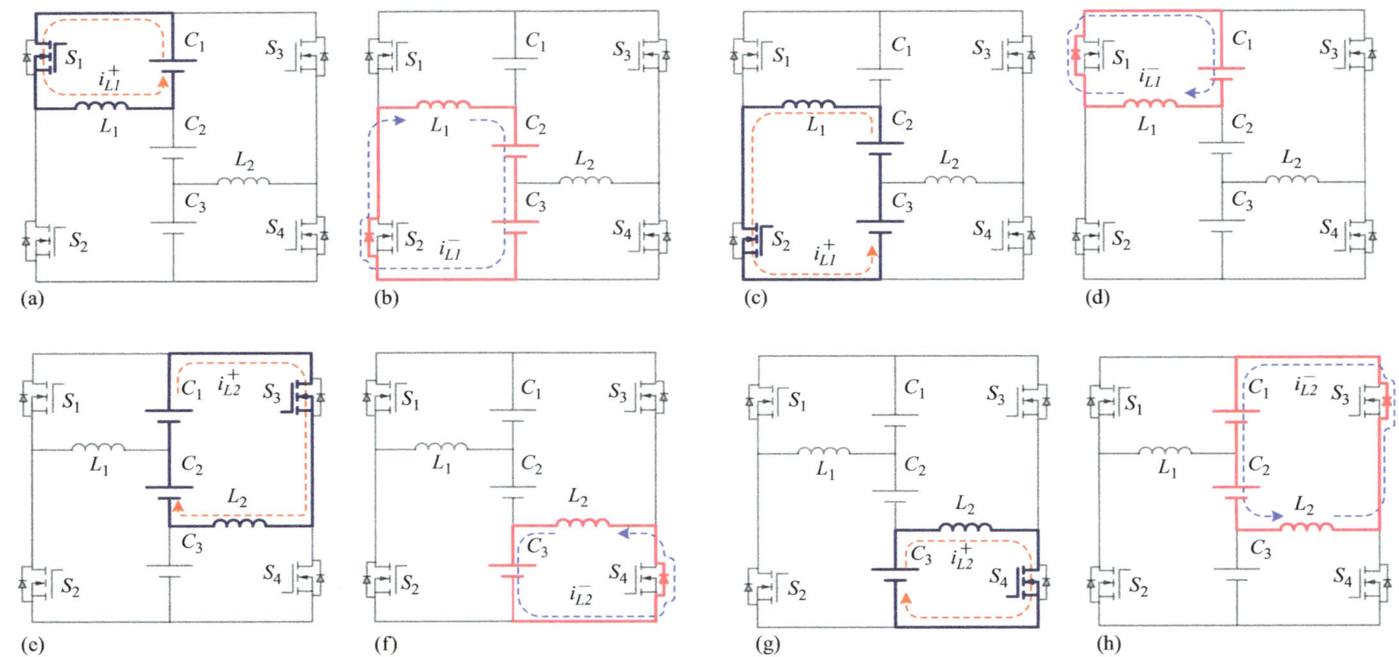

Figure 3.18 The operating principle of the cell balancing circuit

3.10 Summary

Various types of electric energy storage and storage devices with a special mention to battery types have been presented in this chapter for clarity of concepts and understanding their basic behavior. The reader shall be having required concepts for selection of appropriate battery which can be optimum for a particular EV application. The concept of BMS has also been introduced at the end of the chapter and some special techniques have been discussed for basic understanding. The detailed concepts and methodologies for design and selection of appropriate battery management system have been discussed in Chapter 7.

Symbols

V	measured battery voltage
Li^+	lithium ions
e^-	electrons
Li*x*C	lithium-intercalated carbons
H2	hydrogen
H^-	hydrogen ions
MPa	Megapascals
Ah	ampere-hours of a battery
mAh	milli-ampere hours of a battery
OCV	open circuit voltage
$\int idt$	integral of the current over time
***V*max**	*V*max is the maximum voltage observed at a known high SoC
***V*min**	*V*min is the minimum voltage observed at a known low SoC
***f* (OCV)**	*f*(OCV) represents a lookup table or mathematical model that relates the open circuit voltage (OCV) of the battery to the corresponding SoC
***Q*T**	battery's total capacity
***E*0**	cell initial voltage
R_Ω	resistance the battery
***k*1**	constant
vd	denotes the voltage drop across the RC circuit
qd	diffusion charge
qs	stored charge
Cs	voltage across the capacitor
Rsd	self-discharge resistor
L	inductor
S	MOSFET switch

Glossary

Battery capacity	"Battery capacity" is a measure (typically in Amp–h) of the charge stored by the battery, and is determined by the mass of active material contained in the battery.
C rate	C rate is defined as the charge/discharge current divided by the nominally rated battery capacity.
Depth of discharge (DoD)	A battery's DoD indicates the percentage of the battery that has been discharged relative to the overall capacity of the battery.
Electrochemical impedance spectroscopy (EIS)	EIS is one of the most important electrochemical techniques where the impedance in a circuit is measured by ohms (as resistance unit).
Electrolyte	It is the battery component that transfers ions charge-carrying particles back and forth between the battery's two electrodes, causing the battery to charge and discharge.
Fuel cell vehicles (FCEVs)	FCEVs are powered by hydrogen. They are more efficient than conventional internal combustion engine vehicles.
Hybrid electric vehicles (HEVs)	HEVs are powered by an internal combustion engine and one or more electric motors, which uses energy stored in batteries.
Open circuit voltage (OCV)	OCV is the difference of electrical potential between positive and negative terminals of a battery without any load connected.
Power density	Power density is the amount of power (time rate of energy transfer) per unit volume.
Proton exchange membrane fuel cell (PEMFC)	PEM fuel cells, also called proton exchange membrane fuel cells, use a proton-conducting polymer membrane as the electrolyte. Hydrogen is typically used as the fuel.
Phosphoric acid fuel cells (PAFCs)	PAFCs are a type of fuel cell that uses liquid phosphoric acid as an electrolyte.
Remaining useful life (RUL)	RUL is the difference between the total number of charge–discharge cycles when the actual capacity of the battery drops to the threshold value and the number of charge–discharge cycles of the current battery.
Specific energy	The gravimetric energy density or the specific energy of a battery is a measure of how much energy a battery contains in comparison to its weight, and is typically expressed in Watt–hours/kilogram (W–h/kg).

Solid oxide fuel cells (SOFC) An electrochemical conversion device that produces electricity directly from oxidizing a fuel.

State of charge (SoC) SoC of a cell denotes the capacity that is currently available as a function of the rated capacity.

State of health (SoH) SoH is the capability of the battery to retain charge now compared to its rated value.

Uninterrupted power supply (UPS) A device that allows a computer to keep running for at least a short time when incoming power is interrupted.

References

[1] Z. Salameh, *Renewable Energy System Design*, Academic Press, New York, 2014.

[2] N. Ghaeminezhad, Q. Ouyang, X. Hu, G. Xu, and Z. Wang, "Active cell equalization topologies analysis for battery packs: a systematic review," *IEEE Trans. Power Electron.*, vol. 36, no. 8, pp. 9119–9135, 2021.

[3] J. Gallardo Lozano, E. Romero Cadaval, M.I. Milanes Moero, and M.A. Guerrero-Martinez, "Battery equalization active methods", *J. Power Sources*, vol. 246, pp. 934–949, 2014.

[4] A.M. Imtiaz and F.H. Khan, "Time shared fly-back converter based regenerative cell balancing technique for series connected Li-ion battery strings," *IEEE Trans. Power Electron.*, vol. 28, no. 12, pp. 5960–5975, 2013.

[5] K.M. Lee, S.-W. Lee, Y.G. Choi, and B. Kang, "Active balancing of Li-ion battery cells using transformer as energy carrier," *IEEE Trans. Ind. Electron.*, vol. 64, no. 2, pp. 1251–1257, 2017.

[6] C. Lim, K. Lee, N. Ku, D. Hyun, and R. Kim, "A modularized equalization method based on magnetizing energy for a series-connected lithium-ion battery string," *IEEE Trans. Power Electron.*, vol. 29, no. 4, pp. 1791–1799, 2014.

[7] X. Yang, L. Xi, Z. Gao, Y. Li, and J. Wen, "Analysis and design of a voltage equalizer based on boost full-bridge inverter and symmetrical voltage multiplier for series-connected batteries," *IEEE Trans. Veh. Technol.*, vol. 69, no. 4, pp. 3828–3840, 2020.

[8] Y. Chen, X. Liu, Y. Cui, J. Zou, and S. Yang, "A multi winding transformer cell-to-cell active equalization method for lithium-ion batteries with reduced number of driving circuits," *IEEE Trans. Power Electron.*, vol. 31, no. 7, pp. 4916–4929, 2016.

[9] K.M. Lee, Y.C. Chung, C.H. Sung, and B. Kang, "Active cell balancing of Li-ion batteries using LC series resonant circuit," *IEEE Trans. Ind. Electron.*, vol. 62, no. 9, pp. 5491–5501, 2015.

[10] S. Yarlagadda, T.T. Hartley, and I. Husain, "A battery management system using an active charge equalization technique based on a DC/DC converter topology," *IEEE Trans. Ind. Appl.*, vol. 49, no. 6, pp. 2720–2729, 2013.

[11] M. Kim, C. Kim, J. Kim, and G. Moon, "A chain structure of switched capacitor for improved cell balancing speed of lithium-ion batteries," *IEEE Trans. Ind. Electron.*, vol. 61, no. 8, pp. 3989–3999, 2014.

[12] S. Park, K. Park, H. Kim, G. Moon, and M. Youn, "Single-magnetic cell-to-cell charge equalization converter with reduced number of transformer windings," *IEEE Trans. Power Electron.*, vol. 27, no. 6, pp. 2900–2911, 2012.

[13] M. Kim, J. Kim, and G. Moon, "Center-cell concentration structure of a cell-to-cell balancing circuit with a reduced number of switches," *IEEE Trans. Power Electron.*, vol. 29, no. 10, pp. 5285–5297, 2014.

[14] T.H. Phung, A. Collet, and J. Crebier, "An optimized topology for next-to-next balancing of series-connected lithium-ion cells," *IEEE Trans. Power Electron.*, vol. 29, no. 9, pp. 4603–4613, 2014.

[15] Y. Yuanmao, K.W.E. Cheng, and Y.P.B. Yeung, "Zero-current switching switched-capacitor zero-voltage-gap automatic equalization system for series battery string," *IEEE Trans. Power Electron.*, vol. 27, no. 7, pp. 3234–3242, 2012.

[16] F. Baronti, G. Fantechi, R. Roncella, and R. Saletti, "High-efficiency digitally controlled charge equalizer for series-connected cells based on switching converter and super-capacitor," *IEEE Trans. Ind. Inform.*, vol. 9, no. 2, pp. 1139–1147, 2013.

[17] T. Wilberforce, A. Alaswad, A. Palumbo, M. Dassisti, and A.G. Olabi, "Advances in stationary and portable fuel cell applications," *Int. J. Hydrogen Energy.*, vol. 41, no. 37, p. 16509e16522, 2016.

[18] D. Stolten and B. Emonts, *Fuel Cell Science and Engineering, 2 Volume Set: Materials, Processes, Systems and Technology*, Wiley, New York, 2012.

[19] A. Shanian and O. Savadogo, "TOPSIS multiple-criteria decision support analysis for material selection of metallic bipolar plates for polymer electrolyte fuel cell," *J. Power Sources*, vol. 159, no. 2, p. 1095e1104, 2006.

[20] L. Lu, X. Han, J. Li, J. Hua, and M. Ouyang, "A review on the key issues for lithium-ion battery management in electric vehicles," *J. Power Sources* vol. 226, pp. 272–288, 2013, https://doi.org/10.1016/j.jpowsour.2012.10.060.

[21] C. Speltino, A. Stefanopoulou, and G. Fiengo, "Cell equalization in battery stacks through State of Charge estimation polling," *Presented at the American Control Conf.*, Baltimore, MD, June 30–July 2, 2010.

[22] J.S. Goud and K.R.B. Singh, "An online method of estimating state of health of a Li-ion battery," *IEEE Trans. Energy Convers.*, vol. 36, no. 1, pp. 111–119, 2021, doi:10.1109/TEC.2020.3008937.

[23] J.S. Goud, R. Kalpana, B. Singh, and S. Kumar, "A global maximum power point tracking technique of partially shaded photovoltaic systems for constant voltage applications," *IEEE Trans. Sustain. Energy*, vol. 10, no. 4, pp. 1950–1959, 2019, doi:10.1109/TSTE.2018.2876756.

[24] L.H. Saw, A.A.O. Tay and L.W. Zhang, "Thermal management of lithium-ion battery pack with liquid cooling," *2015 31st Thermal Measurement, Modeling and Management Symposium (SEMI-THERM)*, San Jose, CA, USA, 2015 pp. 298–302, doi:10.1109/SEMI-THERM.2015.7100176.

[25] C.C. Pascual and P.T. Krein, "Switched capacitor system for automatic series battery equalization," Paper presented at *12th IEEE Applied Power Electronics Conference and Exposition (APEC)*, Atlanta, GA, 27–27 February 1997.

[26] X. Chen, W.X. Shen, T.T. Vo, Z. Cao, and A. Kapoor, "An overview of lithium ion batteries for electric vehicles," in *10th International Power and Energy Conference (IPEC)*, 2012, pp. 230–235.

[27] X. Han, M. Ouyang, L. Lu, J. Li, Y. Zheng, and Z. Li, "A comparative study of commercial lithium ion battery cycle life in electrical vehicle: aging mechanism identification," *J. Power Sources*, vol. 251, pp. 38–54, 2014.

[28] A.S. Subburaj and S.B. Bayne, "Analysis of dual polarization battery model for grid applications," in *Proceedings of the IEEE 36th International Telecommunications Energy Conference*, Vancouver, BC, Canada, 28 September–2 October 2014, pp. 46–53.

[29] G.H. Min and J.I. Ha, "Active cell balancing algorithm for serially connected Li-ion batteries based on power to energy ratio," in *Proceedings of the IEEE Energy Conversion Congress and Exposition (ECCE)*, Cincinnati, OH, USA, 1–5 October 2017, pp. 1550–1558.

[30] Z. Xi, M. Dahmardeh, B. Xia, Y. Fu, and C. Mi, "Learning of battery model bias for effective state of charge estimation of lithium-ion batteries," *IEEE Trans. Veh. Technol.*, vol. 68, pp. 8613–8628, 2019.

[31] K. Manjunath, R. Kalpana, B. Singh, and Kiran R, "A two-stage module based cell-to-cell active balancing circuit for series connected lithium-ion battery packs," *IEEE Trans. Energy Convers.*, vol. 38, no. 4, pp. 2282–2297, 2023, doi:10.1109/TEC.2023.3283424.

Chapter 4

Power electronic essentials in electric vehicle

Siddhant Gudhe[1] *and Sanjeev Singh*[1]

4.1 Power electronic circuits in electric vehicles

Presently available electric vehicles (EVs) such as hybrid-EVs (HEVs), plug-in hybrid EVs (PHEV), fuel cell EVs (FCVs), and battery EVs (BEVs) require power electronics and control systems to operate the electrical machines, i.e., the traction motor as per need on road.

The battery EVs (BEVs) or all EVs (AEVs) have a battery as the only power source for all needs. Since the present market is flooded by internal combustion engine (ICE)-based vehicles, therefore, the survival of these EVs depends on two factors. One is the initial cost of the vehicle as compared to the ICE-based vehicles and the other is its range of running in one charge. Both these factors are dependent on the availability, economics, and efficiency of the components used in the EV.

To drive the EV, an electric motor is used which may require the desired power with a good quality to deliver the desired torque and speed at good efficiency. There are a variety of electric motors available for use in EVs but all of them require conversion and conditioning of electric power to deliver the desired output. In this process, the role of power electronics becomes very important, and many power-electronic components are required between the source, i.e., the battery and the load. i.e., the electric motor [1,2].

A battery EV or all EVs essentially have a motor, a battery, and power electronic circuitry for effective and efficient control. An electric motor (usually AC type, either induction motor or permanent magnet motor) is used for the propulsion of the vehicle which receives power from an onboard source of electricity, typically a rechargeable battery or fuel cell [3,4]. For obtaining better efficiency in some EVs, ultracapacitors are employed to store the energy received during regenerative braking. This energy is used during the acceleration of EVs for overtaking or hill climbing. The EV may or may not have the gear mechanism in it for power transmission. Even differential axles may also be absent in some designs. Therefore, power electronic circuits for control of all such systems are essential in the EV, which can be broadly categorized into three groups as machine control unit

[1]Electrical Engineering Department, MANIT Bhopal, India

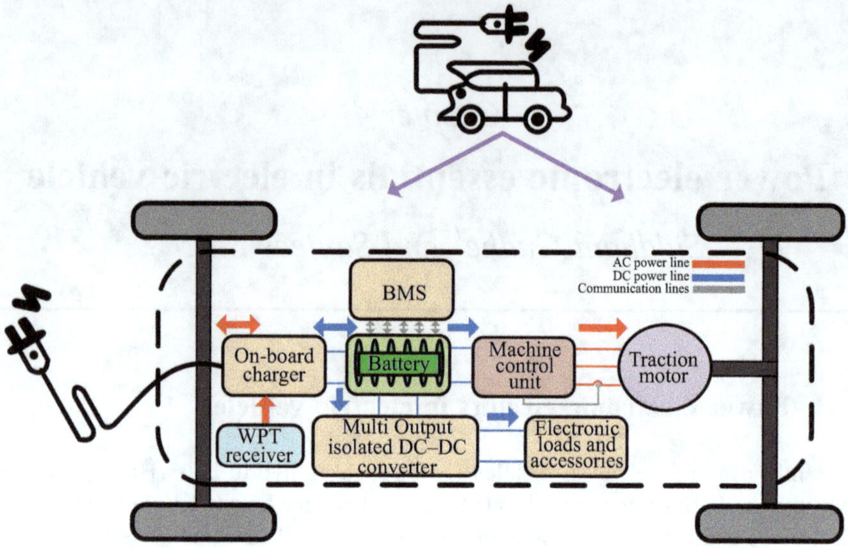

Figure 4.1 Layout of a typical battery EV

(MCU), switched mode power supply (SMPS), and battery management system (BMS).

The layout of a typical battery EV is shown in Figure 4.1 with its components. It has an electric drive system consisting of a power electronic converter as MCU to drive the traction motor. The traction motor (electric motor) transfers the power to the wheels for propulsion. To operate various accessories of the vehicle (ACs, fans, lights, power windows, audio system, power steering, horns, power braking, etc.), a multi-voltage power supply is required which is catered by an SMPS.

To look after the cell health, charging/discharging cycles, and damaged cells in a battery, a BMS is essentially employed in EVs. These power electronic essentials in an EV are discussed in the following sub-sections for a basic understanding of the required power electronic components and circuits for EVs.

4.2 MCU

Present EVs are mostly driven by induction motors, permanent magnet brushless (PMBL) DC motors, or permanent magnet synchronous motors (PMSMs). There are many other motors being researched as a traction motor for EVs such as switched reluctance motor (SRM), synchronous reluctance motor (SyRM), axial flux, and magnetless motors [5,6]. These motors require a three-phase AC supply to run but the EV uses a DC source, i.e., the battery. Therefore, a power electronic converter known as a voltage source converter (VSC) is essentially required to run these motors.

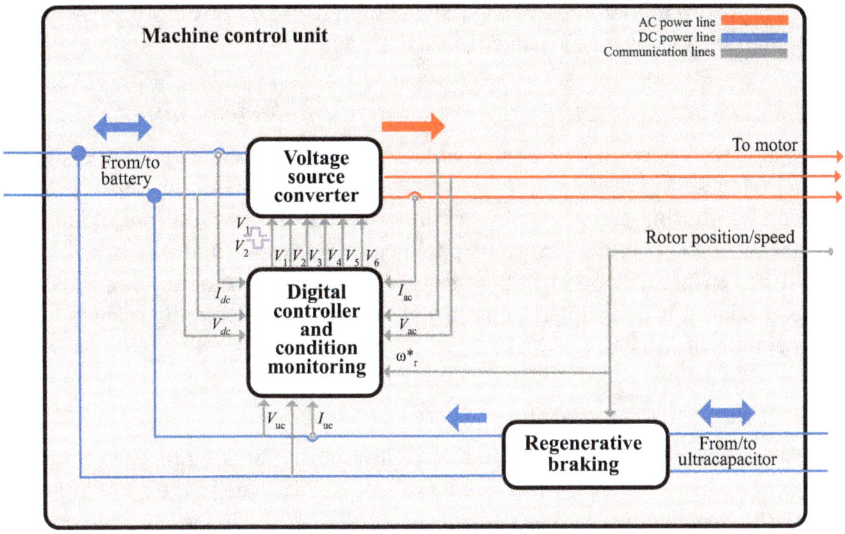

Figure 4.2 Schematic block diagram for the MCU

Moreover, variable speed and load (number of passengers) of the vehicle during its use require precise and efficient control of current and voltage supplied to the motor. There are many transient modes of operation for an EV such as sudden braking, city driving with frequent braking and acceleration, hill climbing, and constant speed cruising on highways [6]. Therefore, the motor control unit (MCU) is designed to have multi-functionality as per the requirement of the EV. The schematic block diagram for the MCU is shown in Figure 4.2. Various components of the MCU used in an EV are discussed below.

4.2.1 VSC

The EV is capable of delivering the power to the electric motor in accordance with the driver's accelerator pedal. To achieve this, the MCU uses a three-phase VSC, consisting of either MOSFETs or IGBTs as power switches, operated on field-oriented control (FOC) algorithm or some other advanced control techniques.

The MCU draws the power from the battery and converts it to the desired three-phase AC supply for the traction motor using the feedback control algorithm. The motor's mechanical and electrical parameters are sensed and feedback to the controller for modification in the switching pulses for the VSC as per the desired speed and torque from EV.

The three-phase VSC used for EVs can have various topologies with two-level and multi-level (more than two-level) voltage outputs [7]. The control algorithms can also be numerous depending on the type of traction motor (induction or PM), the requirement of control parameters, and their accuracies. There is scope for low-cost controllers, fault-tolerant VSC topologies along with improved efficiency converters incorporating zero voltage switching (ZVS) or zero current switching (ZCS), and SiC or advanced material-based switches.

4.2.2 Condition monitoring and control of electric motor

The traction motor for an EV is selected to have high efficiency and power density by design, but during operation, its effective and efficient operation depends on the condition monitoring and preventive maintenance of the traction motor and associated system [8]. Therefore, various parameters are sensed and monitored by the controller to ensure the good health of the EV system. In case of any abnormalities, the motor control is interrupted through additional power electronic circuits for the safe operation of the EV.

4.2.3 Regenerative braking of traction motor

The EV traction motor uses regenerative control during braking to have fast braking with recovery of energy which would otherwise dissipate in the form of heat in brakes. This regenerative braking uses a specific DC–DC converter circuit coupled with a rectifier to store the braking energy in a fast storage device "ultracapacitor" [9–12]. This energy is used during the time of overtaking or hill climbing as a top-up with battery.

4.3 DC–DC converter

The EV requires an additional DC–DC converter to drive various accessories of the EV such as lights, horns, HVAC unit, fans, power windows, audio system, power steering, power braking, and wipers. This DC–DC converter uses battery voltage available at 48 V (for a small low-powered EV) or 800 V (for buses or trucks) and converts to 12 V/24 V for the accessories of the EV. The electrical connection of these accessories is generally connected to the chassis ground which floats with respect to the high-voltage battery pack ground. Hence, an isolated DC–DC converter topology with high-frequency transformer galvanic isolation and multiple secondary windings known as a multi-output DC–DC converter is used [13].

4.3.1 Multi-output DC–DC converter

A multi-output DC–DC converter consists of a high-frequency transformer (HFT) with a single winding for DC–DC converter topology on the input side and multiple windings as secondaries on the output side of the transformer. The schematic block diagram for a multi-output DC–DC converter with galvanic isolation is shown in Figure 4.3.

The control circuit closely tracks the voltage of only one winding and switches the DC–DC converter on the input side for precise control of output voltages in each winding. The accuracy of control depends on the design of the transformer as well as the controller [14]. There are other options reported in the literature as multi-source converters for EV applications. The multi-source converter uses the concept of various batteries combined with other types of energy storage units such as ultracapacitors and fuel cells.

4.3.2 Multi-source converter

In an EV, the supply of power to the traction motor at high DC voltage from the battery eliminates the use of DC–DC boost converter, while reducing conversion stages, thereby improving efficiency. But this shall require a DC–DC converter for supplying the auxiliary load of the EV. Therefore, the use of a single multi-source converter where multiple sources at various voltages are connected through this converter to support all auxiliary loads in EV from another battery or energy storage unit with low voltage. This converter also facilitates the transfer of charge from one battery to the other [13–16]. The schematic block diagram for a dual source converter is shown in Figure 4.4.

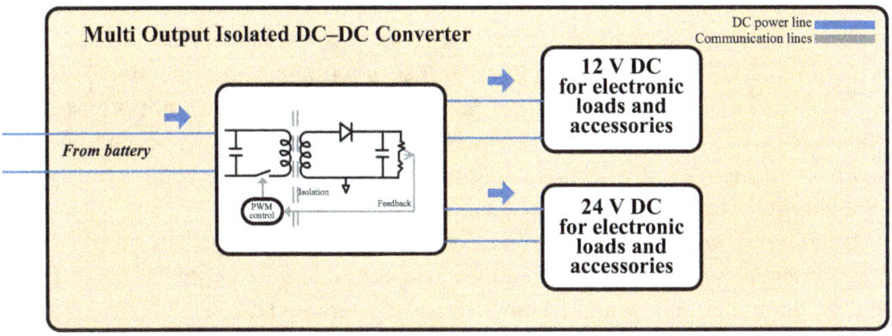

Figure 4.3 Schematic block diagram for the multi-output DC–DC converter

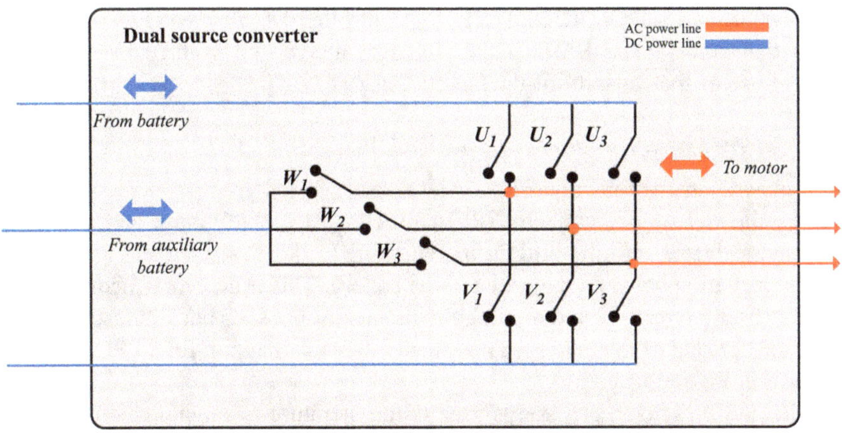

Figure 4.4 Schematic block diagram for the dual-source converter

The scheme presented in Figure 4.4 shows the use of two batteries which can be charged from the single power source or from each other in either direction. The operation of this converter is controlled for the flow of power from the source or batteries as per the need for charge in the respective batteries. This concept is being explored for more than two sources for power transfer in either direction for any number of energy storage units or sources.

4.4 BMS

The battery of any EV needs charging again and again after running the EV to a certain distance depending on the capacity of the battery and loading of the EV. Based on the location of chargers, EVs are classified into onboard chargers, off-board chargers, and integrated chargers. Onboard chargers are located on the vehicle itself and the charger's power electronic parts are included in the design of the vehicle. The onboard charger increases the overall weight of the vehicle and raises the overall cost of the vehicle. Due to weight and space limits on the vehicle, the power rating is constrained. The EV with an onboard charger can be charged at any outlet that is in residential garages or places with ground protection [12].

Off-board chargers are unique pieces of equipment that are not a part of the vehicle; they remain outside the vehicle. The EV battery may be charged at a specific charging station with fast charging capability, while the vehicle must travel to the charger's location to recharge the battery.

Usually, off-board chargers and onboard chargers are rectifiers supplied from three-phase and single-phase AC sources, respectively, with controlled voltage and currents. Fast chargers are off-board chargers, supplied from a three-phase AC source as they require a higher amount of power for fast charging but controlled voltage and current as per battery capacity.

The BMS of an onboard charger has a power management unit that consists of a rectifier with a power factor correction (PFC), an inrush current limiter, protection devices, and a DC–DC converter for controlled DC link voltage and current fed to the battery. The BMS essentially uses active cell voltage balancing for efficient use of the capacity of the battery pack [17–21].

4.4.1 Power factor correction (PFC)

The onboard charger in any BMS uses a rectifier fed from a single-phase AC supply followed by power factor (PF) correction at the AC input and control of harmonics injected into the grid during charging [12]. These chargers also require galvanic isolation between the grid and the battery. This is accomplished by a DC–DC converter having isolated topology with buck or boost or buck–boost operation as per the demand of the battery. The schematic block diagram for a PFC converter is shown in Figure 4.5.

The power factor correction is used in the literature as a synonymous term for power quality converter. The power quality converter is operated for the improvement of various power quality parameters such as PF, crest factor (CF), and

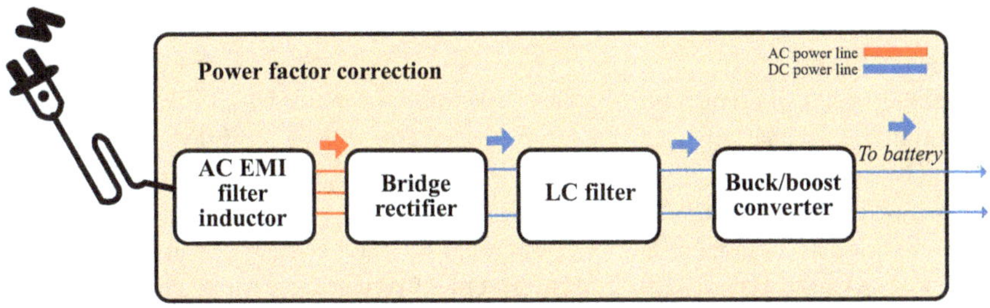

Figure 4.5 Schematic block diagram for the PFC converter

total harmonic distortion (THD) of current drawn. A detailed discussion on PFC and power quality control of battery chargers is included in Chapter 10.

4.4.2 Cell balancing system

The electrical properties of different cells in a battery pack may have variations due to manufacturing as well as operational environment variations. This leads to terminal voltage variations between the cells during charging and/or discharging of the battery pack. This may lead to overcharging or depletion of specific cells, which will reduce usable battery capacity and lifetime. Thus, the battery pack must be provided with a cell balancing system to reduce the cell voltage variations and to enhance the performance of the battery pack [17–21].

There are two techniques of cell balancing: passive and active. A resistance is used in parallel to each cell in passive balancing techniques to dissipate extra energy of the overcharged cell as heat. This is the main drawback of the passive techniques. The passive cell balancing technique is mainly employed in low-power applications due to its low cost and easy control.

On the other hand, the active cell balancing technique uses the transfer of charge between cells to balance the voltage through switches (IGBTs or MOSFETs) and a combination of inductors, transformers, and capacitors. The details of BMS topologies, control, and implementation are discussed in Chapter 9.

4.5 Other applications

The use of power electronics in an EV is not restricted to the above-discussed applications only. There are many other applications such as wireless charging, MPPT converter for fuel cell vehicles, solar PV-based chargers, and bidirectional charterers have come up, which require controlled voltage, current, and power, power electronics shall be used inevitably. The detailed discussions on wireless charging of batteries, bidirectional power flow between the battery and the grid known as vehicle-to-grid (V2G) and grid-to-vehicle (G2V) with extended concepts of vehicle-to-anything (V2X) are discussed in further chapters of this book.

4.6 Summary

The power electronic circuits that are essentially required in an EV have been presented in this chapter with the fundamental concept of the presently used systems. The major essential systems are MCU, battery charging unit (wireless or onboard or off board), DC–DC converter for auxiliary equipment, and BMS. With the technological advancements in EVs and associated systems, many applications shall evolve in the future with a requirement of power electronic converters. The basic concept of these applications shall remain the same as discussed in this chapter. Therefore, all such applications may be understood or explained on the basis of the discussions presented in this chapter.

Symbols

V_{dc}	DC voltage
I_{dc}	DC current
V_{ac}	AC voltage
I_{ac}	AC current
ω^*_r	reference speed of motor in rad/s
$V_1 \ldots V_6$	gating voltage signals
U_1, U_2, U_3	upper switches in different phases of dual-source converter
V_1, V_2, V_3	lower switches in different phases of dual-source converter
W_1, W_2, W_3	switches in three phases for auxiliary battery of dual-source converter

Glossary

BMS	battery management system
PF	power factor
CF	crest factor
THD	total harmonic distortion
PQ	power quality

References

[1] C.C. Chan and K.T. Chau, "An overview of power electronics in electric vehicles," *IEEE Trans. Ind. Electron.*, vol. 44, no 1, pp. 3–13, 1997.

[2] K.T. Chau, *Electric Vehicle Machines and Drives – Design, Analysis and Application*, Singapore: John Wiley & Sons, 2015.

[3] S. Verma, S. Mishra, A. Gaur, *et al.*, "A comprehensive review on energy storage in hybrid electric vehicle," *J. Traffic Transp. Eng. (English Edition)*, vol. 8, no. 5, pp. 621–637, 2021.

[4] T.D. Atmaja and Amin, "Energy storage system using battery and ultra-capacitor on mobile charging station for electric vehicle," *Energy Proc.*, vol. 68, pp. 429–437, 2015.

[5] F. Un-Noor, S. Padmanaban, L. Mihet-Popa, M.N. Mollah, and E. Hossain, "A comprehensive study of key electric vehicle (EV) components, technologies, challenges, impacts, and future direction of development," *Energies*, vol. 10, no. 8, 71pp., 2017.

[6] P.S. Jamwal, *Multilevel Inverter fed Induction Motor Drive for Battery Electric Vehicle*, *PhD Thesis*, SLIET Longowal, India, July 2023.

[7] A. Poorfakhraei, M. Narimani, and A. Emadi, "A review of multilevel inverter topologies in electric vehicles current status and future trends," *IEEE Open J. Power Electron.*, vol. 2, pp. 155–170, 2021.

[8] Motor Control Reference Guide, ST Microelectronics, pp. 1–76, November 2022.

[9] M.R. Khalid, I.A. Khan, S. Hameed, M.S.J. Asghar, and J.S. Ro, "A comprehensive review on structural topologies, power levels, energy storage systems, and standards for electric vehicle charging stations and their impacts on grid," *IEEE Access*, vol. 9, pp. 128069–128094, 2021.

[10] Y. Tahir, I. Khan, S. Rahman, *et al.*, "A state-of-the-art review on topologies and control techniques of solid-state transformers for electric vehicle extreme fast charging," *IET Power Electron.*, vol. 14, no. 9, pp. 1560–1576, 2021.

[11] C. Capasso, S. Riviera, S. Kouro, and O. Veneri, "Charging architectures integrated with distributed energy resources for sustainable mobility," *Energy Proc.*, vol. 105, pp. 2317–2322, 2017.

[12] R. Dwivedi, *Investigations on Fast Charging Converter for Electric Vehicle Application*, PhD Thesis, SLIET Longowal, India, May 2023.

[13] S. Gudhe, S. Singh, M. Rezkallah, and A. Chandra, "Dynamic control of traction motor for EV fed via dual source inverter with a two battery system," *Energies*, vol. 16, p. 1754, 2023.

[14] Ned Mohan, T.M. Undeland, and W.P. Robbins, *Power Electronics: Converters Applications and Design*, 3rd ed. *(An Indian Adaptation)*, Wiley, 2022.

[15] S. Gudhe and S. Singh, "Single stage multiple source bidirectional converter for electric vehicles," In *Flexible Electronics for Electric Vehicles; Lecture Notes in Electrical Engineering*; Singapore: Springer, 2022, vol. 852, pp. 567–574.

[16] S. Gudhe and S. Singh, "Charging of multiple batteries using single-stage multi-source converter with bidirectional power flow," In *Recent Advances in Power Electronics and Drives; Lecture Notes in Electrical Engineering*; Singapore: Springer, 2022, vol. 863, pp. 207–216.

[17] H.R. Eichi, U. Ojha, F. Baronti, and M.Y. Chow, "Battery management system: an overview of its application in the smart grid and electric vehicles," *IEEE Ind. Electron. Mag.*, vol. 7, no. 2, pp. 4–15, 2013.
[18] L. Lu, X. Han, J. Li, J. Hua, and M. Ouyang, "A review on the key issues for lithium-ion battery management in electric vehicles," *J. Power Sour.*, vol. 226, pp. 272–288, 2013.
[19] M.A. Hannan, M.M. Hoque, A. Hussain, Y. Yusof, and P.J. Ker, "State-of-the-art and energy management system of lithium-ion batteries in electric vehicle applications: issues and recommendations," *IEEE Access*, vol. 6, pp. 19362–19378, 2018.
[20] J. Lu, Y. Wang, and X. Li, "Isolated bidirectional DC–DC converter with quasi-resonant zero-voltage switching for battery charge equalization," *IEEE Trans. Power Electron.*, vol. 34, no. 5, pp. 4388–4406, 2019.
[21] D. Roosevelt, *Battery Management System for Li-Ion Batteries for Electric Vehicle Application*, M. Tech. Thesis, MANIT Bhopal, India, April 2023.

Chapter 5

Design, modeling, simulation, and control of electric vehicle

Peng Guan[1]

5.1 Introduction

The history of electric vehicles (EVs) can be traced back to the mid-nineteenth century when EVs were first introduced, as toy-like small-scale cars. The first EV is widely considered to be the electric carriage built by the Hungarian inventor Ányos Jedlik in 1828. It was powered by a small electric motor and used non-rechargeable primary cells. However, the first practical EV was likely built by the Scottish inventor Robert Anderson in the 1830s. Anderson's electric carriage was powered by a rechargeable battery and could reach speeds of up to 4 mph [1]. Throughout the nineteenth and early twentieth centuries, EVs were used primarily for short-distance travel, such as commuting to work or running errands. They were popular because they were quiet, had no emissions, and required less maintenance than gasoline-powered vehicles. However, the development of the internal combustion engine and the availability of cheap gasoline eventually led to the decline of EVs.

During the late 1960s to early 1970s, soaring oil prices and gasoline shortages created a growing interest in lowering the US's dependence on foreign oil and finding homegrown sources of fuel. Congress took note and passed the Electric and Hybrid Vehicle Research, Development, and Demonstration Act of 1976, authorizing the Energy Department to support research and development in electric and hybrid vehicles.

Around this time, many automakers began exploring options for alternative power sources for vehicles, including using electricity. However, vehicles developed and produced during this time suffered from a huge drawback compared to gasoline-powered ones, as they have limited performance, top speeds are below 45 mph, and the range is below 50 miles. Also, the advanced internal combustion engine technology and mass production made gasoline-powered cars more affordable and convenient.

[1]Department of Mechanical, Aerospace, and Biomedical Engineering, University of Tennessee, USA

In recent years, EVs have experienced a resurgence in popularity due to concerns about air pollution and climate change. The passage of the 1990 Clean Air Act Amendment and the 1992 Energy Policy Act—plus new transportation emissions regulations issued by the California Air Resources Board—helped create a renewed interest in EVs in the United States. Many automakers are now producing EVs, and they are becoming more affordable and available, resulting significant increase in the number of EVs on the market, as well as a growing infrastructure of charging stations. Today, EVs are being developed and manufactured by several major automakers and are becoming more widely available to consumers.

5.2 EV modeling

Vehicle modeling is the process of creating a mathematical representation of vehicle behavior and performance. By using vehicle modeling, engineers can optimize vehicle control systems, also simulate, and predict vehicle performance in different scenarios. Vehicle models can be categorized by fidelity levels: low-fidelity vehicle models, which are based on vehicle properties, and high-fidelity vehicle models, which are based on vehicle design parameters like hardpoints, bushing rates, and spring and damper rates. Recently, most automotive OEMs are shifting from developing internal combustion engine vehicles to electrical vehicles; however, the modeling and simulation methods have not changed much, with only more emphasis on aerodynamics and control due to different critical aspects between traditional internal combustion engine vehicles and electrical vehicles.

Nowadays, computer software is widely used to create vehicle models and simulate vehicle performance under various conditions. There are many different aspects of EV modeling and simulation, including:

Control systems modeling: This involves modeling the control systems of a vehicle, such as the steering and braking systems, and simulating their response to different inputs. Here are some key aspects of control systems modeling in EVs:

- Powertrain control: Control systems modeling is used to develop and optimize the control algorithms for the EV's powertrain, which includes the electric motor, battery, and power electronics. By simulating the powertrain control system, designers can evaluate factors such as torque distribution, energy management strategies, regenerative braking, and thermal management. This modeling approach helps optimize powertrain efficiency, range, and performance.
- Vehicle dynamics control: Control systems modeling enables the development of algorithms for vehicle dynamics control, including traction control, stability control, and anti-lock braking systems (ABS). By simulating the vehicle's dynamic behavior, designers can evaluate control strategies that enhance vehicle stability, improve handling characteristics, and ensure safe and predictable performance in various driving conditions.
- Energy management system: EVs require sophisticated energy management systems to optimize the use of electrical energy from the battery, manage power flow, and control charging and discharging processes. Control systems

modeling helps analyze and optimize energy management algorithms, including power allocation between the motor and other vehicle systems, charging strategies, and energy regeneration. This modeling approach ensures efficient utilization of energy and extends the vehicle's range.

- Safety systems: Control systems modeling is used to develop and evaluate safety systems in EVs, such as collision avoidance, adaptive cruise control, and emergency braking. By simulating the behavior of these systems, designers can assess their effectiveness, response times, and integration with other vehicle control functions. This modeling approach helps enhance the safety and reliability of EVs by identifying potential issues and optimizing control algorithms.
- Cybersecurity and fault diagnosis: Control systems modeling plays a role in assessing cybersecurity measures and developing algorithms for fault diagnosis and detection in EVs. By simulating potential cyber threats and system faults, designers can evaluate the robustness of control systems, develop countermeasures, and optimize fault diagnosis algorithms. This modeling approach helps ensure the security and reliability of EV control systems.
- Hardware-in-the-loop (HIL) testing: Control systems modeling allows for HIL testing, where the virtual control algorithms are tested in real time using physical hardware components. HIL testing provides a realistic simulation environment to validate control algorithms, evaluate their performance, and ensure proper integration with the actual vehicle systems.

Overall, control systems modeling in EVs enables the development and optimization of control algorithms for powertrain, vehicle dynamics, energy management, safety systems, and fault diagnosis. By leveraging these modeling techniques, designers can optimize the efficiency, performance, range, safety, and reliability of EVs.

Thermal modeling: This involves modeling the temperature distribution within a vehicle and its components, as well as the heat transfer between different parts. Here are some key aspects of thermal modeling in EVs:

- Battery thermal management: Battery temperature greatly affects the performance, efficiency, and lifespan of EV batteries. Thermal modeling helps assess heat generation, heat transfer, and temperature distribution within the battery pack. By simulating different operating conditions and thermal management strategies, designers can optimize cooling systems, assess the effectiveness of cooling fluids, and ensure that the batteries operate within the desired temperature range.
- Power electronics cooling: Power electronics, including inverters, motor controllers, and DC–DC converters, generate significant heat during operation. Thermal modeling enables designers to analyze the heat dissipation, temperature profiles, and cooling requirements of these components. By optimizing cooling systems, such as heat sinks, fans, or liquid cooling, designers can ensure efficient heat removal and prevent overheating, which can degrade the performance and reliability of power electronics.

- Electric motor cooling: Electric motors in EVs also require effective cooling to manage the heat generated during operation. Thermal modeling helps evaluate the cooling airflow, temperature distribution, and heat transfer within the motor. By optimizing motor cooling systems, such as cooling jackets or oil-based cooling, designers can maintain optimal motor temperature, prevent overheating, and ensure motor efficiency and longevity.
- Cabin thermal comfort: Thermal modeling is used to assess the thermal comfort of the vehicle cabin. By simulating the airflow, temperature distribution, and heat transfer within the cabin, designers can optimize the heating, ventilation, and air conditioning (HVAC) system. This modeling approach helps ensure efficient heating and cooling, even battery-powered heating and cooling while maintaining passenger comfort and minimizing energy consumption.
- Thermal runaway analysis: Thermal modeling assists in analyzing the risk of thermal runaway events in EVs. By simulating abusive conditions, such as short circuits or extreme temperatures, designers can assess the thermal behavior of the battery and identify potential safety risks. This modeling approach helps optimize safety features, such as thermal insulation, cooling channels, or thermal barriers, to prevent and mitigate thermal runaway events.
- Integration and co-simulation: Thermal modeling is often performed in conjunction with other simulations, such as electrical, mechanical, or control system simulations. Integrated modeling and co-simulation enable a comprehensive analysis of the EV's thermal behavior, considering the interactions between various components and systems. This approach allows for more accurate predictions and optimization of the overall thermal management system.

Overall, thermal modeling in EVs helps optimize thermal management, prevent overheating, enhance performance, extend component lifespan, ensure passenger comfort, and improve safety. By leveraging advanced simulation tools and techniques, designers can make informed decisions to achieve efficient thermal control and maximize the overall performance and reliability of EVs.

Structural modeling: This involves modeling the structural behavior of a vehicle and its components, such as the frame and body, under different loads and conditions. Here are some key aspects of structural modeling in EVs:

- Finite element analysis (FEA): FEA is a widely used technique for structural modeling in EVs. It involves discretizing the vehicle structure into finite elements and analyzing the stresses, strains, and deformations under different loading scenarios. FEA helps identify areas of high-stress concentration, evaluate the strength and stiffness of components, and optimize the design to meet structural requirements.
- Crashworthiness and safety: Structural modeling is crucial for evaluating the crashworthiness and safety of EVs. By subjecting the virtual model to crash simulations, designers can assess the behavior of the structure during frontal, side, or rear impacts. FEA allows for the analysis of energy absorption, deformation patterns, and occupant safety. It helps optimize the design of crumple zones, impact-absorbing structures, and reinforcements to enhance crashworthiness and protect occupants.

- Lightweighting and material optimization: EVs benefit from lightweight structures to maximize energy efficiency and range. Structural modeling enables the evaluation of different materials, such as high-strength steels, aluminum alloys, composites, or hybrid materials. By analyzing the weight, strength, and stiffness characteristics of various materials, designers can optimize the vehicle's structural composition to achieve a balance between weight reduction and structural integrity.
- Durability and fatigue analysis: Structural modeling helps assess the durability and fatigue life of EV components subjected to repeated loading cycles. By simulating dynamic loads, such as road vibrations, pothole impacts, or driving maneuvers, FEA can predict the fatigue life of critical components, such as suspension parts, chassis structures, and battery mounts. This allows for optimization of the design, selection of appropriate materials, and identification of potential failure points.
- Noise, vibration, and harshness (NVH) and structural resonance: Structural modeling also plays a role in evaluating the NVH characteristics of EVs. By analyzing the dynamic behavior of the structure and its response to various excitation sources, designers can identify resonant frequencies, modes of vibration, and potential NVH issues. This information helps optimize the design, select appropriate damping materials, and minimize unwanted vibrations and noise transmission.
- Virtual testing and validation: Structural modeling enables virtual testing and validation of EV designs, reducing the need for extensive physical prototyping and testing. By simulating real-world operating conditions and load cases, designers can evaluate structural performance, make design improvements, and ensure compliance with safety and regulatory requirements.

Overall, structural modeling in EVs helps optimize the design, evaluate crashworthiness, enhance safety, reduce weight, and ensure the structural integrity and durability of the vehicle. By leveraging advanced FEA techniques, designers can make informed decisions, reduce development time, and improve the overall performance and reliability of EVs.

NVH modeling: This involves modeling processing to simulate and analyze the NVH characteristics of automotive vehicles. NVH stands for noise, vibration, and harshness, which are important factors in vehicle design and performance. NVH modeling helps engineers understand and predict the behavior of a vehicle's NVH attributes under various operating conditions in the early design stage without a real prototype being produced. Here are some aspects of NVH modeling in electrical vehicles:

- Electric motor noise modeling: Electric motors in EVs can generate noise due to electromagnetic forces, mechanical vibrations, and aerodynamic effects. Modeling techniques such as FEA and boundary element methods (BEM) can be employed to simulate and analyze the structural dynamics and acoustic behavior of the motor. This helps identify potential noise sources, optimize motor design, and evaluate the effectiveness of noise mitigation measures.

- Battery system noise modeling: EV batteries can contribute to the overall noise levels in the vehicle. Battery packs contain cooling systems, fans, and pumps, which generate noise during operation. Computational fluid dynamics (CFD) simulations can be used to analyze the airflow and acoustic behavior within the battery system. By modeling and optimizing the design of the cooling system, noise generated by the battery pack can be minimized.
- Structural dynamics modeling: EVs have unique structural characteristics compared to internal combustion engine vehicles. The absence of a conventional engine and the presence of heavy battery systems can affect the overall vehicle dynamics and vibration behavior. Finite element modeling (FEM) techniques can be utilized to simulate the structural dynamics of the vehicle and identify potential NVH issues related to vibration, resonance, and structural integrity.
- Cabin noise modeling: Cabin noise is a critical aspect of NVH in any vehicle, including EVs. Electric powertrains produce less noise compared to combustion engines, but other noise sources, such as wind, road, and tire noise, become more noticeable. NVH modeling techniques, such as boundary element methods (BEM) or statistical energy analysis (SEA), can help analyze the transmission paths of noise into the cabin and optimize insulation, damping materials, and sealing to achieve desired noise levels.
- Virtual prototyping and testing: NVH modeling allows for virtual prototyping and testing of EV designs, reducing the need for physical prototypes and costly testing. By simulating different configurations and scenarios, designers can assess the impact of various design choices on NVH performance and make informed decisions to optimize the overall vehicle refinement.

Ultimately, NVH modeling in EVs helps engineers understand and address NVH issues during the development stage. By leveraging advanced simulation tools, designers can improve the overall driving experience, enhance comfort, and meet customer expectations for quiet and refined EVs.

Multibody dynamics (MBD) modeling: This involves modeling the interactions between different components of a vehicle, such as the suspension and tires, and simulating their behavior as a system. Here's how MBD modeling is applied in the context of EVs:

- Suspension analysis: MBD modeling enables the analysis of the EV's suspension system, including control arms, springs, dampers, and anti-roll bars. By simulating the motion and forces in the suspension system, designers can assess factors such as ride comfort, handling characteristics, and stability. This modeling approach helps optimize suspension geometry, evaluate different suspension setups, and validate the performance of active suspension systems in EVs.
- Chassis and body dynamics: MBD modeling allows for the analysis of the overall behavior of the EV's chassis and body structure. By considering the interaction between the chassis, body panels, and suspension system, designers can assess factors such as structural integrity, vehicle stiffness, and vibrations. This modeling approach helps optimize the chassis design, evaluate the impact of weight distribution, and ensure proper rigidity and durability of the vehicle.

- Drivetrain and powertrain analysis: MBD modeling helps analyze the behavior of the EV's drivetrain and powertrain components, including the electric motor, transmission, and differential. By simulating the interaction between these components, designers can evaluate factors such as torque distribution, power delivery, and driveline vibrations. This modeling approach helps optimize the drivetrain design, assess the impact of different motor configurations, and validate the performance of regenerative braking systems.
- Vehicle dynamics and handling: MBD modeling allows for the analysis of the EV's overall dynamic behavior, including vehicle handling, stability, and maneuverability. By simulating the vehicle's response to steering inputs, road disturbances, and different driving conditions, designers can assess factors such as cornering performance, stability control systems, and yaw dynamics. This modeling approach helps optimize the vehicle's weight distribution, evaluate the impact of different suspension settings, and ensure safe and predictable handling characteristics.
- Virtual prototyping and testing: MBD modeling enables virtual prototyping and testing of EV designs, reducing the need for physical prototypes and costly testing. By simulating different scenarios and configurations, designers can evaluate the impact of design choices on vehicle dynamics, performance, and safety. This approach helps identify potential issues early in the development process and enables iterative design improvements to achieve desired dynamic characteristics.

In summary, MBD modeling in EVs helps optimize suspension systems, assess chassis and body dynamics, analyze drivetrain behavior, and evaluate overall vehicle dynamics and handling. By leveraging these modeling techniques, designers can make informed decisions to enhance ride comfort, improve handling characteristics, and ensure the safety and performance of EVs.

Aerodynamics modeling: This is very important for EVs, as aerodynamics can affect range significantly, due to the cost, weight, and space limitation, the number of battery cells that can be installed on EVs is limited, aerodynamics performance of EVs can be optimized using these models to maximize the range. Here are some key elements and techniques involved in aerodynamics modeling for EVs:

- Computational fluid dynamics (CFD): CFD is a powerful tool used to simulate and analyze the flow of air around a vehicle. By creating a virtual model of the EV and defining its surroundings, CFD software can calculate airflow patterns, pressure distribution, and drag coefficients. This allows designers to optimize the vehicle's shape, surface contours, and aerodynamic features to minimize drag and improve efficiency.
- Drag reduction: Drag is a significant factor affecting EV range and energy consumption. Aerodynamic modeling helps identify areas of high drag, such as sharp edges, turbulent flow regions, or poorly designed body features. By analyzing CFD results, designers can make iterative improvements to the vehicle's shape, including smoothing body contours, optimizing front and rear-end designs, and minimizing frontal area. These modifications help reduce drag and improve overall efficiency.

- Underbody aerodynamics: The underbody of an EV is often complex, with battery packs, electric motors, and other components affecting airflow. CFD modeling assists in analyzing the underbody airflow patterns and optimizing the design to minimize turbulence, reduce lift, and enhance overall stability. The addition of underbody panels, diffusers, or air curtains can help manage airflow and improve the vehicle's aerodynamic performance.
- Cooling system optimization: Efficient thermal management is crucial in EVs to maintain optimal operating temperatures for batteries, power electronics, and electric motors. Aerodynamic modeling helps design effective cooling systems by assessing the airflow through heat exchangers, radiators, and air ducts. By optimizing the position, size, and geometry of these components, CFD analysis can ensure proper cooling while minimizing aerodynamic losses.
- Wind noise and windshield wiper aerodynamics: In EVs, where traditional engine noise is reduced, wind noise can become more noticeable. Aerodynamic modeling helps analyze and reduce wind noise by optimizing the design of side mirrors, door seals, window frames, and other components that can generate noise-inducing turbulence. Additionally, CFD simulations can assess the aerodynamic performance of windshield wipers, ensuring they do not create excessive drag or lift.
- Virtual wind tunnel testing: Aerodynamics modeling allows for virtual wind tunnel testing, reducing the need for physical prototypes and expensive testing. By simulating different driving conditions, including various speeds, yaw angles, and crosswinds, designers can evaluate the vehicle's stability, lift, and drag characteristics. This enables them to make informed design decisions early in the development process, optimizing aerodynamic performance and improving overall efficiency.

Overall, aerodynamics modeling in EVs helps optimize vehicle efficiency, range, and performance by reducing drag, managing airflow, and enhancing cooling systems. By leveraging CFD simulations and virtual testing, designers can achieve improved aerodynamic performance without the need for extensive physical prototyping and testing.

Electromagnetic interference (EMI) modeling: This is a very essential aspect of electrical vehicle design and development since electrical vehicles rely on electrical and electronic systems, including power electronics, motor controllers, charging systems, and various communication systems. All of these systems generate electromagnetic fields that can potentially interfere with each other and external devices, which can lead to performance degradation and safety issues. To model EMI in EVs, several approaches and techniques can be employed. Here are some commonly used methods:

- Circuit-level modeling: This approach involves creating detailed circuit models of the EV's electrical and electronic systems. These models take into account the characteristics of the components, such as resistors, capacitors, inductors, and semiconductor devices. By simulating the interactions between these components, designers can assess potential EMI issues and identify measures to mitigate them.

- Electromagnetic field simulation: In this method, electromagnetic field simulation software, such as FEA or finite difference time domain (FDTD), is used to analyze the propagation of electromagnetic fields within the EV and its surroundings. This helps identify areas where EMI could occur and allows for optimization of the vehicle's layout and shielding to reduce interference.
- Coupling path analysis: Coupling paths refer to the mechanisms through which electromagnetic energy is transferred from one part of the EV to another. By analyzing these paths, designers can identify potential sources of interference and susceptible components or subsystems. Common coupling paths include conducted emissions through power and signal cables, radiated emissions through antennas or harnesses, and capacitive or inductive coupling through common conductive structures.
- EMC standards and regulations: Compliance with electromagnetic compatibility (EMC) standards and regulations is crucial for EV manufacturers. These standards define acceptable levels of EMI and provide guidelines for testing and measurement. Modeling EMI can help ensure that EV designs meet these requirements and facilitate the certification process.

Overall, EMI modeling in EV development helps optimize the design, reduce interference issues, and ensure compliance with EMC standards, leading to improved performance, reliability, and safety of EVs.

5.3 Critical aspects of EV design

To illustrate the critical aspects of EV design, we need to step back and check the differences between ICE vehicles and EVs. From the names, we know that ICE vehicles ignite and combust fuel within an internal combustion engine (ICE). EVs are powered by electricity from a rechargeable battery instead. Distinct differences are: ICE vehicles need complex gear systems, whereas EVs only need one gear system which makes it relatively simple. The weight of fuel is very low and the fuel tank only takes relatively little space in ICE vehicles, whereas the weight of batteries in EVs is very high and they take up a lot of space in EVs. Also, for ICE vehicles, there are ample refilling infrastructures across the world and the refilling time is very quick which is usually less than 5 min depending on the fuel tank size, whereas, for EVs, there are not many charging infrastructures available, and the charging time is relatively long, usually around 0.5 hours with fast charging technology, 8 h without it. The noises generated from ICE vehicles are mainly from the ICE, and other noises are immersed by the engine noises, whereas the EVs are much quieter, however, other noises are more easily captured by the driver and passengers.

Since most of the EV customers have experience driving ICE vehicles, therefore, they expect EVs to have similar or better performance than ICE vehicles, for example, EVs should have roughly 300 miles range, and NVH performance should be comfortable for both driver and passengers. The vehicle should be safe and easy to handle, the ride should be comfortable, and the charging time should not be too long compared to ICE vehicles.

Therefore, due to these distinct differences between EVs and ICE vehicles, and the customers' expectations, the critical aspects of EV design can be summarized:

1. **Battery pack**
   - Most modern EVs use a lithium-ion battery pack to store energy and power all accessories of the EV. However, lithium-ion batteries are expensive, and their capacity needs to be properly designed together with vehicle curb weight, and aerodynamic efficiency to make sure EVs can perform similarly to ICE vehicles.
   - The onboard battery management is also critical since it manages the power usage for all accessories like AC, stereo, lighting, etc. It will affect the longevity of the battery
   - Though the chances are low, batteries have the potential to overheat and catch fires. Therefore, for the safety of drivers and passengers of EVs, the battery needs to be designed carefully.
   - Extreme temperatures affect the charging and discharging of the batteries. For example, customers expect EVs will have consistent range performance during the year, they do not want EVs to drop the range in winter and summer.
   - All EVs sold today include a battery warranty of at least eight years and 100,000 miles, given the cost of replacing a battery pack, no OEM wants to be stuck with the bill due to the fact they overestimated the battery's life span and resiliency.

2. **Aerodynamics performance**
   - Aerodynamics is the way air moves around things, for example, as a vehicle moves through the air, it pushes aside air molecules, the force that prevents the vehicle from moving through is called air resistance force, and it can be calculated as (5.1):

$$F_{air\ resistance} = \frac{\rho C_d A}{2} v^2 \tag{5.1}$$

   Fair resistance = is the force due to air resistance drag (N); v is the velocity of the vehicle (m/s); ρ is the density of the air that the vehicle is moving through (kg/m^3); C_d is the drag coefficient (unitless); A is the area of the vehicle the air presses on (m^2).

   From the equation, aerodynamics is crucially important for EVs, as the air resistance forces increase quadratically with the vehicle speed. So getting aerodynamics right is paramount to increasing the range and reducing weight and also cost. Also from (5.1), the C_d drag coefficient which is determined by the shape affects air resistance. Since EVs are commodities, therefore, they should have their looks, but from an aerodynamics point of view, the most efficient look is pretty much the same look, so it would be a balancing act by designers and engineers.

3. **NVH performance**
   - The NVH performance of an EV is a key design and development consideration, as a noisy drivetrain system will be perceived as poor quality

and tonal noise is especially unwanted. In an ICE vehicle, a certain amount of tonal noises may be masked by engine noises. Whereas in an EV, there will be minimum masking noise due to the replacement of ICE by e-motors. Therefore, the EV NVH design challenges may be greater than ICE vehicles [2].

4. **EV safety**
   - Vehicle safety is always one of the most important key factors in vehicle performance. It refers to the measures that are taken to ensure the safety of passengers and drivers while operating a vehicle. Besides all the active and passive safety technology that ICE vehicles have, EVs have some special safety considerations:
     - The EV voltage is far higher than the safety voltage the human body can take, therefore, all wirings need to be insulated securely, they should stand for all different kinds of friction, collision, extrusion, or corrosion.
     - Battery explosion danger due to mechanical shock, high temperature, or accidents. EVs cause more damage in collisions than ICE vehicles, partly due to their incredible acceleration ability, and also due to the battery safety issues. A battery pack in an EV typically consists of thousands of cells connected in parallel or series to increase the power density requirement. This significantly increases the energy stored in the batteries which leads to a severe safety issue when accidents happen. Additionally, the batteries in an EV always face harsh working conditions, and they may overcharge, overheat, short circuit, vibration, shock, collision, or nail penetration conditions [3]. If not designed properly, EVs can be very dangerous after a collision, not only to the drivers in EVs but also to the other parties like pedestrians, for example, EVs can be burst into blames after collisions with unsafe designed batteries. Therefore, a battery health-monitoring system is also critical to EVs.

5.4 Tools and techniques for modeling and simulation of EVs

There are many different tools and software packages available for vehicle modeling and simulation, including commercial software packages like ADAMS, SIMPACK, and CarSim, as well as open-source tools like OpenSim and VehicleSim.

Starting from the 1990s, the richer version of the simulation—digital twins concept is introduced [4]. A digital twin is a virtual representation of an object or system that spans its lifecycle, is updated from real-time data, and uses simulation, machine learning, and reasoning to help decision-making [5]. Nowadays, more and more OEMs are adopting or planning to implement digital twins in their development cycle, as digital twin includes a richer virtual environment for different simulation scenarios. By using digital twins, OEMs can be more efficient in engineering, getting a better understanding of what happens to the product, avoiding potential problems, saving costs, and correcting errors earlier in the development cycle [6].

To create an EV model, similar to creating ICE vehicle models, all the design parameters of each subsystem need to be achieved. Usually for OEMs, these can be obtained from suppliers, for example, controller suppliers can provide the controller model, and tyre suppliers can provide tyre models, and so on. These models can be exported through a functional mock-up interface (FMI). FMI is a free standard that defines a container and an interface to exchange dynamic simulation models using a combination of XML files, binaries, and C code, distributed as a ZIP file. By using FMI, different models can be integrated for simulation. Depending on the fidelity requirement of the model, model complexity will vary.

To use the modern simulation software, a detailed CAD model needs to be created first, then if a FEM is used, the CAD needs to be properly meshed. The FEM of each component can treat and integrated into an MBD model for system simulation. If properly treated, they can also be used in real-time applications to increase model accuracy.

Once the MBD model is created and verified, the controllers can be verified on hardware in the loop system (HILS) in real-time, we will discuss these techniques and tools in later sections in this section.

5.4.1 Aerodynamics

Aerodynamics is especially important for EVs as it directly affects the efficiency of the EV. OEMs spend millions of dollars on wind tunnels and simulation software to squeeze out the last bits of aerodynamic performance for EVs. EV drivetrains feature specific packaging needs, but they are more flexible than ICE vehicles since e-motors are much smaller than engines and batteries are often placed on the floor. This enables new shapes that are inherently much more aerodynamic.

Before simulation is used in the automotive industry, aerodynamic performance is evaluated by wind tunnel tests, which are large tubes with air blowing through vehicles to replicate the interaction between air and the vehicle moving through the air, as shown in Figure 5.1. However, wind tunnels are very expensive to do as a high-end wind tunnel usually costs over 100 million dollars to build, or more than 5,000 dollars per hour to rent [7], and this procedure takes a long time since it involves building prototypes, scheduling test facilities, personnel, etc.

Figure 5.1 Typical wind tunnel diagram

To reduce development costs and facilitate the process, virtual wind tunnels are used through using computational fluid dynamics (CFD). Engineers can estimate the vehicle's aerodynamic performance via CFD simulation in the early design stage before a prototype is built. After several iterations in simulation, the design can be optimized, or even using optimizers, a physical prototype can be built, and perform final tests in the wind tunnel for verification purposes. Nowadays, some OEMs are even trying "zero prototypes" development, once proven effective, may be physical prototypes can be reduced to zero or minimum.

Nowadays, there is a lot of commercial CFD software that can help engineers do from modeling to simulation for aerodynamic performance:

1. PowerFLOW CFD is a solution from Dassault Systems that uses the Lattice Boltzmann method to predict real-world conditions. PowerFLOW imports fully complex model geometry and accurately and efficiently performs aerodynamic, aeroacoustics, and thermal management simulations.
2. STAR-CCM+ is a multiphysics CFD software solution from Siemens that includes everything from CAD, automated meshing, multiphysics CFD, sophisticated postprocessing, and design exploration. This allows engineers to efficiently explore the entire design space to make better design decisions faster.
3. Ansys Fluids is a CFD solution from Ansys, it reduces simulation solve time and power consumption using a multi-GPU solver in FLUENT and automates processes that improve the product's performance and safety.
4. OpenFOAM is a leading open-source software for CFD maintained by CFD Direct, the acronym OpenFOAM stands for Open Source Field Operation and Manipulation. OpenFOAM constitutes a C++ CFD toolbox for customized numerical solvers (over 60 of them) that can perform simulations of basic CFD, combustion, turbulence modeling, electromagnetics, heat transfer, multiphase flow, stress analysis, and even financial mathematics modeled by the Black–Scholes equation [8]. Now, with the open-source libraries in OpenFOAM, one does not have to spend one's whole career writing CFD codes or be forced to buy commercial software. Many other users of OpenFOAM have developed relevant libraries and solvers that are either posted online or may be requested for free (illustration Figure 5.2).

5.4.2 Finite element

The finite element method (FEM) is a method used to simulate structurally with strength visualizations, production and weight determination, proper management of materials and costs, and numerical predict how a part or assembly behaves under certain conditions FEA [10]. The FEM was first invented in 1956 for stress analysis on airplane frames and then started to be used in all kinds of different engineering solving. For FEA, to obtain parameters in the equation and calculation process, the drawing geometry (CAD) needs to be done first, and then material and boundary conditions need to be applied. Before running the calculation, the geometry needs to be meshed, which means all the nodes are connected to some other nodes and no nodal point is independent of each other (illustration Figure 5.3).

Figure 5.2 Aerodynamics simulation in OpenFOAM [9]

Figure 5.3 Vehicle body in white (BIW) with mesh [11]

The FEM usually is used in stress analysis, durability analysis, and NVH analysis.

- Stress analysis evaluates vehicle components under different working conditions to predict whether the component will fail or not. Usually, it applied static loads to mimic the real loads to the component.
- Durability analysis evaluates failures under repeated loads. By using fatigue algorithms or software, like FE-safe, engineers can predict the life span of certain parts under normal or extreme conditions.
- NVH analysis is defined as the study of the noise and vibration characteristics of vehicles, especially in EVs, NVH is especially important since there are fewer environmental sounds to blend noises. Interior NVH deals with noise and vibration experienced by cabin occupants like drivers and passengers, whereas exterior NVH is largely related to vehicle-emitted noise and vibration.

Nowadays, there is a lot of FE software to help solve complex FEMs, the most famous software includes the following: Abaqus, Nastran, LS-DYNA, Ansys, etc.

5.4.3 MBD

MBD have grown in the past decades to become an important analysis tool in vehicle development. MBD is the science of studying the motion of complex mechanical systems under the application of mechanical forces. With MBD simulation, engineers can judge the vehicle at the system level and reduce development time and cost. Traditionally, MBD simulation is more to simulate the interaction between rigid components however, with the computation power increases, MBD simulation can now solve complex systems with flexible bodies in the system.

EVs have similar suspension designs but with different drivetrains, and nowadays with advanced driving assist systems being used always in every EV, MBD is used to help controller design and verification.

MBD simulation software can help engineers to create high-fidelity vehicle models. With all the design parameters like hardpoints, mass and inertia properties, and spring and damper rate, EV MBD models can be created in weeks in the simulation software environment. This model then can be used in different simulation scenarios, even control prototyping in the real-time environment [12]. Such MBD software includes Simpack, Adams, VL-Motion, Dymola, etc. Since MBD software can be run in real-time, some OEMs are using it on driver-in-loop simulators (DILS) for vehicle subjective evaluation (illustration Figure 5.4) [13].

5.4.4 EV control simulation and verification

Control simulation and verification are important for EVs as these vehicles are controlled by wire. Not many mechanical valves like ICEs have. To speed up the verification process and reduce the physical prototyping, software in loop (SIL) and hardware in loop (HIL) are used.

Figure 5.4 VI-grade DiM250 DYNAMIC driving simulator [14]

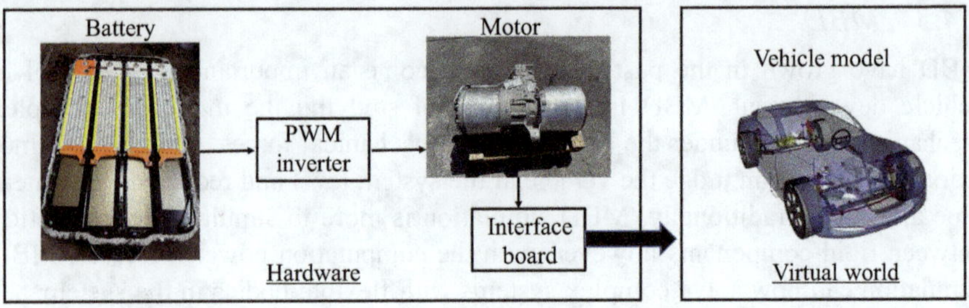

Figure 5.5 HILS testing structure

SIL is a method of testing and validating code in a simulation environment to quickly and cost-effectively catch bugs and improve the quality of the code. Typically, SIL testing is conducted in the early stages of the software development process, while the more complex, costlier HIL testing is done in later stages.

Each new software program has thousands of requirements, and it is not practical to perform manual testing to make sure the software does what it is supposed to do. It is prohibitively expensive and time-consuming to physically load software under development into an actual vehicle and test-drive it for the potentially hundreds of thousands of miles needed to make sure the software works in all types of driving conditions. Also, engineers develop new codes every day, these must be continuously tested.

SIL has many features that make it advantageous for testing in the automotive industry:

- SIL simulations can be run on any standard desktop computer without requiring the special equipment or test benches needed for HIL testing. This makes it cost-effective to deploy SIL testing across many instances, which reduces testing bottlenecks and speeds up the development process.
- Because the simulation is being performed entirely in software, the testing can go faster than it would in real time.
- Simulation programs deliver flexibility and repeatability creating a more effective feedback loop with software developers [15].

HIL testing is a technique where real signals from a controller are connected to a test system that simulates reality, tricking the controller into thinking it is in the assembled product. Test and design iteration takes place as though the real-world system is being used. You can easily run through thousands of possible scenarios to properly exercise your controller without the cost and time associated with actual physical tests (illustration Figure 5.5) [16].

5.5 EV motor control

5.5.1 Control modules

Electric motors in EVs are controlled through a combination of hardware and software systems. The control system ensures the motor operates efficiently, delivers the desired torque and power, and responds to driver inputs. Motors are controlled by multiple control modules in EVs, including:

Motor control unit (MCU): The motor control unit, also known as the inverter or motor controller, is a key component responsible for controlling the operation of the electric motor. It typically consists of power electronics, such as insulated gate bipolar transistors (IGBTs), gate drivers, and sensors. The MCU receives input signals from various sources, including the driver's commands, vehicle sensors, and a battery management system.

Torque control: One of the primary objectives of motor control in EVs is to regulate the torque output of the electric motor. The MCU adjusts the current supplied to the motor windings based on the driver's pedal position or other inputs. This torque control ensures that the motor delivers the desired acceleration, deceleration, or regenerative braking based on the driving conditions.

Field-oriented control (FOC): FOC is a common technique used in motor control for EVs. FOC aligns the stator magnetic field with the rotor magnetic field, allowing for precise control of torque and speed. By controlling the amplitude and phase of the motor currents, FOC maximizes motor efficiency, minimizes losses, and provides smooth operation.

Pulse-width modulation (PWM): To control the magnitude of the motor currents, PWM is employed. PWM involves rapidly switching the power electronics (IGBTs) on and off to regulate the average current supplied to the motor windings. By varying the duty cycle of the PWM signal, the motor controller adjusts the average voltage and current, thereby controlling the torque and speed of the motor.

Regenerative braking: Electric motors in EVs can act as generators during deceleration and braking, allowing for regenerative braking. The motor controller detects the driver's braking input or deceleration, and instead of dissipating the energy as heat through the braking system, it converts the kinetic energy into electrical energy. The motor controller adjusts the motor operation to act as a generator, which charges the battery and provides energy recuperation.

Motor temperature and protection: The motor control system monitors the temperature of the electric motor to ensure it operates within safe limits. Sensors embedded in the motor or external temperature sensors provide feedback to the controller. If the motor temperature exceeds the predetermined threshold, the control system may reduce the torque output or adjust the cooling system to prevent overheating and protect the motor from damage.

Communication and integration: The motor control system is integrated into the overall vehicle control architecture and communicates with other systems, such as the battery management system, vehicle control unit, and safety systems. This integration ensures coordination among various vehicle functions, such as power distribution, energy management, thermal management, and vehicle stability control.

In summary, the control of electric motors in EVs involves the use of MCUs that receive input signals, regulate the torque output, employ FOC, utilize PWM, and enable regenerative braking. These control systems play a critical role in optimizing motor performance, efficiency, and integration within the overall vehicle control architecture.

5.5.2 Classic motor control model

A classic DC motor control could help to illustrate how to model the EV system motor control plant equation of motion, since technically, the EV is using a similar system, but a more complex version.

In the classic control theory, as Figure 5.6, the actuator in the system is the DC motor, with electric resistance, *R*, and electric inductance, *L*. The moment of inertia, *J*, of the rotor which can be treated as the combination of wheels and motor rotor in an EV scenario, motor viscous friction constant is also included here, as b. Motor constant such as electromotive force constant, Ke; motor torque constant, Kt; the battery has voltage, V.

Now, we can derive the equation of motion of the system, in general, the torque generated by the motor is proportional to the current, *i*, and the strength of the magnetic field, which is related to Kt:

$$T = K_t i \tag{5.2}$$

The back emf, *e*, is proportional to the rotation speed of the shaft times electromotive force constant:

$$e = K_e \dot{\theta} \tag{5.3}$$

After applying Newton's Second law and Kirchoff's law, we can have:

$$J\ddot{\theta} + b\dot{\theta} = K_t i \tag{5.4}$$

$$L\frac{di}{dt} + Ri = V - K_e\dot{\theta} \tag{5.5}$$

Applying Laplace transform to (5.4) and (5.5):

$$s(Js + b)\theta(s) = K_t I(s) \tag{5.6}$$

$$(Ls + R)I(s) = V(s) - K_e s\theta(s) \tag{5.7}$$

The transfer function of the DC motor plant is:

$$P(s) = \frac{\dot{\theta}(s)}{V(s)} = \frac{K_t}{(Ls + R)(Js + b) + K_e K_t} \tag{5.8}$$

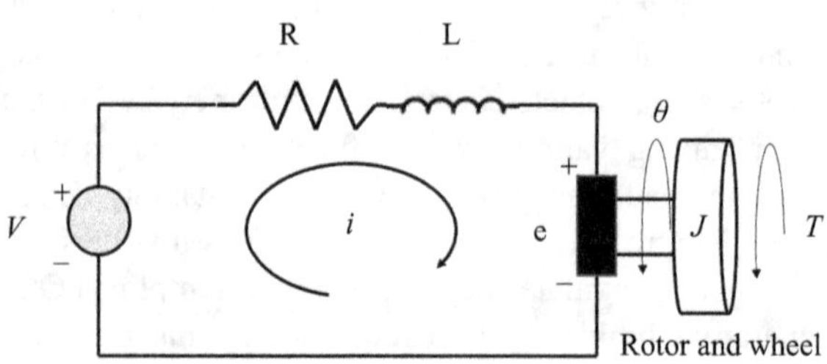

Figure 5.6 Classic DC motor control diagram

The equation of motion in state space is

$$\frac{d}{dt}\begin{bmatrix}\dot{\theta}\\ i\end{bmatrix} = \begin{bmatrix}-\frac{b}{J} & \frac{K_t}{J}\\ -\frac{K_e}{L} & -\frac{R}{L}\end{bmatrix}\begin{bmatrix}\dot{\theta}\\ i\end{bmatrix} + \begin{bmatrix}0\\ \frac{1}{L}\end{bmatrix}V \tag{5.9}$$

$$y = \begin{bmatrix}1 & 0\end{bmatrix}\begin{bmatrix}\dot{\theta}\\ i\end{bmatrix} \tag{5.10}$$

5.6 EV control optimization and condition monitoring

This section will briefly provide general concepts of the control optimization and condition monitoring of EVs and hope to give a brief introduction to what they are.

5.6.1 EV control optimization

The control and optimization of an EV involve managing and optimizing various modes of operation to achieve desired performance, efficiency, and range. General concepts for control and optimization of an EV under different modes, including the following:

Electric drive mode: In the electric drive mode, the EV operates solely on electric power, utilizing the electric motor for propulsion. The control system manages the power output of the motor based on driver inputs, such as accelerator pedal position, to provide the desired torque and speed. The optimization in this mode focuses on maximizing efficiency and range while ensuring a smooth and responsive driving experience.

Regenerative braking mode: During deceleration and braking, the EV enters the regenerative braking mode, where the electric motor acts as a generator, converting kinetic energy into electrical energy to recharge the battery. The control system modulates the regenerative braking force based on driver inputs, vehicle speed, and other factors. The optimization in this mode aims to maximize energy recuperation while maintaining vehicle stability and brake feel.

Hybrid mode (parallel hybrid): In a parallel hybrid configuration, the EV can operate in hybrid mode, combining the power of the electric motor and an internal combustion engine (ICE). The control system manages the power distribution between the two power sources based on various factors, such as driving conditions, battery state of charge, and system efficiency. The optimization in this mode focuses on achieving the best balance between electric and ICE operation to optimize fuel efficiency and overall performance.

Hybrid mode (series hybrid/range extender): In a series hybrid configuration or when using a range extender, the EV operates primarily on electric power, while an onboard generator (usually an ICE) charges the battery to extend the range. The control system manages the power flow between the generator and the electric motor, optimizing the generator's operation and battery charging strategy. The optimization in this mode focuses on maximizing the electric range and efficiency while ensuring sufficient power and range extension through the generator.

Charge-sustaining mode: In charge-sustaining mode, the EV primarily relies on the onboard generator or ICE to provide power to the electric motor while simultaneously maintaining the battery charge level. The control system manages the power flow between the generator, battery, and motor to optimize fuel efficiency and maintain a desired battery state of charge.

Energy management and optimization: In all modes of operation, the control system performs energy management and optimization. It considers factors such as vehicle speed, terrain, driver behavior, battery state of charge, and environmental conditions to determine the most efficient power distribution, throttle control, and charging strategies. The optimization aims to maximize overall efficiency, range, and performance while ensuring safety and comfort.

The control and optimization of an EV under different modes involve a complex interplay between the powertrain components, control algorithms, driver inputs, and external conditions. Advanced control strategies, such as predictive control, machine learning, and adaptive algorithms, are continuously being developed to further enhance the efficiency and performance of EVs across different operational modes.

5.6.2 EV condition monitoring

EV condition monitoring refers to the continuous monitoring and analysis of various parameters and operating conditions of an EV to assess its overall condition and performance. It involves collecting and analyzing data from various sensors and systems in real-time to detect abnormalities, identify potential issues, and optimize the vehicle's operation. Some key aspects of EV condition monitoring, include:

Sensor-based data collection: Condition monitoring relies on sensors embedded within the EV to collect data on various parameters. These sensors may include temperature sensors, current sensors, voltage sensors, vibration sensors, pressure sensors, and more. The data collected from these sensors provide insights into the performance, health, and operating conditions of different EV components.

Data analysis and diagnostics: The collected data are analyzed using algorithms and techniques to identify patterns, anomalies, and trends. Data analysis can be performed in real-time or through periodic diagnostics. The goal is to detect deviations from normal behavior and identify potential issues that may require attention or maintenance.

Performance monitoring: EV condition monitoring includes assessing the performance of key components such as the battery pack, electric motor, power electronics, and charging systems. Parameters such as battery state of charge (SoC), motor efficiency, charging speed, and energy consumption are monitored to ensure optimal performance and identify any degradation or inefficiencies.

Fault detection and predictive maintenance: Condition monitoring helps detect and diagnose faults or malfunctions in real-time or proactively through predictive maintenance. By analyzing sensor data and comparing it against pre-defined thresholds or models, condition monitoring systems can identify deviations or warning signs of component failure or degradation. This allows for timely maintenance or repair, minimizing downtime, and optimizing the vehicle's reliability.

State of health (SoH) monitoring: EV condition monitoring involves monitoring the SoH of critical components, particularly the battery pack. SoH monitoring assesses the overall health, capacity, and degradation of the battery over time. It provides insights into the battery's remaining usable life and helps determine the optimal charging and usage patterns to maximize its lifespan.

Driver behavior and usage analysis: Condition monitoring systems can also incorporate driver behavior and usage analysis. By monitoring parameters such as acceleration patterns, braking behavior, speed profiles, and energy consumption, condition monitoring can provide feedback to drivers on how to optimize their driving habits for better efficiency and performance.

Data logging and reporting: Condition monitoring systems typically log and store data collected from sensors for future analysis, reporting, or system optimization. These data can be used for long-term performance tracking, warranty claims, diagnostics, and improving the design of future EV models.

EV condition monitoring enhances the operational efficiency, reliability, and maintenance of EVs. It allows for proactive maintenance, early fault detection, optimized performance, and improved driver awareness, ultimately resulting in better overall vehicle performance and ownership experience.

5.7 Summary

EV modeling and simulation is a very important topic, by using modeling and simulation, EVs are designed to be more efficient, safe, and less costly. Also, design issues can be found early even without a physical prototype being made. With simulation, EV aerodynamics can be optimized to make the air drag minimum in a short development process. EVs need to be updated quickly due to the nature of EVs; therefore, SIL and HIL are important, without modeling and simulation, SIL testing and HIL testing are impossible to do. Nowadays, more and more OEMs are committing to modeling and simulation, and some of them already started to propose a zero-prototype roadmap to their R&D, aiming to reduce the time to market time and cost. Shortly, it would be possible to reduce the development time and cost even further with the help of developing arcuate vehicle models and running them in real-time simulation.

References

[1] R. Matulka, *The History of the Electric Car, Department of Energy*, 2014, https://www.energy.gov/articles/history-electric-car.

[2] R. Holehouse, A. Shahaj, M. Michon, and B. James, "Integrated approach to NVH analysis in electric vehicle drivetrains," *The Journal of Engineering*, vol. 2019, no. 17, pp. 3842–3847, 2019, doi:https://doi.org/10.1049/joe.2018.8247.

[3] J. Zhang, L. Zhang, F. Sun, and Z. Wang, "An overview on thermal safety issues of lithium-ion batteries for electric vehicle application," *IEEE Access*, vol. 6, pp. 23848–23863, 2018.

[4] M.W. Grieves, "Virtually intelligent product systems: digital and physical twins." In *Complex Systems Engineering: Theory and Practice*, American Institute of Aeronautics and Astronautics, 2019, pp. 175–200.

[5] IBM, https://www.ibm.com/topics/what-is-a-digital-twin.

[6] R. Tara, "Are we ready for digital twins?," *Engineering.com Audience Survey of Perceptions and Readiness*.

[7] e-motec, *Optimizing EV Aerodynamics to Increase Range*, September 2, 2021, https://www.e-motec.net/optimizing-ev-aerodynamic.

[8] G. Chen, Q. Xiong, P.J. Morris, E.G. Paterson, A. Sergeev, and Y. Wang, "OpenFOAM for computational fluid dynamics," *Notices of the AMS*, vol. 61, no. 4, pp. 354–363, 2014.

[9] T. A. o. OpenFOAM, "OpenFOAM HPC with AWS EFA," https://cfd.direct/cloud/openfoam-hpc-aws-efa/.

[10] F. Ozcan and S. Ersoy, "Analysis of the vehicle: applying finite element method of 3D data," *Mathematical Models in Engineering*, vol. 7, no. 4, pp. 63–69, 2021.

[11] R. Jobava, F. Bogdanov, A. Gheonjian, and S. Frei, "Application of adaptive scheme for the method of moments in automotive EMC problems," in *Proceedings of 16th International Zurich EMC Symposium, Zurich, Switzerland*, 2005, pp. 131–136.

[12] A. Fox, "Simulation of AWD performance with multibody chassis model and hardware in the loop system (HILS)," in *SIMULIA Great Lakes UGM 2016*, 2016.

[13] I. Mula, G. Tosolin, and X.C. Akutain, "Enhanced ride comfort evaluation on the driving simulator with real-time multibody models," in *12th International Munich Chassis Symposium 2021*, 2022, New York, NY: Springer, pp. 43–60.

[14] "VI-grade Announces Installation of DiM250 DYNAMIC Driving Simulator at CEVT," https://newsroom.notified.com/cevt/posts/pressreleases/vi-grade-announces-installation-of-dim250-dyn.

[15] Apriv, *What Is Software-in-the-Loop Testing*, 2022, https://www.aptiv.com/en/insights/article/what-is-software-in-the-loop-testing.

[16] N. Instruments, *What Is Hardware-in-the-Loop*, 2022, https://www.ni.com/en-us/solutions/transportation/hardware-in-the-loop/what-is-hardware-in-the-loop-.html.

Chapter 6

Design, modelling, simulation and control of electric machines and drives used in electric vehicle

Faz Rahman[1]

6.1 Introduction to motor drives for electric vehicles

The adoption of electric vehicles (EVs) for automotive traction is gathering pace at breakneck speed lately (see Figure 6.1). Passenger vehicle makers are announcing new EV models virtually every day, some with announcements of even discontinuing petrol/diesel vehicles entirely in a few years! At the higher power spectrum, electrification of medium and large power trucks of >500 kW capacity is also a gathering pace, especially for trucks with well-defined regular routes that can take advantage of suitably located charging stations where batteries can be changed in a short time. The change from IC engine-driven vehicles to EVs is akin to the way the world changed over from steam engines to IC engines about a century or so ago. However, battery-electric motor-driven vehicles are not new; one, depicted in Figure 6.2, was announced as far back as 1832, when battery and motor technologies were very primitive. These succumbed by 1920 when IC engines came of age and cheap gasoline became easily available. The tremendous rise in battery energy density and motor power density and associated improvements in power electronic converters for efficient battery charging and control of electric machines in recent years have fundamentally altered the scene of EVs. This chapter will attempt to cover some of the motor designs and control systems that are at the heart of traction drives of EVs.

6.2 Torque–speed capability requirements for EVs

The maximum torque that a traction motor is required to deliver to the wheels of the typical vehicle over the full speed range from *0* to $\omega_{\max}$ is indicated by the black trace in Figure 6.3. The base speed ω_b refers to the speed that is attained with rated (or maximum) voltage at the base frequency (say 50 Hz) applied by the

[1]School of Electrical Engineering and Telecommunications, The University of New South Wales, Australia

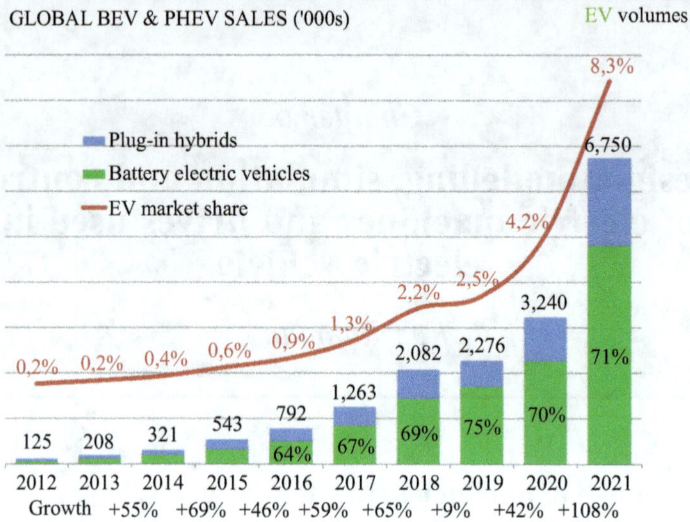

Figure 6.1 Growth in EVs since 2012. Source: ev-volumes.com.

Figure 6.2 An EV in 1832, developed by the Scottish inventor named Robert Anderson

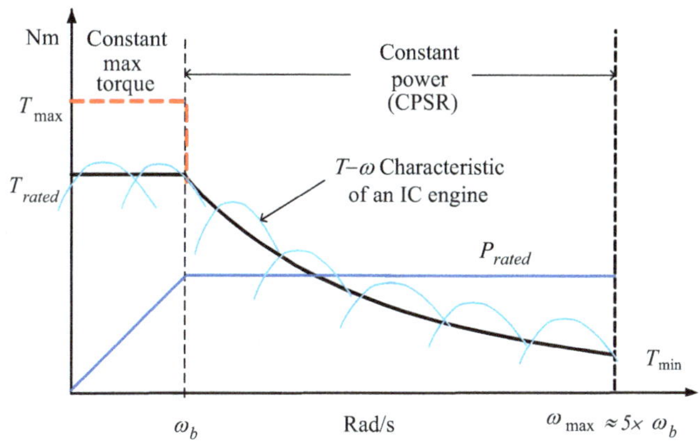

Figure 6.3 Torque–speed characteristic requirement for a vehicle

inverter to the motor. The air-gap flux of the motor up to this speed is the rated or nominal value for the motor. Beyond ω_b and up to ω_{max}, the air-gap flux is reduced inversely with speed, to maintain the back-emf in the stator phase at the nominal value. This is the so-called field weakening region over which the maximum torque that the motor produces falls inversely with speed. Because the maximum developed power remains constant over the speed region $\omega_b - \omega_{max}$ with this field weakening (FW) control, this speed range is known as a constant power speed range (CPSR). A simplified analysis for a reasonable power rating of the motor over this speed range shows that the ω_{max}/ω_b ratio should be about 5 or more to minimize the size of the motor. For an IC engine-driven vehicle, the engine T–ω characteristics (indicated by the faint blue curves) are progressively shifted up or down by gear changes. An automated gearbox for this is a substantial part of the engine-traction drive system.

Modern interior permanent-magnet (IPM) motors are capable of meeting the traction T-ω profile without requiring any gear changes, i.e., just one gear stage is normally required to match the motor torque to the requirement of the traction system, as indicated in Figure 6.4. This advantage with the IPM machine also allows the motor to be designed with the minimum volume, i.e., the highest torque/power density, compared to other available machines. The motor also delivers the short-time constant maximum torque requirement (indicated by the red broken line) by exploiting the short-time capacities of the motor and the inverter switches.

A typical control system of an EV is indicated in Figure 6.5 in which the accelerator pedal produces the torque reference for the vehicle. This reference is then vectored into torque references for left and right motors on the axle with an electronic differential gear-box. The steering angle δ and speeds of the left and wheels are used for calculating the torque references for the two motors. These motors are then controlled by their individual inverter and torque controllers. This chapter is mainly concerned with the motor and its control system.

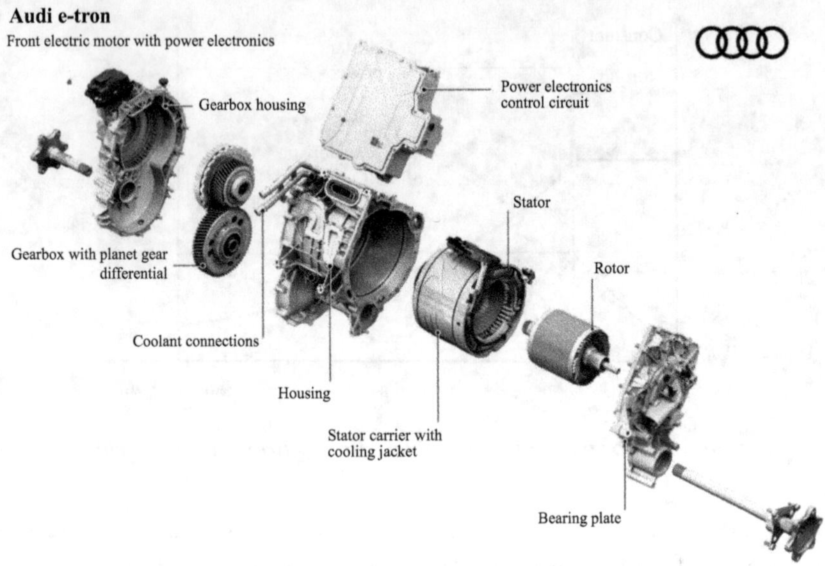

Figure 6.4 A typical motor-gear and housing arrangement of a vehicle motor. Courtesy: Audi e-tron.

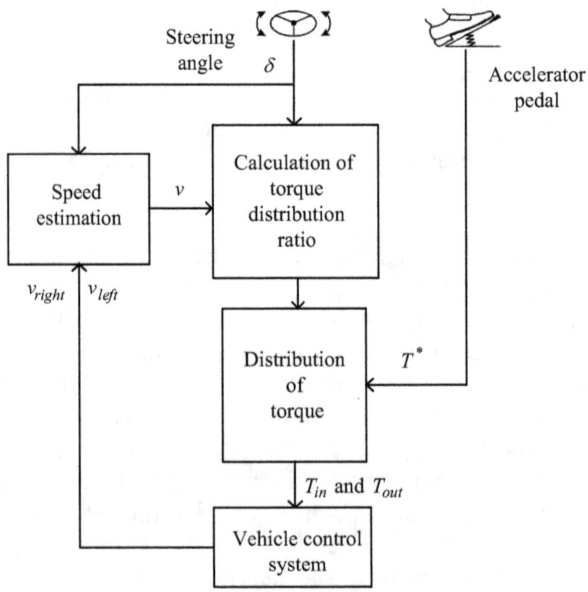

Figure 6.5 A typical supervisory control system for an EV

6.3 The evolution of IPM machines for EV application

Many contenders as electric motors for the current generation of EVs include induction motors (IMs), wound rotor synchronous motors (WRSMs), axial-flux PM motor (AFPMM), switched reluctance motors (SRMs) and the interior permanent-magnet synchronous motors (IPMSMs). The IMs were used in the Tesla 2017 EV Model S, however, an IM has limited CPSR and low efficiency at high speed. As a result, recent models of Tesla EVs are using the IPMSM. The WRSM has the disadvantage of slip rings for supplying current to the rotor for rotor excitation and control of the rotor field. Its torque/power–volume ratio also does not compare favourably with the IPMSM. The AFPMM has a few advantages when the power rating of the motor is low (a few KWs) and it does not lend easily to high-power design due to the increased diameter and consequent unbalanced axial forces and torque at large diameter. The SRM also loses out to IPMSMs in terms of increased power converter size, audible noise, and low power factor operation, to name a few. Considering these, the IPMSM, with rare-earth PM materials (usually sintered/bonded NdFeB in conjunction with other materials that extend the operating temperature and corrosion resistance) embedded within the rotor iron, has become the motor of obvious choice for virtually all EV manufacturers as of now. Table 6.1 includes a few hybrid, plug-in-hybrid and pure EVs up to 2017, which indicates the overwhelming usage of the IPMSM for EVs.

Table 6.1 IPM motor power and driving range data for a few hybrid and electric vehicles

	Vehicle	**Electric motor**		**ECE data**	**Battery**		
		Type	**Power**		**Type**	**Capacity**	**Range (electric)**
Hybrid	Honda Insight, 2010	PMSM	9.7 kW	1.31 petrol	Li-ion	0.6 kW	
	Toyota Prius, 2017	PMSM	23 kW 53 kW	1.8 L gasoline	Li-ion	8.8 kWh	40 km
	GM Volt hybrid, 2016	PMSM	55 kW (MG) 111 kW(M)	1.4 L petrol	Li-ion	18.4 kWh	85 km
	Hyundai loniq, 2017	PMSM	32 kW	1.6 L petrol	Li-ion	1.56 kWh	
	BMW i3, 2014	PMSM	125 kW	647 cc petrol	Li-ion	33 kWh	290 km
	Ford Fusion, 2015	PMSM	88 KW	2.5 L petrol	Li-ion	7.6 KW	31 KM
	Hyundai Sonata, 2016	PMSM	50 KW	2.0 L petrol	Li-ion	9.8 KW	43 KM
All-electric	Tesla Model S, 2017	IM	581 kW (two motors combined)	$m > 1$	Li-ion	100 kWh	550 km
	Chevrolet Bolt, 2017	PMSM	150 kW	–	Li-ion	60.0 kWh	380 km
	Ford Focus, 2017	PMSM	107 kW		Li-ion	33.5 kWh	185 km
	Renault Fluence, 2014	PMSM	70 kW		Li-ion	22 kWh	185 km

6.3.1 The IPM rotor

The design of the rotor of IPM machines for EVs has developed much since the first incorporation PM rotors in Toyota hybrid vehicles (Prius I) [1–7]. Early IPM machines such as in Figure 6.6(a) and (c) did not exhibit adequate field weakening until the V-shape magnet pole with iron bridges between magnets was incorporated in each pole. Many other aspects of minimizing cogging torque, torque ripple, and magnet volume and losses in the rotor, and maximizing the efficiency were incorporated in the design of these and other still emerging structures. A few examples of rotor structures adopted by several EV and PHEV suppliers are indicated in Figure 6.7.

6.3.2 The fractional-slot concentrated winding (FSCW) stator winding of the IPM machine

Another development that has taken place for the IPM machine is the fractional-slot, concentrated windings [8–22] which have several desirable attributes. The FSCW, in contrast with distributed windings, concentrates whole per-pole and per-phase windngs on one or two adjacent stator tooths. This is indicated in the simple illustration of

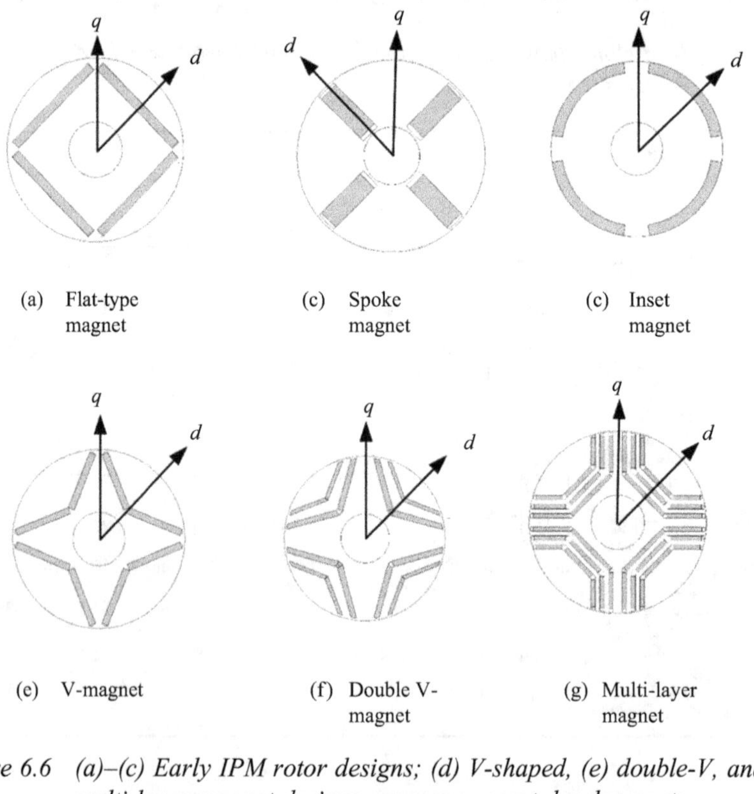

Figure 6.6 (a)–(c) Early IPM rotor designs; (d) V-shaped, (e) double-V, and (f) multi-layer magnet designs are more recent developments

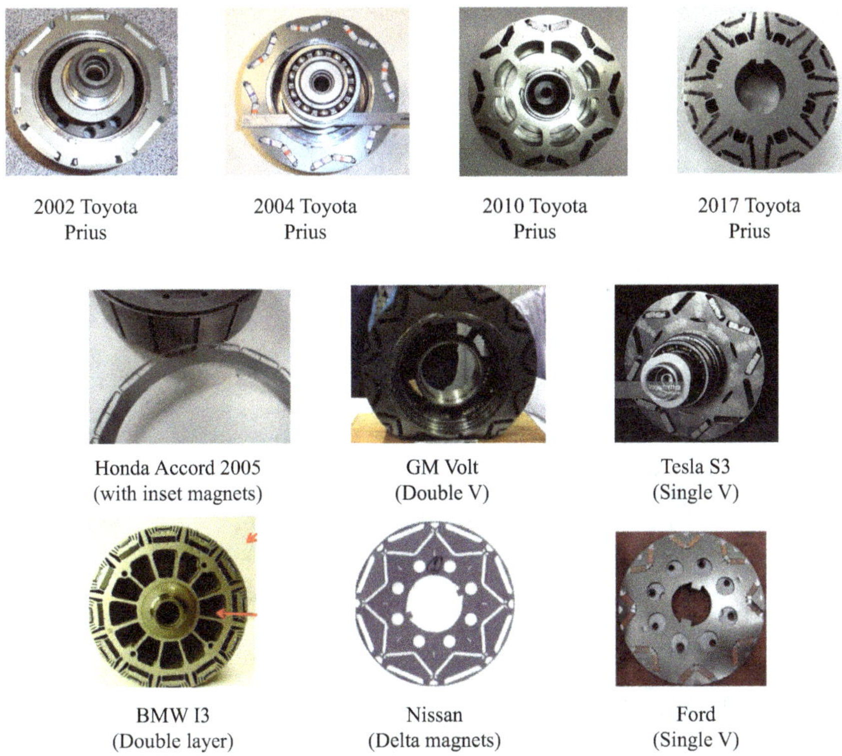

Figure 6.7 Examples of a few rotor structures adopted by the hybrid and EV industry

Figure 6.8 where a single-layer winding is shown. In practice, a double layer winding in each stator slot is preferred. The FSCW offers the following advantages:

- The condition of optimum CPSR is more easily achievable in FSCW machines. Increased field-weakening (or high CPSR ≥ 5) capability due to high d-axis inductance.
- Reduction of cogging torque due to a large least common multiple (LCM) between the number of poles and the number of slots.
- Reduction of copper losses due to shorter end-winding length, and hence reduction of machine total length, leading to increase in torque and power density.
- Reduction of short-circuit current due to higher inductances.
- Increased fault-tolerant capability due to the reduction of mutual inductance and absence of overlapping of windings.
- Compact size and high pole number are also useful for low-speed direct-drive applications.
- Simplified manufacturing, maintenance, and repair of the winding due to its modular structure.
- The FSCW offers the opportunity for integrating motor and inverter in a single unit.

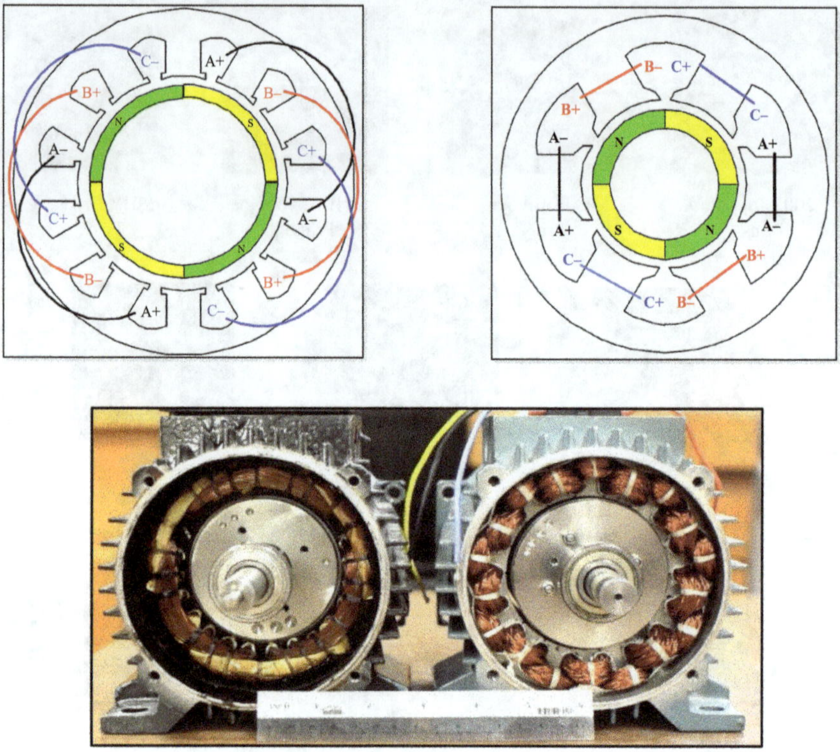

Figure 6.8 The IPM machine with (a) conventional distributed winding and (b) FSCW with slot/pole/phase less than 1

Elaborate designs meeting the optimization goals mentioned above can nowadays be undertaken using a finite-element analysis platform and multi-objective optimization algorithm guiding the analyses to optimize the dimensions in the chosen rotor and stator structure (illustration Figure 6.9) [23–39]. This process is rather time-consuming, requiring high computing resource.

The desirable attributes of the Spoke and V-type rotors, indicated in Figure 6.6(b) and (e) were combined in the Y-type rotor of Figure 6.10 [40]. This rotor with the FSCW in the stator has attributes better than what is possible in the multi-layer and V-type rotors. These attributes include cogging torque, torque ripple, magnet volume and losses in the rotor, operating efficiency, and CPSR.

6.3.3 IPM machine performance from FE analysis

Typical back-emf waveforms, torque ripples, torque–speed characteristics, and efficiency maps of the Spoke, V- and Y-type motors of the same stator and rotor dimensions obtained through FE analysis are shown in Figures 6.11–6.14. These figures are included for the purpose of showing the depth of analysis and design procedures and their efficacies. A fuller description of these procedures is however beyond the scope of this chapter.

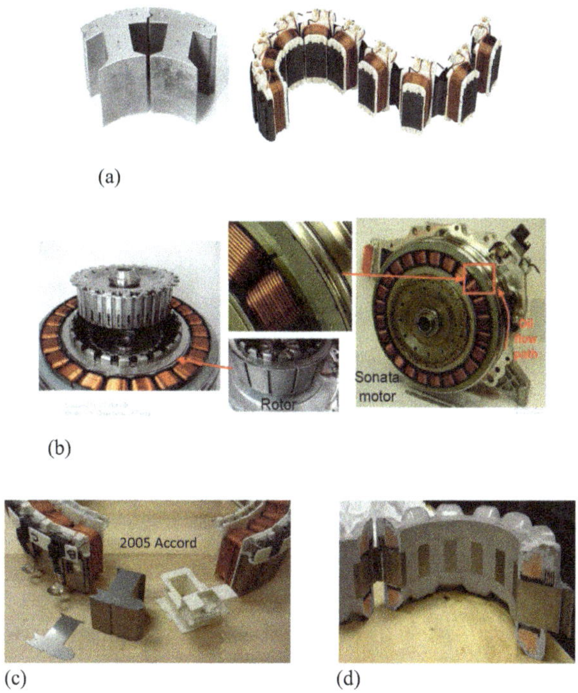

Figure 6.9 *(a)–(d) A few examples of FSCW in hybrid and EVs. (a) Typical FSCW stator poles and windings. (b) Hyundai Sonata 2010 with 24 slots, 16 poles, 14.4 kW, 840 rev/min (c) Stator poles for Honda 2005 Accord IPM motor (d) Toyota Prius 2010 and 2017 FSCW stator.*

Figure 6.10 *The Y-type rotor with FSCW stator [40]*

6.3.4 Steady-state performance from measured stator parameters

The FE analyses as means for design and assessment of an IPM machine are usually beyond the scope of a practicing engineer familiar with phasor analysis of

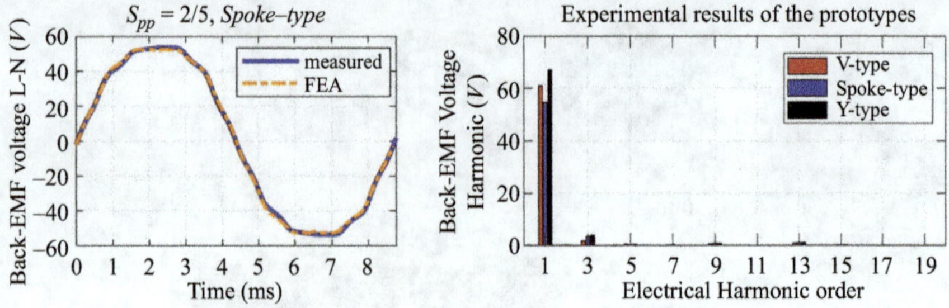

Figure 6.11 (a) Bemf of the spoke type motor. (b) Harmonic spectrum of phase voltages of Spoke, V- and Y-motor

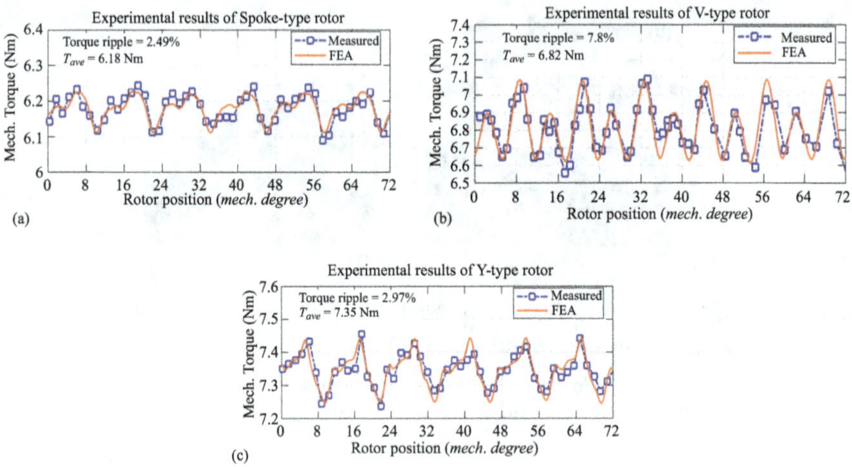

Figure 6.12 Torque ripples in optimized Spoke, V- and Y-motors

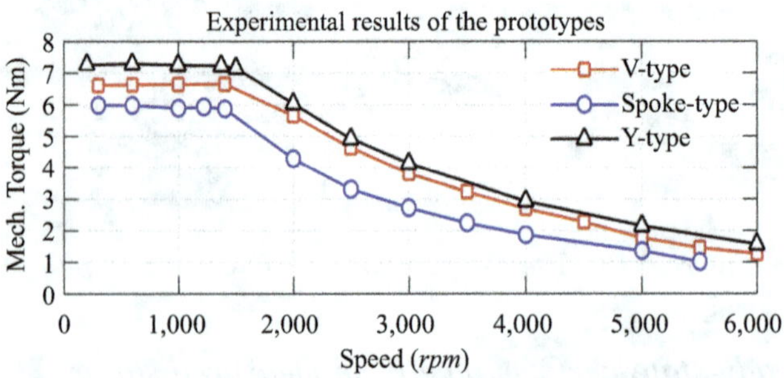

Figure 6.13 Torque–speed characteristics of optimized Spoke, V- and Y-motors

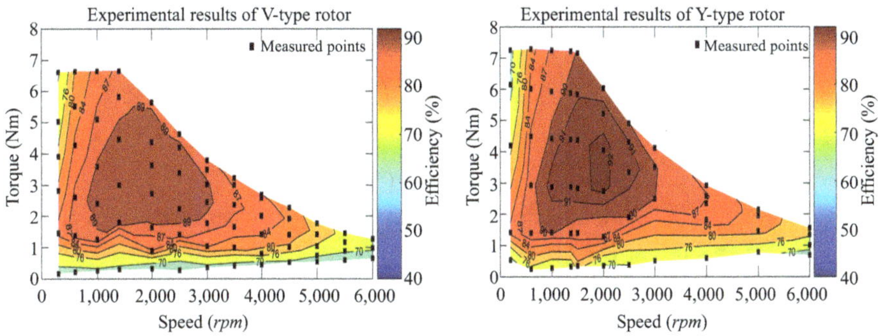

Figure 6.14 Efficiency maps of the optimized Spoke, V- and Y-motors

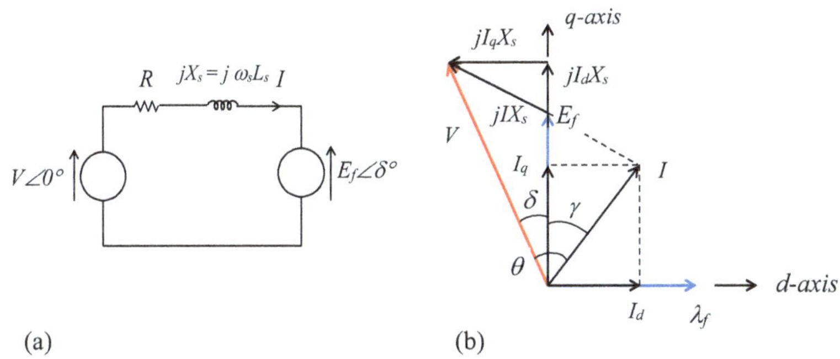

Figure 6.15 (a) IPMSM per-phase equivalent circuit and (b) phasor diagram

AC machines. The IPM machines of the foregoing section are in fact synchronous machines which can also be analyzed using machine steady-state inductance parameters L_d and L_q, stator resistance R and the stator flux linkage parameter λ_f. Machine steady-state performance of torque and efficiency for various applied voltages and operating speeds can be found using phasor algebra. It will be shown in the latter section that the steady-state model can be deduced from the dynamic model in the rotor *dq* frame. The above-mentioned parameters are easily obtained via a self-test routine in which the inverter operates the motor at a few speeds and at zero frequency to determine these parameters. It may be mentioned here that the FOC and DTC controllers of the motor use parameters obtained via such self-tests for running the motor.

The steady-state analysis of an IPMSM starts with representing the per-phase equivalent circuit and the phasor diagram is shown in Figure 6.15(a) and (b), respectively. For the phasor diagram of 6.14(b), the stator resistance/phase R has been neglected.

The torque–speed characteristic of the IPMSM can be obtained from (6.1):

$$T = \frac{3p}{\omega_o}\left[\frac{\frac{V_1^2}{\lambda}R + V_1 E_{fo}\lambda X_{qo}\sin\delta + \frac{V_1^2}{2}\left(X_{do} - X_{qo}\right)\sin 2\delta - V_1 R E_{fo}\cos\delta}{R^2 + \lambda^2 X_{do}X_{qo}}\right] \tag{6.1}$$

where $\lambda = f_1/f_o$; f_1 is the input supply frequency and f_o is the base supply frequency (say, 50 Hz), V_1 is the per-phase input RMS supply voltage to the motor with frequency f_1, p is the number of pole pairs, E_{fo} is the back-emf at the base speed, and X_{do} and X_{qo} are dq synchronous reactances at the base frequency f_o or speed ω_o. The IPMSM operates near base speed, the voltage drop in R can be negligible compared to V_1, so that for $R \sim 0$,

$$T = \frac{3p}{\omega_o}\left[\frac{E_{fo}V_1}{\lambda X_{do}}\sin\delta + \frac{V_1^2}{2\lambda^2}\left(\frac{X_{do} - X_{qo}}{X_{do}X_{qo}}\right)\sin 2\delta\right] \tag{6.2}$$

With an open-loop V–f inverter, as shown in Figure 6.16(a), the torque–speed characteristics from (6.1) and (6.2) are as shown in Figure 6.15(b). At a low speed, the torque capability of the motor drops (equation (6.1)) due to the voltage drop in R as indicated by the curved dotted line of Figure 6.16(b). With closed-loop control

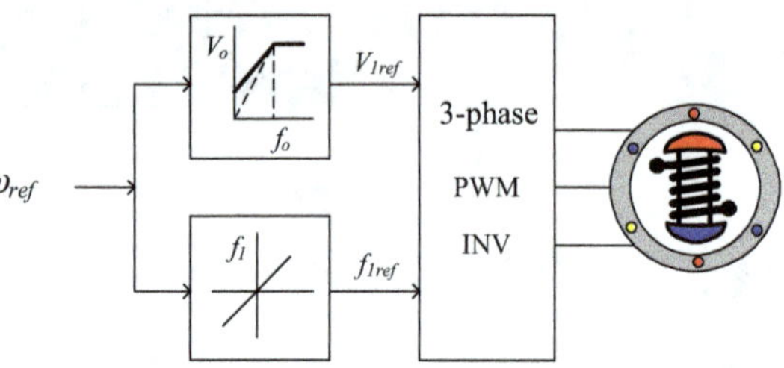

(a)

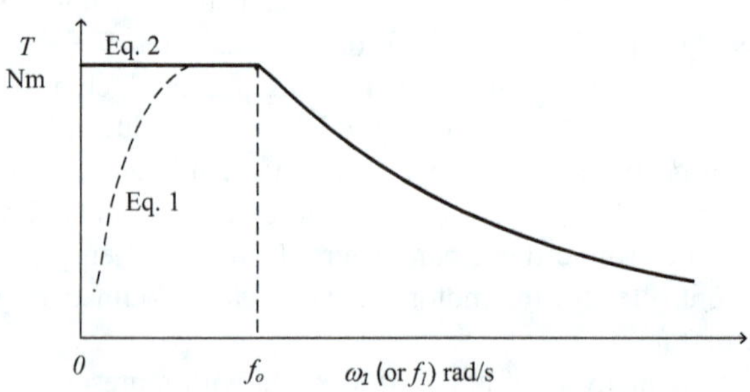

(b)

Figure 6.16 The torque–speed capability of an IPMSM under (a) open v–f control and (b) closed-loop current control which overcomes the voltage drop in stator resistance R. (a) Inverter control for an IPMSM.

of I_d and I_q, the voltage drop in R is compensated, and the motor torque capability is retained right down to zero speed. This production of the rated torque is one important distinguishing capability compared to the IC engine. A single-stage gear is thus only required, as indicated in Figure 6.4 to match the motor T–ω characteristic with the referred torque from the driven wheel. The maximum torque T, the required CPSR, and ω_o for the drive allows the gear ratio indicated in Figure 6.4 to be selected.

6.4 The dynamic model and control of IPMSMs

An EV requires very fast dynamic control of the IPMSM used for the traction drive. The required dynamic performance requires field-oriented controls (FOC) based on the dynamic model of the IPMSM. For this, the IPMSM is represented in the rotor reference frame, via Park's dq-transformation. Equations (6.3)–(6.6) represent the d- and q-axes voltages, flux linkages and the developed torque of the machine.

The dq voltages:

$$v_d = Ri_d + \frac{d\lambda_d}{dt} - \lambda_q \frac{d\theta}{dt} = Ri_d + L_d \frac{di_d}{dt} - \omega L_q i_q \tag{6.3}$$

$$v_q = Ri_q + \frac{d\lambda_q}{dt} + \lambda_d \frac{d\theta}{dt} = Ri_q + L_q \frac{di_q}{dt} + \omega\left(L_d i_d + \lambda_f\right) \tag{6.4}$$

The dq flux linkages:

$$\lambda_d = L_d i_d + \lambda_f \quad \lambda_q = L_q i_q \tag{6.5}$$

The developed torque:

$$T = \frac{3p}{2}\left(\lambda_d i_q - \lambda_q i_d\right) = \frac{3p}{2}\left[\lambda_f i_q - \left(L_q - L_d\right) i_q i_d\right] \tag{6.6}$$

It should be noted that for an IPM machine, $L_q > L_d$, and $\xi = \frac{L_q}{L_d}$, the saliency ratio is an important measure of the machine's reluctance torque and field weakening capability. Also, the flux linkage λ_d, which is responsible for the q-axis bemf E_q to be reduced to zero when

$$i_d = i_{ch} = -\frac{\lambda_f}{L_d} \tag{6.7}$$

The circuit representation of the machine in the rotor dq frame, as indicated in Figure 6.17, follows from (6.3).

Control of flux linkage and torque are based on (6.5) and (6.6), respectively, when the machine is driven from a current source inverter with independent and decoupled control of i_d and i_q. To understand the torque–speed envelope of the machine and some of its control restrictions, it is useful to represent the machine in terms of its steady-state equivalent circuit, in which the dq stator windings are

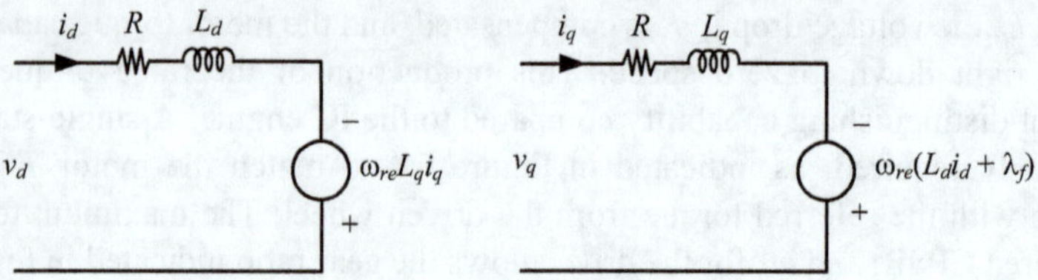

Figure 6.17 Circuit representation of an IPMSM in rotor d- and q-axes

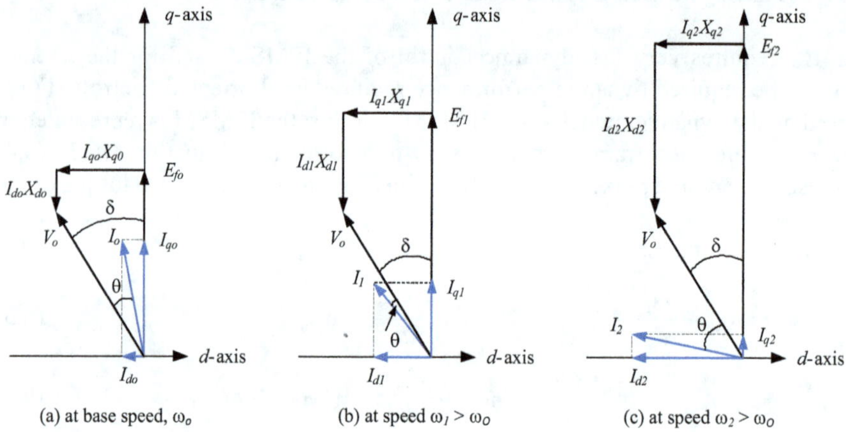

Figure 6.18 Phasor diagrams of an IPMSM at base and higher speeds

stationary, but produces their MMFs along the rotor *dq* axes. In this condition, the *dq* stator windings have AC voltage, current and flux linkage phasors at the frequency corresponding to the speed of rotation. Figure 6.18 shows phasor diagrams of the IPMSM at the base and two other speeds that are higher than the base speed, if the voltage drops in the stator resistances are negligible at these speeds. V_o is the rated phase voltage. E_{fo} is the PM excited phase voltage at the base speed. Angles θ and δ are the power factor and load angles, respectively.

6.4.1 dq *Current controls below base speed and above with field weakening*

It should be clear from (6.2) that the inverter output voltage V should increase proportionately with speed up to the rated voltage at base speed so that the motor develops its maximum torque at the base speed. For operation at higher speeds, the inverter output voltage remains fixed at the rated value corresponding to the DC link voltage, while the speed is increased by increasing the supply frequency ω (= $2\pi f$). The excitation voltage E_f and the *d*- and *q*-axes reactances also increase proportionately with ω. The phasor diagrams of Figure 6.18 show the V, E_f, and the input current I phasors at three speeds at and above the base speed. These figures also indicate that operation above speed can be arranged by appropriately controlling the phase angle of the input current phasor from the excitation voltage E_f, when a current source inverter drive is used. This type of control increases the $-$ve d-axis current, which reduces the flux linkage (see (6.5)) as speed is increased above the base speed, thereby reducing the air-gap field via armature reaction. Equation (6.6) implies that the $-$ve d-axis current $-i_d$ increases the reluctance torque component of an IPM machine (the second term in (6.6)), which is advantageous because the total current $I = \sqrt{i_d^2 + i_q^2}$ of the machine must also be subject to a maximum limit. The increased reluctance torque afforded by increased $-i_d$ allows a reduction in i_q, helping the drive to operate within its maximum current limit while, at the same time, maximizing the total developing torque. It should be clear also from Figure 6.18 that using $-I_d$ for field weakening is equivalent to making input current phasor I_1 advance in phase angle with respect to E_f.

6.4.2 Power converters for EVs

The IPM motor in an EV is driven by an inverter, as indicated in Figure 6.19. The inverter DC voltage is in the range of 500–800 VDC for today's EVs. The inverter DC voltage is maintained using a bidirectional DC–DC converter which operates off the vehicle battery of about 300 V DC. The regenerative energy of the IPM motor charges the inverter capacitor and the LV battery as required for maintaining the inverter DC-bus voltage.

6.4.3 Torque control

The torque control system is shown in Figure 6.20. The torque reference T^* is from the acceleration pedal of the vehicles, or it may also be the output of a speed controller in a cruise control system. The reference torque T^* must be transformed into current references i_d^* and i_q^* by the current reference generator according to the measured or estimated speed of the motor shaft and certain optimum control requirements to be discussed in the following section. Motor currents $i_a - i_c$ are sensed and transformed into i_d and i_q using the measured rotor position θ. The shaft

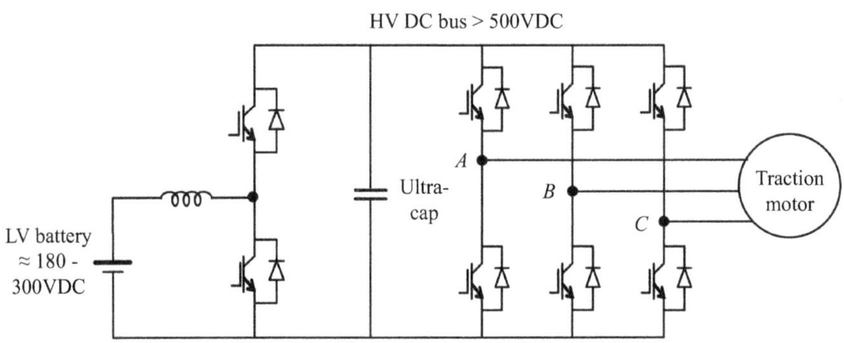

Figure 6.19 Power converters for IPMSM drive in an EV

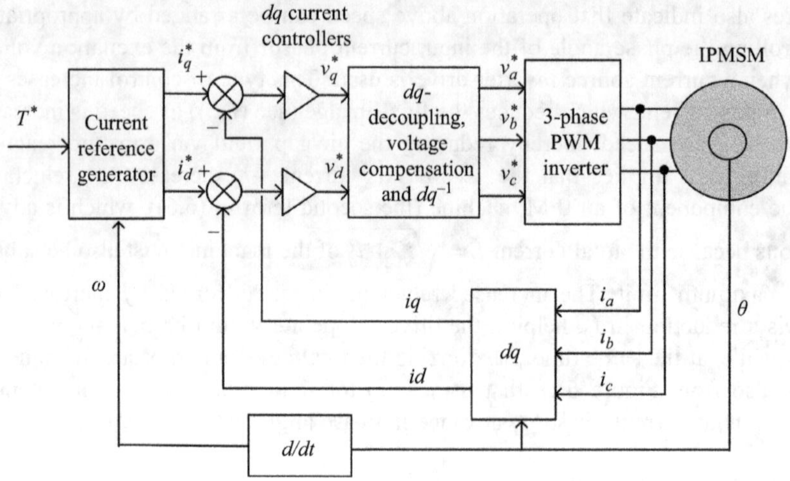

Figure 6.20 Torque control block diagram of the IPMSM

angle θ is obtained using an encoder/resolver of adequate accuracy. The i_d and i_q current controllers (usually PI regulators with a low-pass filter) operate in the rotor dq reference frame, producing voltage references v_d^* and v_q^*.

6.5 Optimum control trajectories

The current reference generator in Figure 6.20 is necessitated by the torque equation (6.6), in which the developed torque T is contributed partly by the product of the rotor flux linkage λ_f and i_q, which may be termed the excitation (or magnet) torque T_e and partly by the product of the reluctance difference $(L_q - L_d)$ and $i_d{*}i_q$, which is known as the reluctance torque T_r. Because i_d and i_q both contributes to the torque, references for i_d and i_q from the torque reference T^* must be found [41–47].

6.5.1 *The condition for maximum torque per ampere (MTPA) characteristic*

Assuming that the input current phasor I will lie on or to the left of the q-axis in the i_d–i_q plane, from Figure 6.15, we may write

$$i_d = -I \sin\gamma \text{ and } i_q = I \cos\gamma \tag{6.8}$$

where $I = \sqrt{i_d^2 + i_q^2}$ is the phase current phasor magnitude and γ is the angle of the I phasor with the q-axis on which the E_f phasor lies. Substituting (6.8) into the torque equation (6.6), we obtain

$$T = \frac{3}{2}p\left\{\lambda_f I \cos\gamma + \frac{1}{2}(L_q - L_d)I^2 \sin 2\gamma\right\} \tag{6.9}$$

The first term in (6.9) is torque T_e due to rotor excitation (flux) and the second term is the reluctance torque T_r. To obtain the fastest transient response and highest torque for a current i, the current phase angle γ must be such as to satisfy the maximum torque condition. The relationship between the amplitude of the stator current and the phase angle γ for the maximum torque can be derived by setting the derivative of (6.9) with respect to γ to zero.

$$\frac{dT}{d\gamma} = \frac{3}{2}p\lambda_f I \sin \gamma + \frac{3}{2}p(L_q - L_d)I^2 \cos 2\gamma = 0 \tag{6.10}$$

By replacing i with i_d and i_q from (6.8) leads to the following relationship between i_d and i_q which procures the MTPA of stator current i:

$$i_d = \frac{\lambda_f}{2(L_q - L_d)} - \sqrt{\frac{\lambda_f^2}{4(L_q - L_d)^2} + i_q^2} \tag{6.11}$$

Equation (6.11) implies that the maximum torque per-ampere can be achieved if i_d is determined by this equation for any i_q. The reference i_q is usually determined by the torque reference T^*. The developed torque components of an IPMSM are shown in terms of angle γ in Figure 6.21, which is a plot of (6.6). Driving the motor with i_d and i_q current references from (6.11) is equivalent to operating the machine with the current angle γ that produces the maximum torque. Hence this mode is referred to as the MTPA control. Note that with this mode of control, the motor develops torque with the least current, thus minimizing the I^2R loss (copper loss) of the machine.

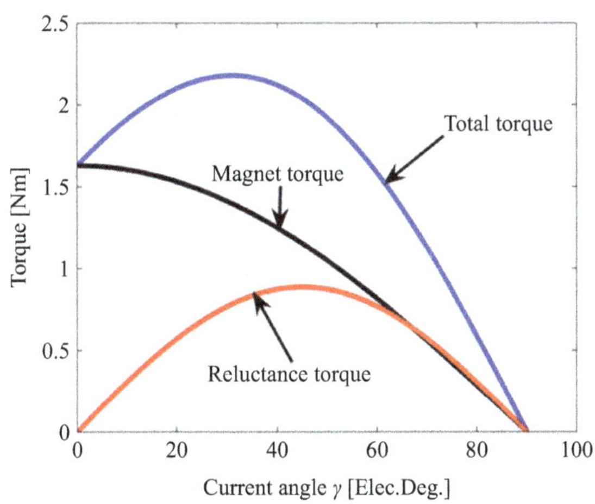

Figure 6.21 Excitation (or magnet), reluctance, and total torque of an IPMSM

6.5.2 Operation under current and voltage limits

When an IPM synchronous motor is run from an inverter, the maximum stator current and the voltage are limited by the inverter/motor current and DC-link voltage ratings, respectively. These constraints can be expressed as

$$I = \sqrt{i_d^2 + i_q^2} \leq I_{sm} \tag{6.12}$$

$$V = \sqrt{v_d^2 + v_q^2} \leq V_{sm} \tag{6.13}$$

where I_{sm} and V_{sm} are the available maximum current and voltage limits of the inverter/motor, respectively.

When an IPMSM runs at a steady speed, i_d and i_q become DC variables and (6.3) and (6.4) can be written as

$$\begin{bmatrix} v_d \\ v_q \end{bmatrix} = \begin{bmatrix} R & -\omega L_q \\ \omega L_d & R \end{bmatrix} \begin{bmatrix} i_d \\ i_q \end{bmatrix} + \begin{bmatrix} 0 \\ \omega \lambda_f \end{bmatrix} \tag{6.14}$$

Substituting (6.13) into (6.14) yields

$$\mathrm{V} = \sqrt{(Ri_d - \omega L_q i_q)^2 + (Ri_q + \omega L_d i_d + \omega \lambda_f)^2} \leq V_{sm} \tag{6.15}$$

If the stator resistance R is neglected, which is applicable when the motor operates near speed and above, (6.15) can be written as

$$(L_q i_q)^2 + (L_d i_d + \lambda_f)^2 \leq \left(\frac{V_{om}}{\omega}\right)^2 \quad \text{where } V_{om} = V_{sm} - RI_{sm} \tag{6.16}$$

Equation (6.12) describes a circle in the i_d–i_q plane with radius I_{sm}. This implies that i_d and i_q currents, or the I trajectory, must always remain within this current limit circle if the motor phase current is not to exceed its rated or maximum value. This circle is centred at the origin of the i_d–i_q plane, as shown in Figure 6.22.

Equation (6.16) describes ellipses with origin at $\left(0, -\frac{\lambda_f}{L_d}\right)$ in the i_d–i_q plane. These ellipses progressively become smaller as the speed ω become higher, as also shown in Figure 6.22. This implies that i_d and i_q currents, or the I trajectory at any speed, must always remain on the voltage limit ellipse for a given speed if the motor phase voltage is not to exceed its rated maximum value. It should be noted that for $i_q > 0$, the IPMSM acts as a motor while for $i_q < 0$, it acts as a generator. Furthermore, $i_d < 0$ applies for field weakening for IPMSMs with $L_q > L_d$. With higher $-i_d$, λ_d is reduced, leading to operation at higher than base speed.

6.5.3 The crossover speed ω_c

Figure 6.23 includes the current-limit circle and a few voltage-limited ellipses for the motor of Table 6.2. From Figure 6.23, the crossover speed (ω_c) may be defined. For $\omega_c = 2{,}400$ rpm, the voltage limit trajectory with zero load torque (or $i_d = i_q = 0$)

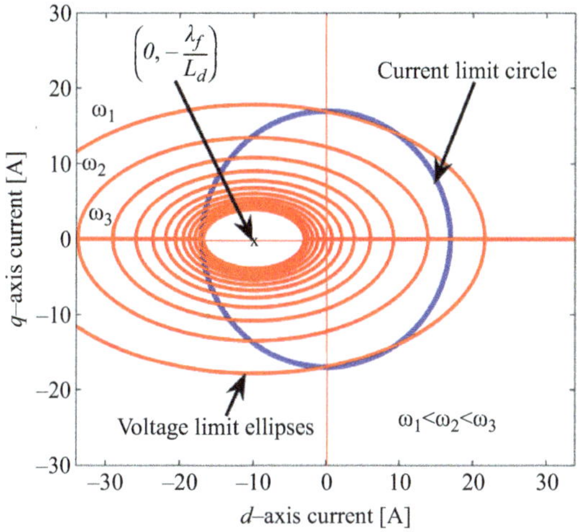

Figure 6.22 Current and voltage limit trajectories in the i_d–i_q plane

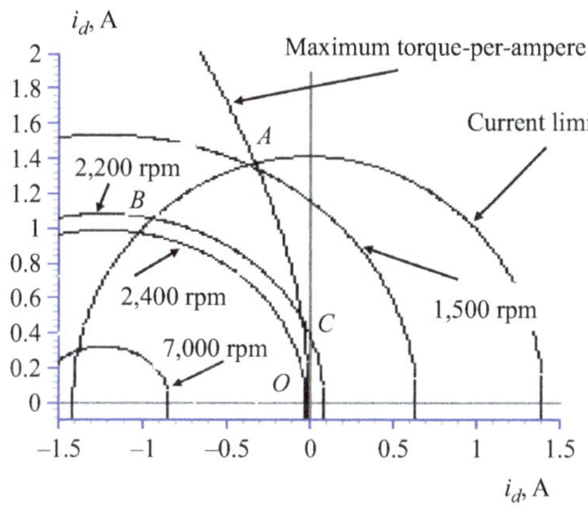

Figure 6.23 Current and voltage limit trajectories in the i_d–i_q plane for motoring

Table 6.2 Parameters of the IPM Motor IPMSM-I

Number of pole pairs	p	2
Stator resistance	R	18.6 Ω
Magnet flux linkage	λ_f	0.447 Wb
d-axis inductance	L_d	0.3885 H
q-axis inductance	L_q	0.4755 H
Rated phase voltage	V	240 V
Rated phase current	I	1.4 A
Base speed	ω_b	1500 rpm
Rated torque	T_b	1.95 Nm
Crossover speed	ω_c	2400 rpm

intersects the MTPA trajectory. When the machine is operated between the base and crossover speeds, the operating mode can be both along the MTPA and FW trajectories, as determined by the load. For example, when the motor runs at 2,200 rpm, the corresponding voltage limit trajectory is *BCO* in Figure 6.23. If the machine is heavily loaded, the i_d–i_q trajectory is along *BC*. When lightly loaded, the trajectory is along *CO* which is on the MPTA trajectory.

6.5.4 Operation with MTPA and field-weakening under maximum current and voltage limits [43,48,50–58]

Figure 6.23 also shows the MTPA trajectory and voltage limit ellipses, for a motor at a few speeds from 1,500 to 7,000 rev/min for a 4-pole IPMSM for which the base speed is 1,500 rev/min. The MTPA trajectory given by (6.11) is independent of rotor speed ω and determined only by three motor parameters and I_{sm}. It should be expected that the MTPA trajectory *OCA* in Figure 6.23 intersects the current limit circle at the base speed ω_b where the motor will develop its maximum torque and rated voltage.

It can also be seen from (6.16) that as the speed increases to infinity, the voltage limit ellipse shrinks to the point $\left(0, -\frac{\lambda_f}{L_d}\right)$, on the $-d$-axis. The d-axis current represented by the centre of the ellipses is the *Characteristic Current*, I_{ch}, defined as.

$$I_{ch} = \frac{-\lambda_f}{L_d} A \tag{6.17}$$

6.5.5 The characteristic current and flux-weakening control conditions

For an IPMSM, three flux-weakening control conditions based on the characteristic current of a machine may exist, as explained below.

(i) $I_{ch} > I_{sm}$
If I_{ch} lies outside the current limit circle, i.e., to the left of the intersection of the current limit circle and the $-d$-axis, voltage limit ellipses above certain high speeds will fall outside the current limit circle, as shown in Figure 6.24(a). At these speeds, the current–voltage limits cannot both be satisfied. As a result, the field-weaking range will be limited and a wide CPSR cannot be achieved. In this case, the maximum speed (ω_{max}) which the IPMSM can be operated is defined when $i_d = -I_{sm}$ and $i_q = 0$. At this point, the torque becomes zero and the machine cannot be operated beyond that speed. The maximum speed is thus given by $\omega_{max} = V_{om}/\left(\lambda_f - L_d I_{sm}\right)$.

(ii) $I_{ch} = I_{sm}$
In this case, the centre of the voltage limit ellipse lies on the cross point of the current limit trajectory and the negative d-axis as shown in Figure 6.24 (b). Thus, there will always be an intersecting point between the current limit trajectory and the voltage limit ellipse, regardless of the shrinkage of the

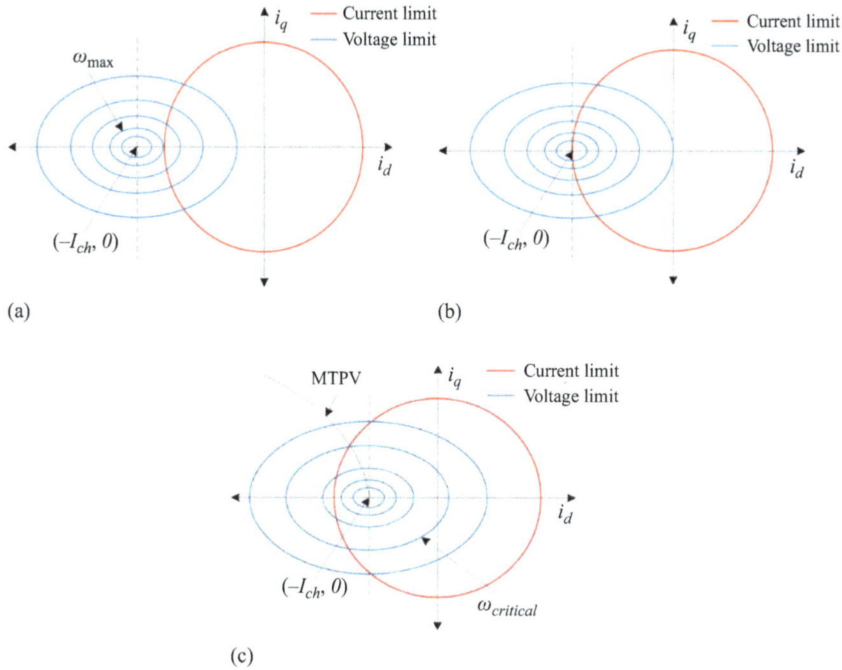

Figure 6.24 The significance of the characteristic current in determining CPSR

voltage ellipse with the increasing speed. In this case, the CPSR is infinite theoretically and it is considered as the optimum flux-weakening condition.

(iii) $I_{ch} < I_{sm}$

If the characteristic current is less than the rated current of the system, voltage limit ellipses tend to shrink at a point inside the current limit circle with increasing speed as shown in Figure 6.24(c). Even though this demonstrates that the machine can run up to an infinite speed, the CPSR is limited to a finite speed which may not cover the required speed range of an EV. For such a machine the CPSR can be extended by using the maximum torque per voltage (MTPV) control [48,49]. This entails the i_d–i_q trajectory moving along the MTPV trajectory as indicated in Figure 6.24(c). It should be noted that the characteristic current of an IPMSM will tend to move to the right due to fall of If with rise of magnet temperature during normal operation. The *d*-axis inductance L_d does not change much with temperature or operating current.

6.6 Selection of control modes

6.6.1 Operation with MTPA below base speed

It is seen in Figure 6.23 that the voltage limit ellipse for operation below the base speed is larger than the one for base speed. Therefore, if the stator current vector is controlled according to the maximum torque-per-ampere trajectory and satisfies the current limit, which is the trajectory *OA*, it must satisfy the voltage limit since the current vector is inside the voltage limit ellipse. Therefore, the maximum torque-per-ampere mode is selected for constant torque operation.

The *d*- and *q*-axis currents, i_{dA} and i_{qA}, with which the maximum torque is produced, are determined by (6.11) and (6.12) when $I_s = I_{sm}$.

$$i_{dA} = \frac{\lambda_f}{2(L_q - L_d)} - \sqrt{\frac{\lambda_f^2}{4(L-L)^2} + I_{sm}^2 - i_{dA}^2} \tag{6.18}$$

Solving (6.18), one gets

$$i_{dA} = \frac{\lambda_f}{4(L_q - L_d)} - \sqrt{\frac{\lambda_f^2}{16(L_q - L_d)^2} + \frac{I_{sm}^2}{2}} \tag{6.19}$$

$$i_{qA} = \sqrt{I_{sm}^2 - i_{dA}^2} \tag{6.20}$$

This is the operating point at which the motor produces the maximum torque and the limit value of the outer control loop for constant torque operation is then determined by (6.20).

6.6.2 Operation between base and crossover speeds

When the rotor speed is in the range from ω_b to ω_c, the control mode is determined by the load. When the motor is under light load, it may operate near the crossover speed with MTPA control. When the motor is fully loaded, it must be controlled according to the voltage limit trajectory above the base speed. For instance, if the motor runs at 2,200 rpm, the corresponding voltage limit trajectory is *BCO* in Figure 6.23. Thus, if the motor is heavily loaded and the current vector is along the trajectory *BC*, it must be controlled according to the voltage limit trajectory. Otherwise, the motor may be still controlled with MTPA control along the trajectory *CO*. The determination of the control mode is based on the calculated i_d from both equations (6.11) and (6.16). If the calculated i_d from (6.11) is smaller than the one calculated from (6.16), the MTPA trajectory is used for constant torque operation. Otherwise, voltage limit trajectory is used to control the current vector for field weakening operation. The limit value for the outer control loop is still determined by (6.12).

6.6.3 Operation with field-wakening above the crossover speed

To run the motor above the crossover speed ω_c, the flux λ_f, which is the flux linkage along the rotor *d*-axis, must be reduced. Although the flux is already reduced with the MTPA control, the voltage limitation is no longer satisfied when the rotor speed is above the crossover speed. The stator current vector is, therefore, controlled according to the voltage limit trajectory instead of the MTPA trajectory. Thus, i_d

and i_q are determined according to the voltage and current limit equations (6.16) and (6.12), respectively. The limit value, i_{dv} and i_{qv}, of the outer control loop for such a field weakening operation is, therefore, determined by these two equations with $I_s = I_{sm}$, which is inversely proportional to the rotor speed:

$$i_{dv} = -\frac{\lambda_f L_d}{a} + \frac{1}{a}\sqrt{\lambda_f^2 L_d^2 - a\,b} \tag{6.21}$$

where

$$a = L_d^2 - L_q^2 \tag{6.22}$$

$$b = I_{sm}^2 L_q^2 + \lambda_f^2 - \frac{V_{sm}^2}{\omega^2} \tag{6.23}$$

and

$$\mathrm{i_{qv}} = \sqrt{\mathrm{I_{sm}^2} - \mathrm{i_{dv}^2}} \tag{6.24}$$

6.6.4 MTPV trajectory control

If $I_{ch} < I_{sm}$, the field-weakening speed range can be extended by using MTPV control after the conventional flux-weakening control. The MTPV trajectory delivers the optimal operating point that develops the maximum torque under the voltage constraint. The relationships between i_d and i_q for MTPV control can be derived as follows.

From the flux linkage Equation (6.5),

$$i_d = \frac{\lambda_d - \lambda_f}{L_d} \tag{6.25}$$

$$i_q = \frac{\lambda_q}{L_q} = \frac{\sqrt{\lambda_s^2 - \lambda_d^2}}{L_q}, \text{ where } \lambda_s = \sqrt{\lambda_d^2 + \lambda_q^2} \tag{6.26}$$

By substituting (6.25) and (6.25) into (6.6),

$$T = \frac{3}{2}p\sqrt{\lambda_s^2 - \lambda_d^2}\left\{\frac{\lambda_d}{L_q} - \frac{(\lambda_d - \lambda_f)}{L_d}\right\} \tag{6.27}$$

By taking $\frac{dT}{d\lambda_d}$ and equating the derivative to 0 for maximum torque,

$$\lambda_d = \frac{-L_q\lambda_f + \sqrt{(L_q\lambda_f)^2 + 8(L_q - L_d)^2\lambda_s^2}}{4(L_q - L_d)}, \text{ where } \lambda_s = \frac{V_{om}}{\omega} \tag{6.28}$$

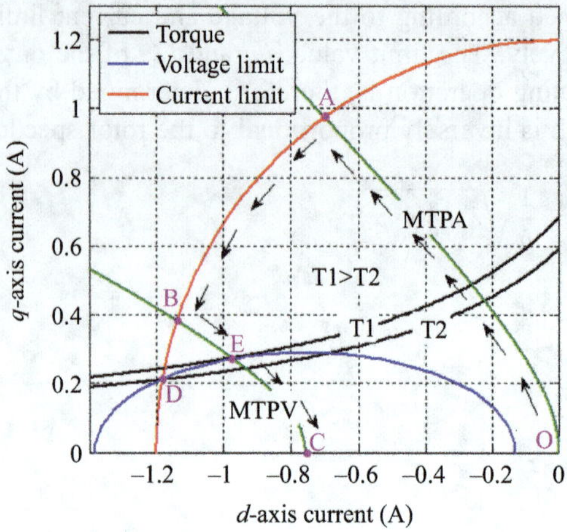

Figure 6.25 MTPA, FW, and MTPV control trajectories

By substituting (6.28) in to (6.25) and (6.26), i_d and i_q for MTPV control are obtained.

$$i_d = -\frac{\lambda_f}{L_d} - \Delta i_d \tag{6.29}$$

$$i_q = \frac{\sqrt{\left(\frac{V_{om}}{\omega}\right)^2 - \left(L_d \Delta i_d\right)^2}}{L_q} \tag{6.30}$$

where

$$\Delta i_d = \frac{-L_q\lambda_f + \sqrt{\left(L_q\lambda_f\right)^2 + 8\left(L_q - L_d\right)^2\left(\frac{V_{om}}{\omega}\right)^2}}{4L_d\left(L_q - L_d\right)}$$

The MTPV trajectory in the i_d–i_q plane is shown in Figure 6.25. If the motor speed is below the critical speed, MTPV control cannot be applied because the intersection of the MTPV trajectory and the voltage ellipse will be outside of the current limit trajectory.

6.7 Controller implementation issues

6.7.1 Voltage compensation for avoiding current controller saturation

Separate i_d and i_q current control are performed in the rotor flux reference frame, as indicated in Figure 6.20. Measured i_d and i_q are fed back to the q- and d-axes

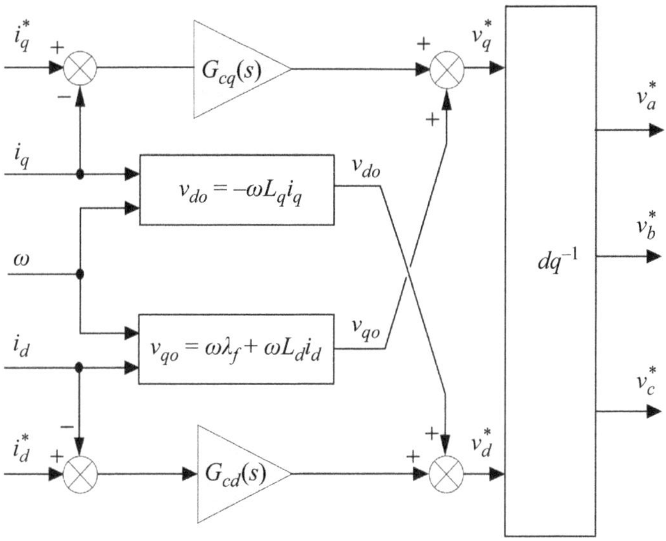

Figure 6.26 The d- and q-axes current controllers with decoupling

current controllers. PI controllers are usually used for fast dynamic performance and high accuracy in the steady state. The de-coupling block of Figure 6.26 helps remove the coupling between the *d*- and *q*-axes voltages and is used for the fast transient response of these controllers. Transient responses of i_d and i_q with and without voltage decoupling based on (6.14) are shown in Figure 6.27(a) and (b).

6.7.2 Prevention of controller saturation during field-weakening

It should be noted that the stator voltage constraint in the previous sections was based on the steady-state equations (6.14) and (6.16). In the flux weakening operation, the stator voltage is kept equal to the maximum stator voltage V_{sm}. The voltage resulting from commanded i_d^* and i_q^* may exceed V_{sm} in transient operations when i_d or i_q are required to change abruptly. As a result, the *dq* current controllers can be saturated and the current control performance then becomes poor. To prevent saturation, current i_d should be controlled prior to i_q current in the case of current controller saturation. The voltage compensation which was indicated in Figure 6.20 and detailed in Figure 6.28 can be used for preventing the current controllers from saturation, in which v_d^* and v_q^* are the output signals from the voltage de-coupling block of Figure 6.26 and v_{dc}^* are the v_{qc}^* voltage references for inverse *dq* transformation (dq^{-1}) and application to the inverter.

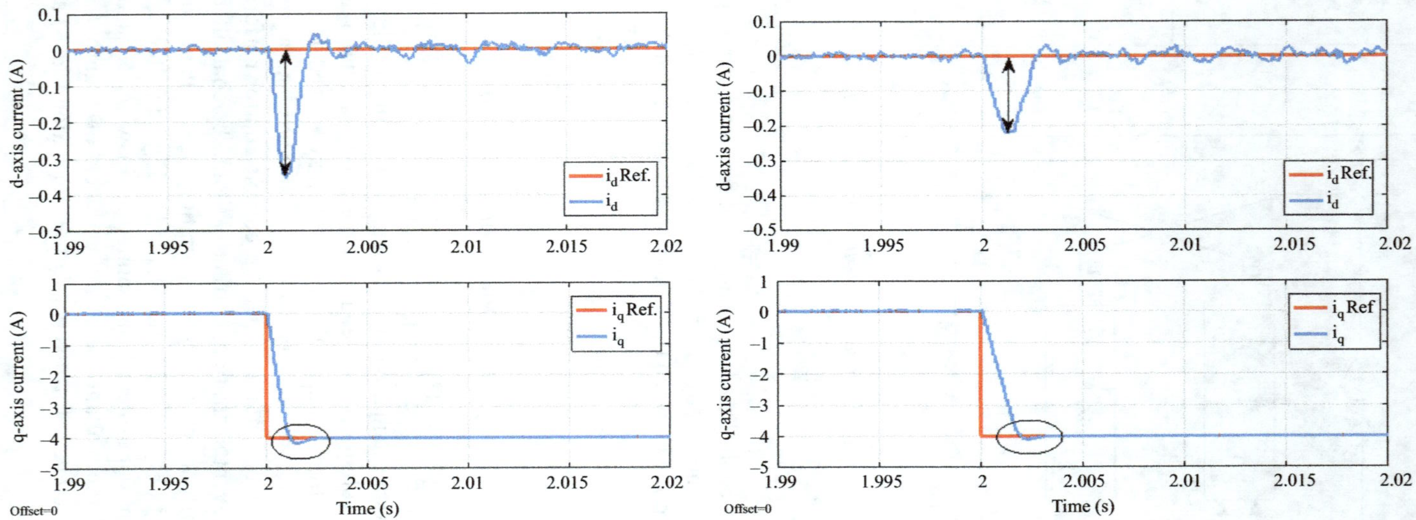

Figure 6.27 (a) i_d and i_q transients without dq decoupling and (b) with decoupling

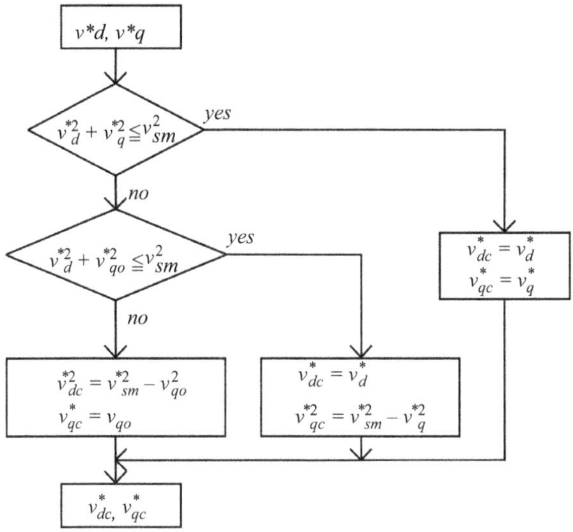

Figure 6.28 Flow-chart of voltage compensation for avoiding controller saturation during field weakening

6.8 Current controller gains for FOC IPMSM drives

Several highly integrated and sophisticated simulation platforms like MATLAB®/Simulink® and PSIM exist that can simulate many steady-state and dynamic performances of a drive system. Many complex drive issues can be addressed at the design stage on such platforms, without having to build a drive at the outset. Tuning of controllers for various control loops, effects of machine, load parameter variations, and drive efficiency map can all be pre-assessed.

To achieve good dynamic performance and operational features, a cascaded structure of closed-loop controls is normally employed in the electric drive system with an outer speed loop Figure 6.29 shows a cascaded control structure for electric drives, which includes speed control (for cruise control only) in the outer loop and current (or torque) controllers in the inner loop comprising of d-axis and q-axis current controllers (see Figure 6.19).

Usually, PI controllers are used for these controllers. Many procedures are available for designing and tuning such controllers. Nevertheless, for the sake of completeness, a short account is included here.

From the dq representation of Figure 6.16, the approximate plant model can be taken as a first order Laplace Equation (6.31):

$$G_i(s) = \frac{K_m}{T_m s + 1} \tag{6.31}$$

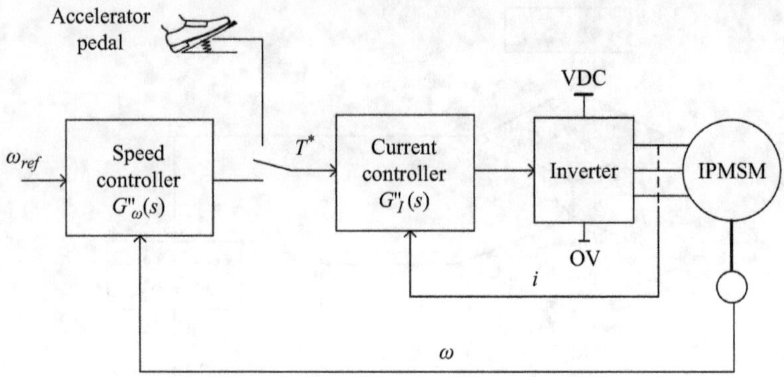

Figure 6.29 Cascaded feedback control loops for electric drives

1. Discretization of the plant model with forward Euler discretization method. Substituting $s = \frac{1-z^{-1}}{T_s z^{-1}}$ into (6.31) yields the discrete plant model:

$$G_i(z^{-1}) = \frac{K_m\left(\frac{T_s}{T_m}\right)z^{-1}}{1+\left(\frac{T_s-T_m}{T_m}\right)z^{-1}} = \frac{b_1 z^{-1}}{1+a_1 z^{-1}} \tag{6.32}$$

where T_s is the sampling time for the discrete current controller.
2. Define a discrete PI controller using the same transform.

$$G_{PI}(z^{-1}) = K_P + K_I\frac{T_s z^{-1}}{1-z^{-1}} = \frac{K_P + (K_I T_s - K_P)z^{-1}}{1-z^{-1}} = \frac{q_0 + q_1 z^{-1}}{1-z^{-1}} \tag{6.33}$$

Combining the discrete transfer function of the plant and PI controller, the closed-loop transfer function is given by

$$G_{CL}(z^{-1}) = \frac{q_0 b_1 z^{-1} + q_1 b_1 z^{-2}}{1+(a_1 - 1 + q_0 b_1)z^{-1} + (-a_1 + q_1 b_1)z^{-2}} \tag{6.34}$$

where the denominator of the transfer function is the characteristic polynomial of the system.

$$P_{CL}(z^{-1}) = 1 + (a_1 - 1 + q_0 b_1)z^{-1} + (-a_1 + q_1 b_1)z^{-2} \tag{6.35}$$

3. The characteristic polynomial for achieving the required performance is defined as

$$P_{cd}(z^{-1}) = 1 + \alpha_1 z^{-1} + \alpha_2 z^{-2} \tag{6.36}$$

where $\alpha_1 = -2e^{-\xi\omega_n T_s}\cos\left(\omega_n T_s\sqrt{1-\xi^2}\right)$ and $\alpha_1 = e^{-2\xi\omega_n T_s}$.
4. To determine the controller parameters, set the characteristic polynomial for the system equal to the characteristic polynomial for the required performance,

i.e., $P_{CL}(z^{-1}) = P_{cd}(z^{-1})$. Solving for the proportional and integral gains of the controller for the first-order system yields

$$\begin{cases} K_P = \dfrac{a_2 - a_1 + 1}{b_1} \\ K_I = \dfrac{a_2 + a_1}{b_1 T_s} + \dfrac{K_P}{T_s} \end{cases} \tag{6.37}$$

The above general equations for the PI controller gains contain two coefficients, the damping factor ξ and natural frequency ω_n. These two coefficients determine two of the main step-response characteristics, overshoot σ and response time t_r:

$$\sigma = e^{\frac{-\pi\xi}{\sqrt{1-\xi^2}}} \tag{6.38}$$

$$t_r \cong \begin{cases} \dfrac{4}{\omega_n \xi}, \text{if}(\xi < 0.7) \\ \dfrac{6\xi}{\omega_n}, \text{if}(\xi \geq 0.7) \end{cases} \tag{6.39}$$

It should be noted that the gain K_m and time constant T_m of the first-order system $G_i(s)$ for an IPMSM are

$$\begin{cases} K_m^{d,q} = 1/R_s \\ T_m^{d,q} = L_{d,q}/R_s \end{cases} \tag{6.40}$$

The proportional and integral gains of the controllers will be different and must be calculated separately.

Given the system requirements such as overshoot and response time, one can now solve for the PI current controller gains in the rotor *dq* frame.

6.9 Dynamic responses and trajectory following

A few results on the following of the MTPA and FW trajectories are included in Figure 6.30 for the distributed-winding IPMSM of Table 6.2. Figure 6.30(a) shows the i_d and i_q along the MTPA line for this motor and Figure 6.30(b) shows their transient values during the MTPA operation. Figure 6.30(c) and (d) includes i_d and i_q current during MTPA and FW operation to 2200 rpm. For this machine, $I_{ch} > I_{sm}$.

For the IPMSM-2 in Table 6.3, $I_{ch} < I_{sm}$. A few experimental results on the trajectory following, which include MTPV, are included in Figure 6.31(a)–(f). Operation with MTPV for this motor leads to some increase in torque at high speed with deep field weakening and a consequent increase in CPSR, as is shown in Figure 6.32(a) and (b).

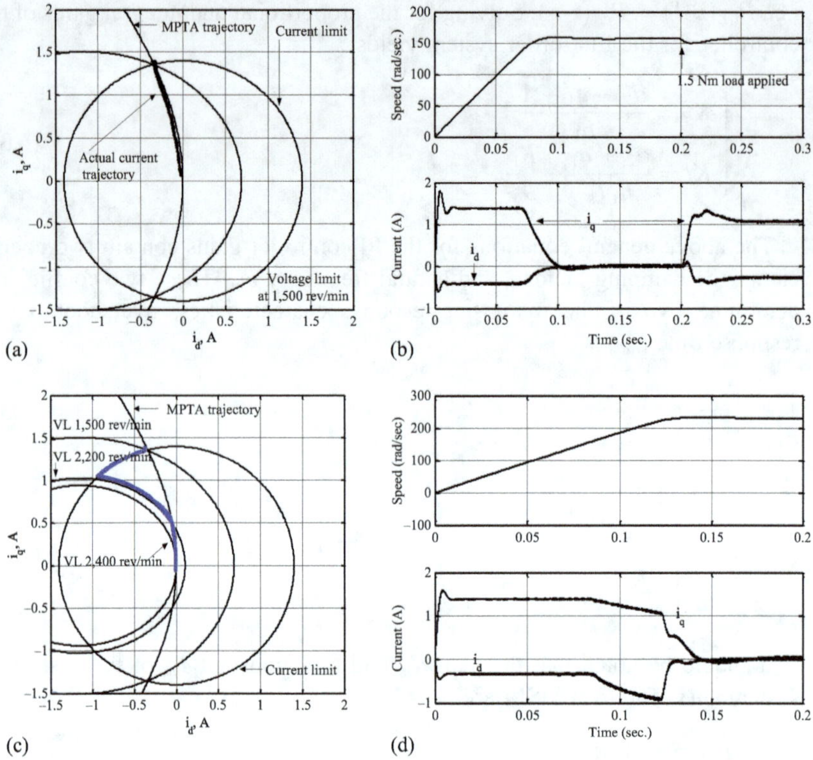

Figure 6.30 i_d *and* i_q *current during MTPA and FW trajectory following for the motor of Table 6.2. (a) Operation with MTPA below base speed (1,500 rev/min), (b) speed and current transients during acceleration to base speed, (c) operation with field weakening under current and voltage limits, and (d) speed and current transients over the whole speed range from 0 to 2,400 rev/min.*

Table 6.3 Parameters of the DW IPMSM-2

Parameter	**Value**
Number of pole pairs, p	2
Stator resistance, R	18.6 Ω
Magnet flux linkage, λ_f	0.18 Wb
d-axis inductance, L_d	0.238 H
q-axis inductance, L_q	0.5128 H
Friction coefficient, T_F	0.00029 Nm/rad/s
Rotor inertia, J	0.001176 kgm^2
Rated phase voltage (peak), V_{sm}	178 V
Rated current (peak) I_{sm}	1.2 A
Characteristic current	0.75 A
Base speed, ω_b	1,500 rpm
Critical speed, $\omega_{critical}$	3,500 rpm
Crossover speed, ω_c	4,200 rpm

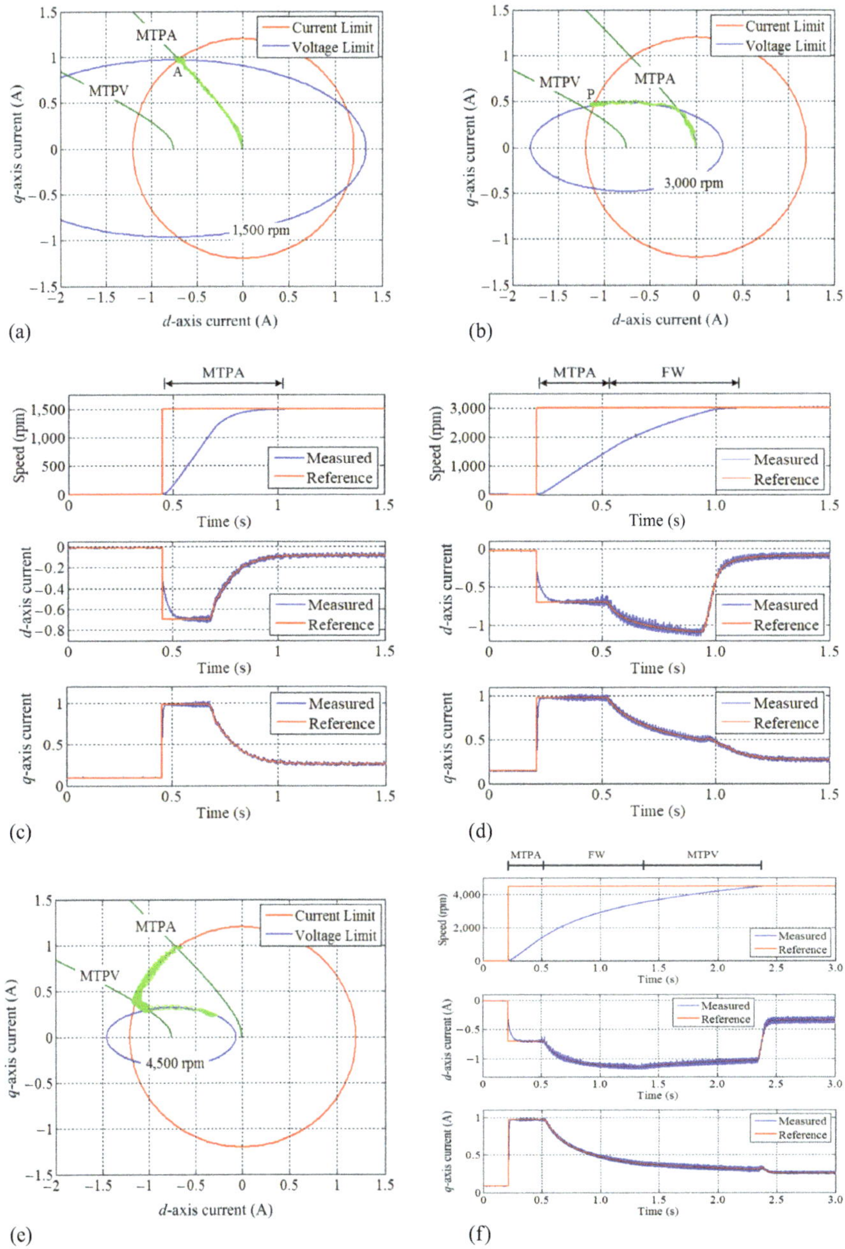

Figure 6.31 Trajectory following of IPMSM-2 with $I_{ch} < I_{sm}$

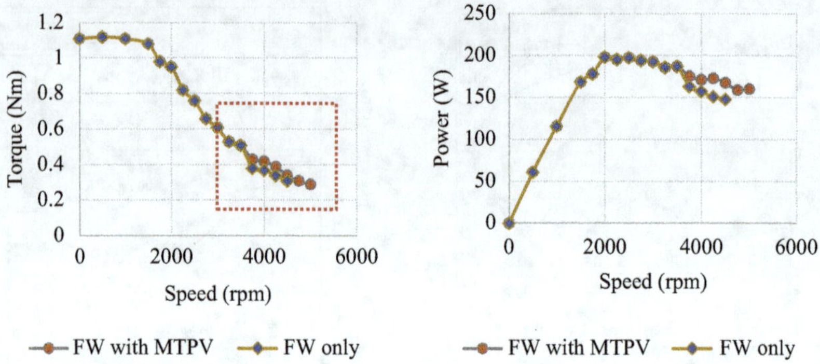

Figure 6.32 Extension of CPSR with MTPV for the IPMSM-2 of Table 6.3

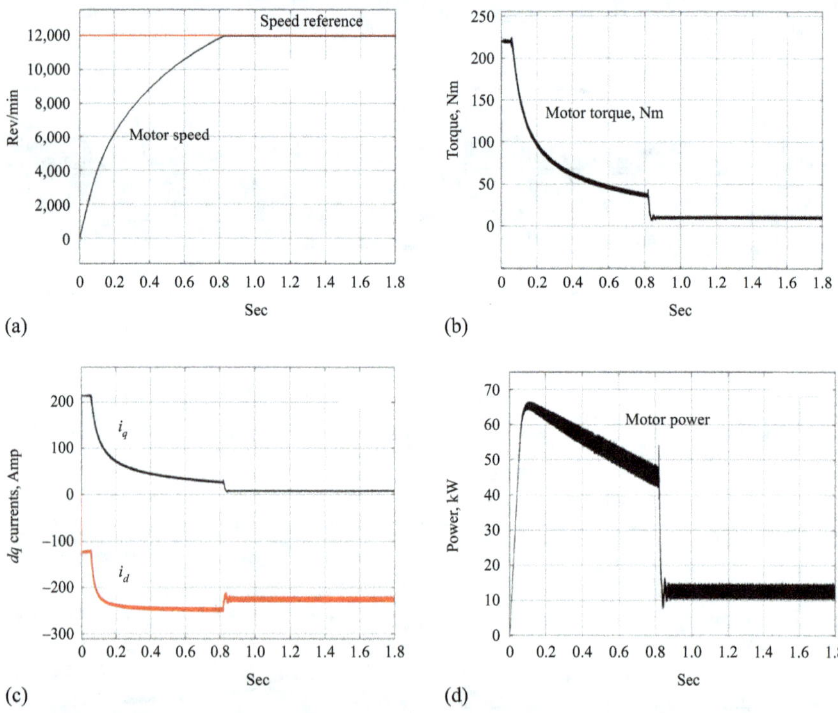

Figure 6.33 Control simulation results of a 40 kW IPM motor for an EV with CPSR = 4

The foregoing simulation and experimental results in Figures 6.30–6.32 are for two low-power IPMSMs-1 and IPMSM-2. Figure 6.33(a)–(d) shows dynamic responses of (a) speed, (b) torque, (c) *dq* currents, and (d) developed power of a 40 kW IPMSM used in an EV. Table 6.4 includes all parameters of this machine for which $I_{ch} \approx I_{sm}$, so that MTPV operation is not required.

Table 6.4 Motor and Drive Parameter Values

Item	Symbol	Value	Unit
DC Bus Voltage	*VDC*	320	V
Stator resistance/phase	*R*	0.04132	Ω
Rotor Flux Linkage	λ_f	0.152	Wb
Pole pairs	*p*	3	
d-axis inductance	L_d	0.46	mH
q-axis inductance	L_q	1.06	mH
Base speed	ω_b	3,000	rpm
Moment of inertia	*J*	0.047565	kg m^2

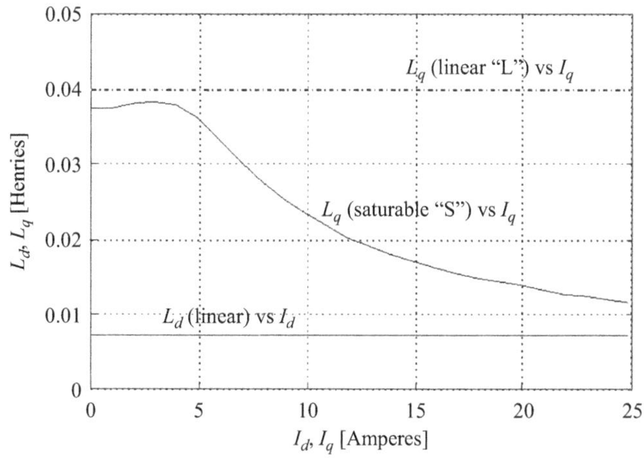

Figure 6.34 Variation of L_q with current

6.10 Variation of machine parameters and impacts

It may be noted that the trajectories described in Section 6.5 have parameter dependency. The magnet flux linkage λ_f decreases by about 0.1%/°C of rise in magnet temperature, however, this change is rather slow and can be obtained by a variety of recursive estimation techniques. The variation of dq inductance with current, especially of L_q, is more pronounced due to magnetic saturation along the *q*-axis, as indicated in Figure 6.34. The dynamics of this change are as fast as current changes and thus more difficult to adapt to. It may be also noted that L_d does not have significant dependency on the current. As a result of variation L_q with current, the MTPA trajectory should ideally follow the trajectory OB in Figure 6.35, rather than OA when saturation of L_q is neglected. The shift of the MTPA obviously has some implications on the value of the base speed, field-weakening trajectories efficiency and CPSR. Look-up tables for determining these

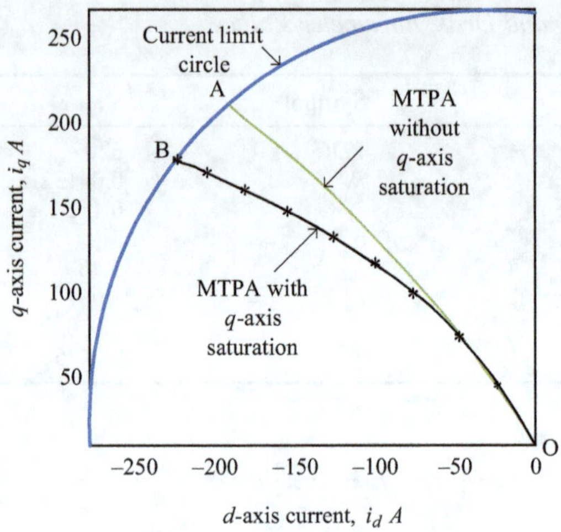

Figure 6.35 MTPA trajectory with q-axis saturation

trajectories thus become the current dependent and rather complex to manage [59–61]. These will not be discussed here. More can be found in [62,63].

6.11 Summary

This chapter has attempted to give an overview of IPM machines for the application in EVs. The progression of this motor over the past several years for the inclusion of high CPSR, power density and efficiency, and low torque ripple have been reviewed. Of particular interest is the rotor configuration that has offered high CPSR, so that these motors could cover the required speed range with the torque requirement of vehicles. This has implications for the number and size of motors in such applications. The control techniques of the IPM machines for exploiting the full CPSR and by meeting the current and voltage limits were discussed by a mix of steady-state and dynamic models in the rotor *dq* reference frame. Controller design and control strategies for effective control of currents (and hence torque) and trajectory following have also been included. Extensive simulation and experimental results on three IPMSMs, one of which is a 40-kW IPM machine for an EV, have been included to demonstrate the suitability and applicability of optimum control boundaries (trajectories) for these motors for high efficiency, high dynamic response, and wide field-weakening.

Symbols

$\omega_{\mathbf{max}}$	full speed range
$\omega_{\mathbf{b}}$	base speed
T–ω	torque–speed

δ_s	steering angle
L_d **and** L_q	machine steady-state inductance parameters
R	stator resistance
λ_f	stator flux linkage parameter
f_1	input supply frequency
f_o	base supply frequency
p	pole pairs
E_{fo}	back-emf at the base speed
X_{do} **and** X_{qo}	*dq* synchronous reactance at the base frequency f_o
I_d **and** I_q	direct and quadrature axis current
E_q	q-axis back-emf
V_o	rated phase voltage
E_{fo}	PM excited phase voltage
δ	load angle
E_f	excitation voltage
V	inverter output voltage
I	input current
T^*	reference torque
i_d^* **and** i_q^*	current references
$i_a - i_c$	Motor currents
θ	rotor position/shaft angle
v_d^* **and** v_q^*	voltage references
T_r	reluctance torque
γ	angle of the current phasor with the q-axis
I_{sm}	radius of i_d–i_q plane
ω_c	crossover speeds
dq^{-1}	inverse *dq* transformation
ξ	damping factor
ω_n	natural frequency
t_r	response time
σ	overshoot
K_m	Gain of first-order system
T_m	time constant of the first-order system

Glossary

AFPMM	axial-flux permanent magnet motor
CPSR	constant power speed range

DTC	direct torque control
EV	electric vehicles
FE analysis	finite-element analysis
FOC	field-oriented control
FSCW	fractional-slot concentrated winding
FW	field weakening
ICE	internal combustion engine
IM	induction motors
IPM	interior permanent-magnet motors
IPMSM	interior permanent-magnet synchronous motors
LCM	least common multiple
MTPA	maximum torque per ampere
MTPV	maximum torque per voltage
NdFeB	neodymium iron boron magnets
PHEV	plug-in hybrid electric vehicles
PI	proportional-integral
SRM	switched reluctance motors
WRSM	wound rotor synchronous motors

References

[1] Y. Honda, T. Higaki, S. Morimoto, and Y. Takeda, "Rotor design optimisation of a multi-layer interior permanent-magnet synchronous motor," *IEE Proceedings – Electric Power Applications*, vol. 145, pp. 119–124, 1998.

[2] R. Dutta and M. F. Rahman, "Design and analysis of an interior permanent magnet (IPM) machine with very wide constant power operation range," *IEEE Transactions on Energy Conversion*, vol. 23, pp. 25–33, 2008.

[3] F. Parasiliti, M. Villani, S. Lucidi and F. Rinaldi, "Finite-element-based multiobjective design optimization procedure of interior permanent magnet synchronous motors for wide constant-power region operation", *IEEE Trans. Ind. Electron*, vol. 59, no. 6, pp. 2503–2514, 2012.

[4] W. Kakihara, M. Takemoto, and S. Ogasawara, "Rotor structure in 50 kW spoke-type interior permanent magnet synchronous motor with ferrite permanent magnets for automotive applications," in 2013 *IEEE Energy Conversion Congress and Exposition*, 2013, pp. 606–613.

[5] J. Cros and P. Viarouge, "Synthesis of high performance PM motors with concentrated windings," *IEEE Transactions on Energy Conversion*, vol. 17, pp. 248–253, 2002.

[6] C. Jae Seok, K. Izui, S. Nishiwaki, A. Kawamoto, and T. Nomura, "Topology optimization of the stator for minimizing cogging torque of IPM motors," *IEEE Transactions on Magnetics*, vol. 47, no. 10, pp. 3024–3027, 2011.

[7] Z. Wenliang, Z. Fei, T. A. Lipo, and K. Byung-Il, "Optimal design of a novel V-type interior permanent magnet motor with assisted barriers for the improvement of torque characteristics," *IEEE Transactions on Magnetics*, vol. 50, no. 11, pp. 1–4, 2014.

[8] A. M. El-Refaie, "Fractional-slot concentrated-windings synchronous permanent magnet machines: opportunities and challenges," *IEEE Transactions on Industrial Electronics*, vol. 57, pp. 107–121, 2010.

[9] R. Dutta, L. Chong, and M. F. Rahman, "Design and experimental verification of an 18-slot/14-pole fractional-slot concentrated winding interior permanent magnet machine," *IEEE Transactions on Energy Conversion*, vol. 28, pp. 181–190, 2013.

[10] N. Bianchi, S. Bolognani, Pre, x, M. D., and G. Grezzani, "Design considerations for fractional-slot winding configurations of synchronous machines," *IEEE Transactions on Industry Applications*, vol. 42, no. 4, pp. 997–1006, 2006.

[11] P. B. Reddy, A. M. El-Refaie, H. Kum-Kang, J. K. Tangudu, and T. M. Jahns, "Comparison of interior and surface PM machines equipped with fractional-slot concentrated windings for hybrid traction applications," *IEEE Transactions on Energy Conversion*, vol. 27, no. 3, pp. 593–602, 2012.

[12] L. Alberti and N. Bianchi, "Theory and design of fractional-slot multilayer windings," *IEEE Transactions on Industry Applications*, vol. 49, no. 2, pp. 841–849, 2013.

[13] J. Yang, G. Liu, W. Zhao, Q. Chen, L. Sun, and X. Zhu, "Quantitative comparison for fractional-slot concentrated-winding configurations of permanent-magnet vernier machines", *IEEE Transactions on Magnetics*, vol. 19, no. 7, pp. 3826–3829, 2013.

[14] P. B. Reddy, A. M. El-Refaie, and H. Kum-Kang, "Effect of number of layers on performance of fractional-slot concentrated-windings interior permanent magnet machines," *IEEE Transactions on Power Electronics*, vol. 30, no. 4, pp. 2205–2218, 2015.

[15] G. Dajaku, X. Wei, and D. Gerling, "Reduction of low space harmonics for the fractional slot concentrated windings using a novel stator design," *IEEE Transactions on Magnetics*, vol. 50, no. 5, pp. 1–12, 2014.

[16] J. K. Tangudu, T. M. Jahns, and A. El-Refaie, "Unsaturated and saturated saliency trends in fractional-slot concentrated-winding interior permanent magnet machines," *in Energy Conversion Congress and Exposition (ECCE)*, 2010 *IEEE*, 2010, pp. 1082–1089.

[17] A. M. El-Refaie, J. P. Alexander, S. Galioto, *et al.*, "Advanced high-power-density interior permanent magnet motor for traction applications,"

IEEE Transactions on Industry Applications, vol. 50, no. 5, pp. 3235–3248, 2014.

[18] S. J. Galioto, P. B. Reddy, A. M. El-Refaie, and J. P. Alexander, "Effect of magnet types on performance of high-speed spoke interior-permanent-magnet machines designed for traction applications," *IEEE Transactions on Industry Applications*, vol. 51, no. 3, pp. 2148–2160, 2015.

[19] E. Carraro, N. Bianchi, S. Zhang, and M. Koch, "Design and performance comparison of fractional slot concentrated winding spoke type synchronous motors with different slot-pole combinations," *IEEE Transactions on Industry Applications*, vol. 54, no. 3, pp. 2276–2284, 2018.

[20] J. Han, J. Lee, and W. Kim, "A study on optimal design of the triangle type permanent magnet in IPMSM rotor by using the Box-Behnken design," *IEEE Transactions on Magnetics*, vol. 51, no. 3, pp. 1–4, 2015.

[21] L. Alberti, M. Barcaro, and N. Bianchi, "Design of a low-torque-ripple fractional-slot interior permanent-magnet motor," *IEEE Transactions on Industry Applications*, vol. 50, no. 3, pp. 1801–1808, 2014.

[22] W. Ren, Q. Xu, and Q. Li, "Asymmetrical V-shape rotor configuration of an interior permanent magnet machine for improving torque characteristics," *IEEE Transactions on Magnetics*, vol. 51, no. 11, pp. 1–4, 2015.

[23] P. G. Alotto, C. Eranda, B. Brandstatter, *et al.*, "Stochastic algorithms in electromagnetic optimization," *IEEE Transactions on Magnetics*, vol. 34, no. 5, pp. 3674–3684, 1998.

[24] S. D. Sudhoff, *Power Magnetic Devices: A Multi-Objective Design Approach*, John Wiley & Sons, New York, 2014.

[25] D. Yao and D. M. Ionel, "A review of recent developments in electrical machine design optimization methods with a permanent-magnet synchronous motor benchmark study," *IEEE Transactions on Industry Applications*, vol. 49, no. 3, pp. 1268–1275, 2013.

[26] J. Legranger, G. Friedrich, S. Vivier, and J. C. Mipo, "Combination of finite-element and analytical models in the optimal multidomain design of machines: application to an interior permanent-magnet starter generator," *IEEE Transactions on Industry Applications*, vol. 46, no. 1, pp. 232–239, 2010.

[27] A. Arkadan, M. ElBsat, and M. Mneimneh, "Particle swarm design optimization of ALA rotor SynRM for traction applications," *IEEE Transactions on Magnetics*, vol. 45, no. 3, pp. 956–959, 2009.

[28] G. Y. Sizov, Z. Peng, D. M. Ionel, N. A. O. Demerdash, and M. Rosu, "Automated multi-objective design optimization of PM AC machines using computationally efficient FEA and differential evolution," *IEEE Transactions on Industry Applications*, vol. 49, no. 5, pp. 2086–2096, 2013.

[29] G. F. Uler and O. A. Mohammed, "Utilizing genetic algorithms for the optimal design of electromagnetic devices," *in Southeastcon'95. Visualize the Future., Proceedings*, IEEE, Piscataway, NJ, 1995, pp. 92–95.

[30] N. Bianchi and S. Bolognani, "Design optimisation of electric motors by genetic algorithms," *IEE Proceedings-Electric Power Applications*, vol. 145, no. 5, pp. 475–483, 1998.

[31] Y. Ahn, J. Park, C.-G. Lee, J.-W. Kim, and S.-Y. Jung, "Novel memetic algorithm implemented with GA (genetic algorithm) and MADS (mesh adaptive direct search) for optimal design of electromagnetic system," *IEEE Transactions on Magnetics*, vol. 46, no. 6, pp. 1982–1985, 2010.

[32] X. Jannot, J. C. Vannier, C. Marchand, M. Gabsi, J. Saint-Michel, and D. Sadarnac, "Multiphysic modeling of a high-speed interior permanent-magnet synchronous machine for a multiobjective optimal design," *IEEE Transactions on Energy Conversion*, vol. 26, no. 2, pp. 457–467, 2011.

[33] G. Y. Sizov, D. M. Ionel, and N. A. O. Demerdash, "Modeling and parametric design of permanent-magnet AC machines using computationally efficient finite-element analysis," *IEEE Transactions on Industrial Electronics*, vol. 59, no. 6, pp. 2403–2413, 2012.

[34] Z. Peng, D. M. Ionel, and N. A. O. Demerdash, "Morphing parametric modeling and design optimization of spoke and V-type permanent magnet machines by combined design of experiments and differential evolution algorithms," *in Energy Conversion Congress and Exposition (ECCE)*, 2013 IEEE, Piscataway, NJ, 2013, pp. 5056–5063.

[35] A. Fatemi, D. M. Ionel, N. A. O. Demerdash, and T. W. Nehl, "Fast multi-objective CMODE-type optimization of PM machines using multicore desktop computers," *IEEE Transactions on Industry Applications*, vol. 52, no. 4, pp. 2941–2950, 2016.

[36] Y. Duan and D. Ionel, "A review of recent developments in electrical machine design optimization methods with a permanent-magnet synchronous motor benchmark study", *IEEE Transactions on Industry Applications*, vol. 49, no. 3, pp. 1268–1275, 2013.

[37] Y. Wang, D. Ionel and D. Staton, "Ultrafast steady-state multi-physics model for pm and synchronous reluctance machines", *IEEE Transactions on Industry Applications*, vol. 51, no. 5, pp. 3639–3646, 2015.

[38] G. Lei, C. Liu, J. Zhu and Y. Guo, "Techniques for multilevel design optimization of permanent magnet motors", *IEEE Transactions on Energy Conversion*, vol. 30, no. 4, pp. 1574–1584, 2015.

[39] A. Fatemi, D. M. Ionel, N. A. O. Demerdash, and T. W. Nehl, "Optimal design of IPM motors with different cooling systems and winding configurations," *IEEE Transactions on Industry Applications*, vol. 52, no. 4, pp. 3041–3049, 2016.

[40] R. Dutta, A. Pouramin, and M. F. Rahman, "A novel rotor topology for high-performance fractional slot concentrated winding interior permanent magnet machine", *IEEE Transactions on Energy Conversion*, vol. 36, no. 2, pp. 658–670, 2021.

[41] R. Dutta and M. F. Rahman, "A comparative analysis of two test methods of measuring *d*- and *q*-axes inductances of interior permanent-magnet

machine," *IEEE Transactions on Magnetics*, vol. 42, no. 11, pp. 3712–3718, 2006.

[42] A. Sneyers, D. W. Novotny, and T. A. Lipo, "Field-weakening in buried permanent magnet AC motor drives," *IEEE Transactions on Industry Applications*, vol. IA-21, pp. 398–407, 1985.

[43] T. M. Jahns, "Flux-weakening regime operation of an interior permanent-magnet synchronous motor drive," *IEEE Transactions on Industry Applications*, vol. IA-23, pp. 681–689, 1987.

[44] S. Morimoto, M. Sanada, and Y. Takeda, "Wide-speed operation of interior permanent magnet synchronous motors with high-performance current regulator," *IEEE Transactions on Industry Applications*, vol. 30, pp. 920–926, 1994.

[45] K. Jang-Mok and S. Seung-Ki, "Speed control of interior permanent magnet synchronous motor drive for the flux weakening operation," *IEEE Transactions on Industry Applications*, vol. 33, pp. 43–48, 1997.

[46] G. Gallegos-Lopez, F. S. Gunawan, and J. E. Walters, "Optimum torque control of permanent-magnet AC Machines in the field-weakened region," *IEEE Transactions on Industry Applications*, vol. 41, pp. 1020–1028, 2005.

[47] K. Ueda, S. Morimoto, Y. Inoue, and M. Sanada, "A novel control method in flux-weakening region for efficient operation of interior permanent magnet synchronous motor," in *Power Electronics Conference (IPEC-Hiroshima 2014 – ECCE-ASIA), 2014 International*, 2014, pp. 312–317.

[48] S. Morimoto, Y. Takeda, T. Hirasa, and K. Taniguchi, "Expansion of operating limits for permanent magnet motor by current vector control considering inverter capacity," *IEEE Transactions on Industry Applications*, vol. 26, pp. 866–871, 1990.

[49] H. Dakai, Z. Lei, and X. Longya, "Maximum torque per volt operation and stability improvement of PMSM in deep flux-weakening region," in *Energy Conversion Congress and Exposition (ECCE)*, 2012 IEEE, Piscataway, NJ, 2012, pp. 1233–1237.

[50] K. H. Shin, J. Y. Choi, and H. W. Cho, "Characteristic analysis of interior permanent-magnet synchronous machine with fractional-slot concentrated winding considering nonlinear magnetic saturation," *IEEE Transactions on Applied Superconductivity*, vol. 26, pp. 1–4, 2016.

[51] S. R. Macminn and T. M. Jahns, "Control techniques for improved high-speed performance of interior PM synchronous motor drives," *IEEE Transactions on Industry Applications*, vol. 27, pp. 997–1004, 1991.

[52] K. Tae-Suk, C. Gi-Young, K. Mu-Shin, and S. Seung-Ki, "Novel flux-weakening control of an IPMSM for quasi-six-step operation," *IEEE Transactions on Industry Applications*, vol. 44, pp. 1722–1731, 2008.

[53] Y. C. Kwon, S. Kim, and S. K. Sul, "Voltage feedback current control scheme for improved transient performance of permanent magnet synchronous machine drives," *IEEE Transactions on Industrial Electronics*, vol. 59, pp. 3373–3382, 2012.

[54] A. Bing and T. R. Tesch, "Torque feedforward control technique for permanent-magnet synchronous motors," *IEEE Transactions on Industrial Electronics*, vol. 57, pp. 969–974, 2010.

[55] J. Faiz and S. H. Mohseni-Zonoozi, "A novel technique for estimation and control of stator flux of a salient-pole PMSM in DTC method based on MTPF," *IEEE Transactions on Industrial Electronics*, vol. 50, pp. 262–271, 2003.

[56] S. Bolognani, S. Calligaro, R. Petrella, and F. Pogni, "Flux-weakening in IPM motor drives: Comparison of state-of-art algorithms and a novel proposal for controller design," in *Proceedings of the 2011-14th European Conference on Power Electronics and Applications (EPE 2011)*, 2011, pp. 1–11.

[57] T. M. Jahns, G. B. Kliman, and T. W. Neumann, "Interior permanent-magnet synchronous motors for adjustable-speed drives," *IEEE Transactions on Industry Applications*, vol. IA-22, pp. 738–747, 1986.

[58] S. Wang, J. Kang, M. Degano, A. Galassini, and C. Gerada, "An accurate wide-speed range control method of IPMSM considering resistive voltage drop and magnetic saturation," *IEEE Transaction on Industrial Electronics*, vol. 67, no. 4, pp. 2630–2641, 2020.

[59] S. Jung, J. Hong and K. Nam. "Current minimizing torque control of the IPMSM using Ferrari's method," *IEEE Transactions on Power Electronics*, vol. 28, no. 12, pp. 5603–5617, 2013.

[60] H. Eldeeb, C. M. Hackl, L. Horlbeck, and J. Kullick, "A unified theory for optimal feedforward torque control of anisotropic synchronous machines," *International Journal of Control*, vol. 91, no. 10, pp. 2273–2302, 2018/10/03 2018.

[61] K. Choi, Y. Kim, K. S. Kim, and S. K. Kim, "Real-time optimal torque control of interior permanent magnet synchronous motors based on a numerical optimization technique," *IEEE Transactions on Control Systems Technology*, vol. 29, no. 4, pp. 1815–1822, 2021.

[62] S. M. Showybul Islam Shakib, D. Xiao, R. Dutta and M. F. Rahman, "An analytical approach to direct torque and flux control of interior permanent magnet synchronous machine for deep field weakening without using pre-calculated lookup tables," *2019 10th International Conference on Power Electronics and ECCE Asia (ICPE 2019 – ECCE Asia)*, 2019, pp. 3196–3202.

[63] S. M. Showybul Islam Shakib, D. Xiao, R. Dutta and M. F. Rahman, "Online deadbeat predictive direct torque and active flux control for IPMSM drive,", International Power Electronics Conference (IPEC-Himeji 2022 – ECCE Asia), 15–19 May 2022, Japan.

Chapter 7

Battery management system for electric vehicle

Khare Mangesh[1], Mandhana Abhishek[1], Gudhe Siddhant[2], Singh Sanjeev[2] and D Giribabu[2]

7.1 Introduction

Lead-acid batteries had been an integral part of internal combustion engine (ICE) vehicles supplying power for cranking, ignition and supporting auxiliary loads like lights and horns. Increasing expectations on improving vehicle efficiency, reducing emissions, and enhancing safety and comfort lead to higher demands on vehicle battery. This triggered the need for better management of the battery as well as development of newer battery technology.

During the 1970s to early 2000, pressure to reduce the emissions and fuel consumption, forced several automotive original equipment manufacturers (OEMs) and tier1 suppliers to introduce new technologies to tackle these challenges. One such technology was "Automatic Start Stop" system. This was the first trigger for the introduction of an intelligent battery sensor (IBS) for actively managing then state of art lead-acid batteries. Leading German Automotive tier1 company, HELLA was among the first few companies to launch IBS in the year 2000 [1] and has sold more than 30million units of it till date. The picture of Hella IBS generation II is shown in Figure 7.1. Bosch [2] and Continental [3] also became the leading suppliers of IBS. The schematic block diagram of present IBS is shown in Figure 7.2.

IBS provided measurement of voltage, current and temperature of the 12V Lead Acid Battery. It also gave an indication of battery status and thus marked the beginning of Battery Management Systems (BMS) for automotive applications [4]. With the advent of hybrid electric and battery electric vehicles, the need for batteries for traction application became evident. Lead-acid batteries were no longer suitable for traction application due to their inherent limitations, hence NiMH and Li-ion batteries got entry into Automotive.

Li-ion batteries had distinct advantages in terms of higher energy density, longer cycle life, lighter weight, and higher charge–discharge capacities; however,

[1]Hella India Automotive Private Limited, Pune, India
[2]Department of Electrical Engineering, MANIT Bhopal, India

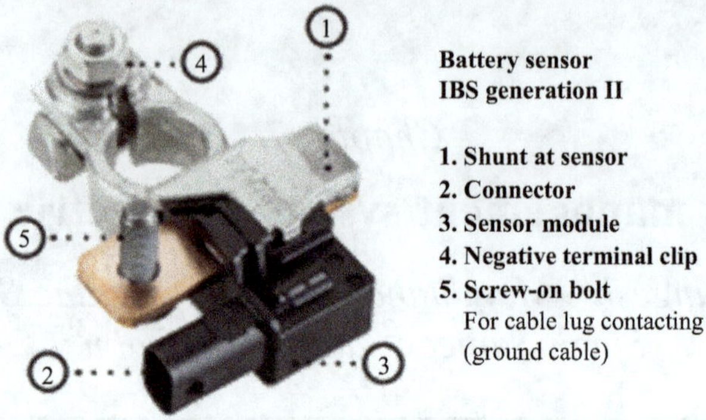

Figure 7.1 Picture of Hella IBS (courtesy: Hella)

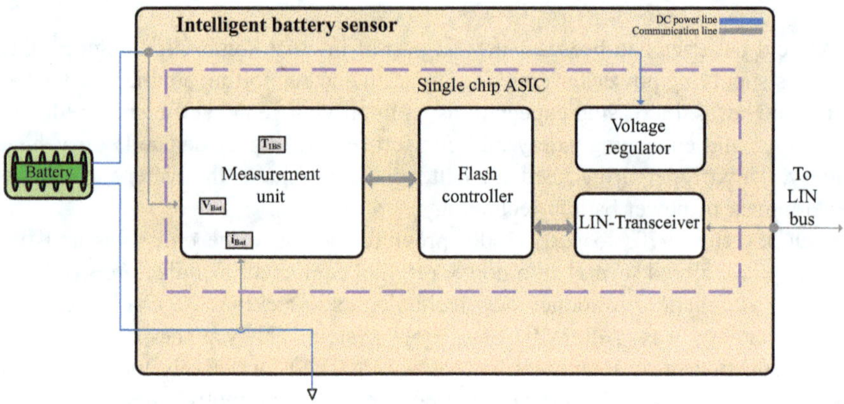

Figure 7.2 IBS schematic block diagram

they required highly controlled operating conditions for safe operations, and this led to the evolution of modern-day BMS [5].

Today Li-ion battery pack along with BMS forms the heart of a battery electric vehicle (BEV). Ensuring safe operation, optimizing performance, prolonging battery life and providing accurate information regarding battery's state are the core functions of BMS.

This chapter discusses various aspects of BMS along with the future trends of battery management system for EVs.

7.2 BMS overview

Figure 7.3 shows a system overview of BMS through a schematic block diagram. The main functions of BMS include sensing various battery parameters, analyzing them to control the load engagement or disengagement, informing the thermal

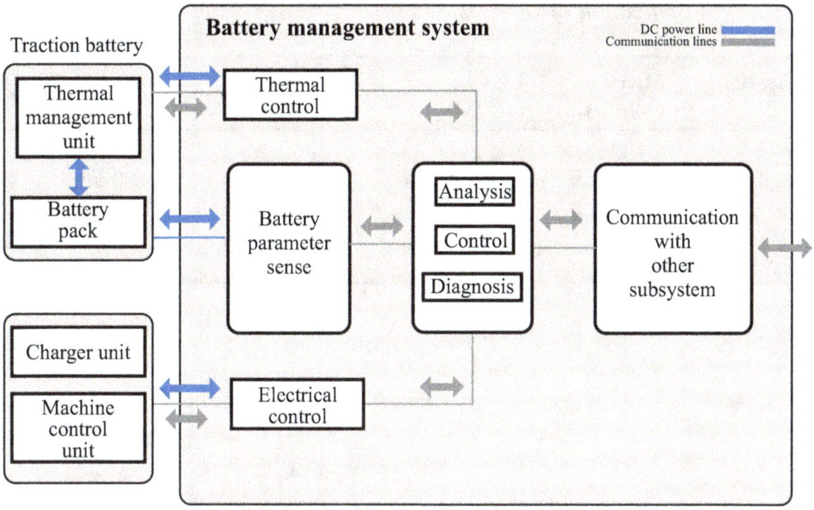

Figure 7.3 BMS schematic block diagram

control units to maintain battery temperatures, estimating the different battery parameters and communicating this information to other subsystem in the vehicle.

7.2.1 Common concepts in BMS

The following concepts are important for understanding BMS functionalities.

7.2.1.1 Low voltage and high voltage automotive systems

Low voltage automotive systems are those which operate at 60 V DC and below. Systems operated above 60 V DC are referred to as high voltage automotive systems.

7.2.1.2 Vehicle classification

Vehicle classification influences some of the design criteria for the BMS. Typical vehicle classification with reference to BMS is given in Table 7.1.

7.2.1.3 Li-ion battery pack structure

Li-ion battery pack structure as shown in Figure 7.4 has significant impact on the selection of suitable BMS architecture.

Cell: It is the fundamental unit of a battery pack. Its electrical specifications would vary based on its chemistry, form factor and size.

Module: It is a group of cells connected together to generate the specific voltage and currents. In the low voltage packs, these are usually parallel-connected cells.

Pack: It is a group of modules connected together to form a single entity of the battery which can be used to provide necessary power and energy to the vehicle.

Table 7.1 Classification of vehicles

Specifications	Micro mobility	LEVs (2W/3W)	Passenger vehicles	Heavy commercial vehicles
Power	<1 kW	3–6 kW	100 kW	>300 kW
Range	20–40 km	100–150 km	300–450 km	>500 km
Speeds	10–30 km/h	80–100 km/h	120–180 km/h	100–120 km/h
Battery voltages	12 V, 24 V	48 V, 60 V, 72 V	>300 V	>800 V
No. of wheels	2 or 3W	2 or 3W	4W	>4W

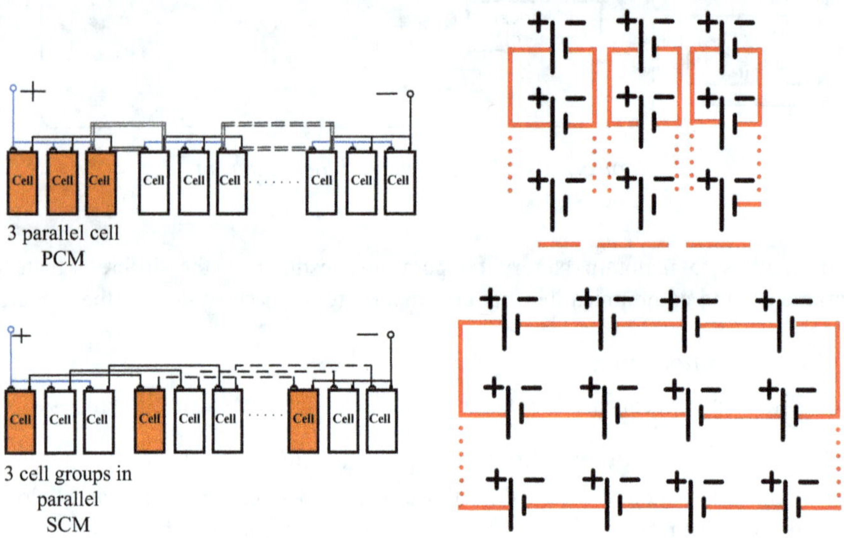

Figure 7.4 Block diagram of Li-ion battery pack structure

PCM approach for a battery pack: The parallel-cell-module (PCM) approach wires cells in parallel to make modules, then wires modules in series to create a battery pack.

SCM approach for battery pack: The series-cell-module (SCM) approach wires cells in series to make modules, then wires modules in parallel to make a battery pack.

In automotive applications, commonly PCM approach is used to fabricate battery packs.

7.2.1.4 Common Li-ion cell chemistries

Commonly used Li-ion battery chemistries for automotive traction application are lithium nickel manganese cobalt oxide ($LiNiMnCoO_2$), abbreviated as NMC and lithium iron phosphate ($LiFePO_4$) or lithium ferro phosphate abbreviated as LFP.

Table 7.2 Li-ion cell operating voltage range and nominal voltage

Battery chemistry	Operating voltage range (V)	Nominal voltage (V)
NMC	3.0–4.2	3.7
LFP	2.5–3.6	3.20

The typical operating voltage range and nominal voltage of Li-ion cells are shown in Table 7.2.

7.2.1.5 Battery capacity

It is the amount of electrical energy (charge) that a battery can store or deliver to various electrical devices or system over a specific period of time. It is typically measured in ampere-hours (Ah).

A battery's capacity is an important characteristic as it directly relates to its runtime or how long it can power a device or system before it needs to be recharged. Every cell in a battery is marked with an Ah rating called as nominal capacity. For example, a cell with a rating of 5 Ah means it can deliver a continuous current of 5 A for 1 h, before its charge is depleted.

7.2.1.6 C-rate

C-rate specifies the speed at which the battery is charged or discharged [5,6]. For a 5 Ah fully charged cell, at 1 C discharge rate, the battery discharges completely in 1 h. Similarly at 0.5 C discharge rate a fully charged cell discharges completely in 2 h. The same interpretation can be applied for charging cases. Sometimes C rates are also represented in fraction form. For example, C/2 rate means 0.5 C.

7.3 Measured parameters

Battery parameter measurement plays a crucial role in BMS. Measurement resolution, accuracy and sampling frequency are some of the important considerations.

7.3.1 Voltage, current, and temperature measurement

These are the most crucial directly measured battery parameters. Some important considerations for voltage (V), current (I) and temperature (T) sensing are discussed below.

7.3.1.1 Synchronous sampling

Current drawn from battery and its voltage are interdependent. Also, traction batteries are often subjected to high load/current dynamics. So synchronous sampling of voltage and current is important. Typically sampling rate is in milliseconds.

7.3.1.2 Voltage

The battery pack consists of several cells connected in the application-specific configuration. Each cell has a direct impact on the overall battery performance. Ideally BMS should measure each cell voltage levels. However, with system considerations and to reduce the overall complexity, voltage measurements are usually done at the module level.

7.3.1.3 Current

Current sensing becomes very crucial as it forms the basis of many algorithms for the state estimation of the battery [7,8]. The higher the accuracy of measurement is, the better the algorithm accuracy is. Hall sensor or shunt-based sensor is typically used in BMS as shown for illustration in Figure 7.5. Both the technologies have its merits and demerits. Hall sensor is better in terms of thermal management; however, it is bulky and less accurate. Usually, it is used in very high current measurement requirements. The shunt-based sensing is very accurate and compact; however, it generates heat which needs to be managed separately. As discussed earlier, the accuracy expectations are higher in BMS automotive applications, usually shunt-based measurements are preferred. Shunt sensors in combination with multiple amplifiers provide the best resolutions and accuracy for a wider range of current measurement.

7.3.1.4 Temperature

Lithium-ion batteries are sensitive to temperatures. Hence BMS is expected to measure the temperatures of each cell in the battery pack accurately to ensure the safety and performance of the pack [7,8]. However due to wiring and cost considerations, the number of sensors used is limited. Thermal simulations of the battery are performed to identify critical areas of heating and accordingly sensors are placed inside the battery pack. Usually, thermistors with negative temperature coefficient (NTC) are used for measurement.

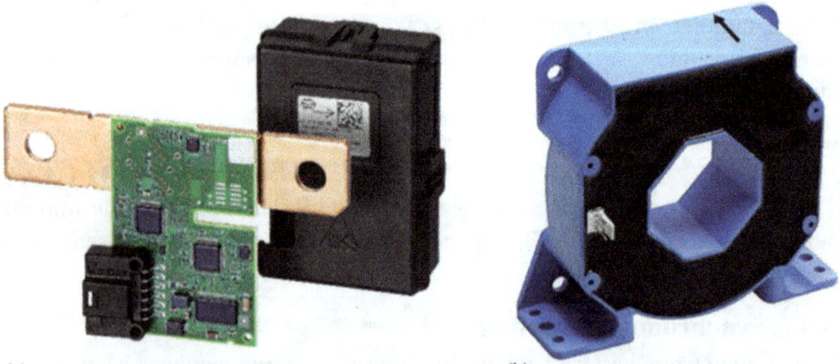

(a) (b)

Figure 7.5 (a) Shunt-based current sensor (courtesy: Hella). (b) Hall-based current sensor (courtesy: Aliexpress.com)

7.3.2 Gas sensors

It is used primarily as an early warning mechanism for any abnormal chemical reaction within the battery pack. Analysis of gases emitted helps in analyzing different internal faults and thereby enables timely maintenance of affected battery cells/packs. Phenomena like thermal runaway can be detected well in advance and accordingly user can be intimated of the imminent hazard. Gas sensors are still not used commonly due to cost and system integration-related considerations (Figure 7.6).

7.3.3 Inferred parameters

There are several battery parameters that are inferred based on the measured parameters and the battery operational condition.

7.3.3.1 State of charge (SoC)

SoC is an indicator of the remaining capacity of the battery during discharging or charging scenarios. It is one of the most important indicators which translates to the remaining distance a vehicle can travel (called distance to empty, DTE). SoC is a cyclic parameter that is measured during each charge and discharge cycle. It is usually represented in %. The equation used to represent the SoC is given as,

$$\mathrm{SoC}(t) = \frac{C_{\text{remaining}}}{C_{\text{actual}}} * 100\%$$

$C_{\text{remaining}}$ is the available capacity that can be extracted from battery at time t; C_{actual} is the actual capacity is the maximum possible charge (in Ah) that can be extracted or accepted by the battery during initial complete charge/discharge cycle at room temperature and C/30 rate.

SoC is sometimes also represented as the depth of discharge (DoD)

DoD = 100% − SoC

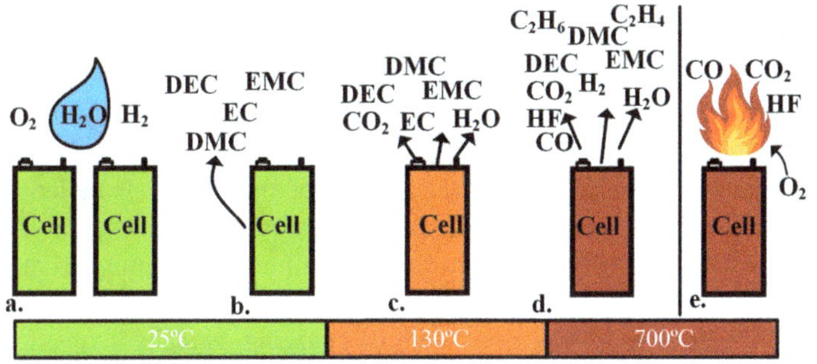

Figure 7.6 Investigated battery failure cases, which involve gas emissions. (a) Unwanted electrolysis, (b) vaporizing electrolyte of damaged cells, (c) the first venting of a failing cell, (d) the thermal runway (TR), and (e) battery fire not investigated.

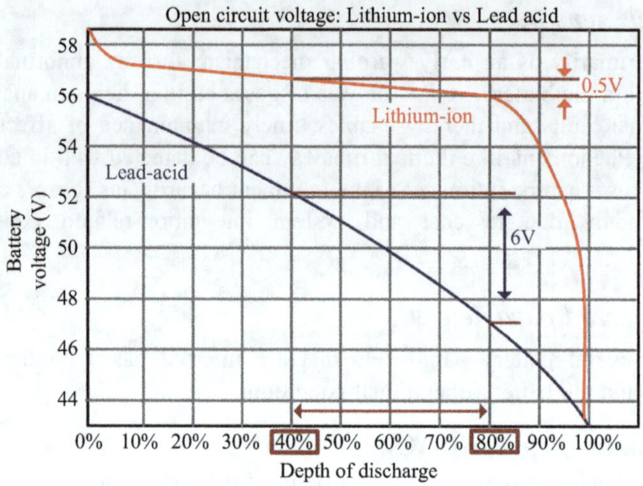

Figure 7.7 DoD versus OCV characteristics of different batteries

SoC estimation is one of the important tasks of BMS. However, it is also one of the complex tasks because of the non-linear nature of SoC the battery. Hence, advanced algorithms are used to estimate SoC of lithium-ion batteries [9].

SoC or DoD measurement in lithium-ion batteries is complex compared to the lead-acid batteries because of non-linearity between SOC versus OCV (open circuit voltage graph).

Figure 7.7 describes the difference between DoD versus OCV characteristics of lead acid and lithium-ion batteries.

From Figure 7.7, the characteristics of SOC versus OCV for lithium ion are highly non-linear compared to lead acid. Hence, there is need of complex algorithms to estimate SoC.

7.3.3.2 State of health (SoH)

SoH indicates the reduction in battery capabilities due to the ageing. As the battery is an electrochemical system, it ages with the usage. Due to ageing, there is a reduction in the charge storage capacity of the battery and its ability to provide the instantaneous power to the system [10].

A reduction in the charge storage capacity of the battery is called "capacity fade". Whereas, the reduction in the capability to provide instantaneous power is called "power fade". Battery ages because of continuous charging and discharging events and due to storage without use.

Ageing effect due to charging and discharging is called "cyclic ageing" whereas, the ageing effect due to storage without use is called calendric ageing.

Calendric ageing also intensifies the cell health degradation process. Usually this is ageing due to storage of the cell and mainly depends on the conditions at which the cells are stored. These conditions include SOC at which the cells are stored and ambient conditions like temperature at which cells are stored.

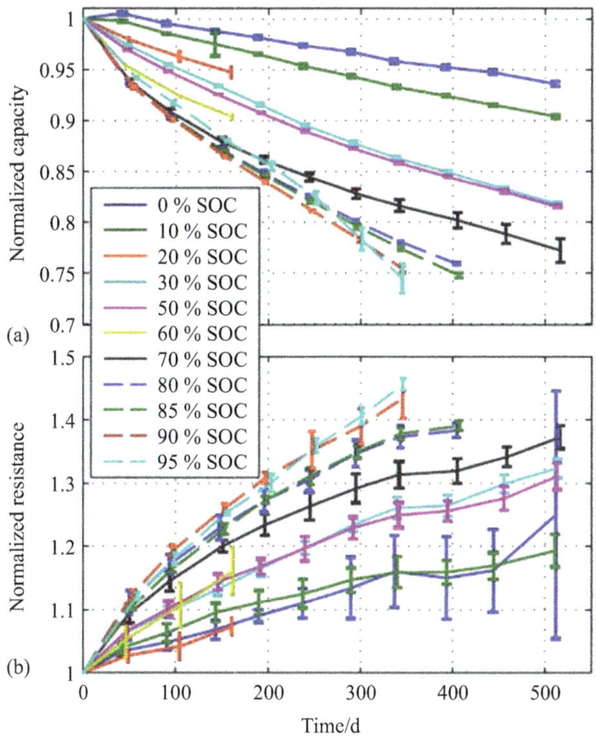

Figure 7.8 (a) Normalized capacity over time and (b) normalized resistance over time for calendar aging tests at 50 C

In Figure 7.8, the cells used are NMC cells stored at 50°C at different SOCs.

From the figure, it can be concluded if cells are stored at higher SOC, calendric ageing happens faster.

Cyclic ageing is dependent on rate at which batteries are charged and discharge, the temperature range of SOC at which charge, and discharge is happening and chemistry of the battery.

In Figure 7.9, the cells used are NMC cells cycled at 1 C charge discharge rates at 35°C and SoC depth of 10%.

From the figure, it can be concluded that the cell ages faster if they are cycled at extreme SoC ranges and minimum ageing observed when cycled in the range of 45%–55%.

In general, higher temperatures and higher C rates influence ageing faster as it produces electrical stress over battery.

SoH using capacity fade

$$SoH = \frac{C_a}{C_r} * 100\%$$

or

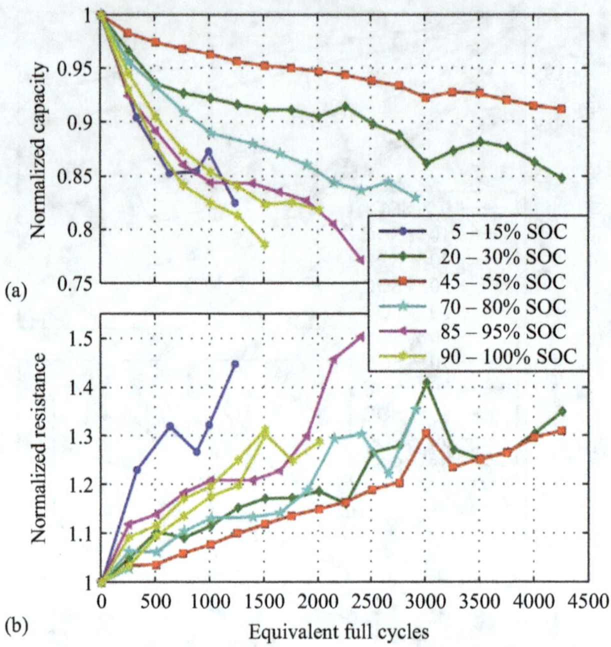

Figure 7.9 (a) Normalized capacity and (b) normalized resistance over equivalent full cycles

where C_a and C_r represent actual and rated capacity.

SoH using power fade

$$SOH = \frac{R_a - R_r}{R_r} * 100\%$$

where R_a and R_r are actual and rated internal resistances.

The SoH representation used in BMS is decided based on the vehicle application's sensitivity to capacity or power fade. The estimation of SoH becomes very crucial as it impacts the range and performance of the vehicle. SoH in combination with SoC is used to provide the accurate DTE. SoH also indicates the remaining life of the battery and plays a key role in making the decision to scrap the battery for specific applications. Hence, from the industry perspective, it addresses very important problems of battery warranties. The SoH measurement is used to define battery warranty conditions; hence, it is important.

7.3.3.3 State of power (SoP)

SoP indicates the capability of battery to supply and absorb power. This indicator is very important to ensure that the charge or discharge power does not exceed certain limits. This helps in using the battery as good as possible to extend its life expectancy.

A battery's SoP is defined as the ratio of peak power to nominal power and is represented in %.

$$\text{SoP} = \frac{P_{\max}(t)}{P_{\text{nominal}}(t)} * 100\%$$

The peak power, based on present battery-pack conditions, is the maximum power that is maintained constant for a predefined time in seconds without violating operational design limits on battery voltage, SoC, power, or current.

The SoP depends primarily on battery chemistry, capacity, voltage and SoC.

7.3.3.4 State of energy (SoE)

SoE indicates the energy left in the battery from initial energy stored under fully charged condition. It is defined as the ratio of the residual energy to the maximum available energy. SoE is represented in % and it gives a wholistic indication of the overall capability of the battery in terms of power and capacity.

$$\text{SoE}(t) = \text{SoE}(t_o) + \frac{\int_{to}^{t} p(\tau)d\tau}{E_N}$$

where, E_N is nominal energy amount and $p(\tau)$denotes power at time τ.

7.4 BMS system architecture

Overall BMS as system consists of the following key subsystems:

- BMS controller unit: This component is responsible for making key decisions, communicating information and controlling the battery engagement with the load. This unit holds all intelligence capabilities for BMS.
- BMS monitoring unit: This component is responsible for collecting battery information including current, voltage, and temperature. This unit must be controlled via the BMS controller.
- BMS power distribution unit (PDU): This component consists of switches that are used to engage or disengage battery from loads and charger.

Based on the combination of these BMS modules, the overall system architecture is classified into the following types.

7.4.1 Centralized architecture

In centralized BMS architecture, BMS controller unit and BMS monitoring unit are integrated into a single BMS unit as shown in Figure 7.10.

7.4.2 Distributed architecture

Distributed Architecture is used in electric vehicles where the battery pack size is big and there are multiple battery modules. In distributed BMS architecture, the BMS monitoring units reside inside the battery modules. As described earlier, the

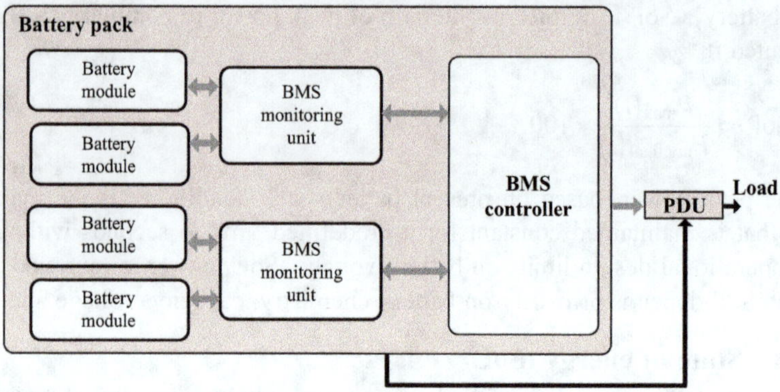

Figure 7.10 Centralized BMS architecture

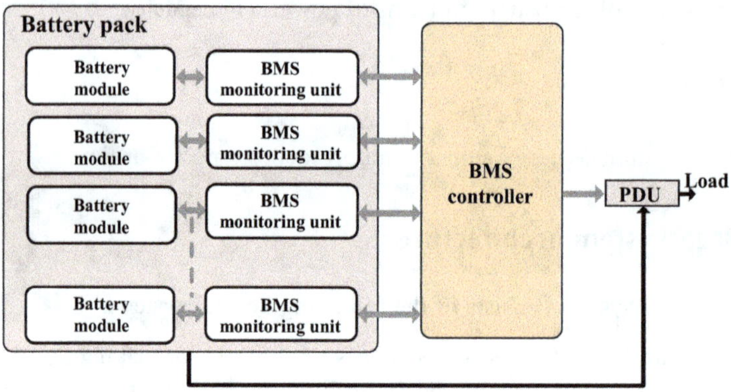

Figure 7.11 Distributed BMS architecture

main function of the BMS monitoring unit is to sense the voltage, current, and temperature information from the battery pack.

There are multiple BMS monitoring units communicating with the BMS controller unit. This BMS controller unit compiles the information received from different BMS monitoring units and relays this information to other ECUs in the vehicle. The BMS controller unit ensures that appropriate decisions are taken during fault conditions. It does it by controlling the switches in PDUs.

Cell balancing is done in combination of BMS monitoring unit and BMS monitoring unit. The BMS monitoring unit senses all the information and communicates to the BMS controller unit. Based on information and type of algorithm implemented, the BMS controller unit identifies the channel needs to be balanced and communicates back this information to BMS monitoring unit. The BMS monitoring unit ensures the specific channels to be balanced and the defined rates.

PDU is outside battery packs. It consists of switches to engage or disengage the battery packs from the load. In some architecture these switches could be specific to modules (Figure 7.11).

7.4.3 Factors influencing choice of architecture

There are various factors influencing the choice of the BMS architecture.

7.4.3.1 Vehicle type

Vehicle Categories →	Low Voltage system		High Voltage system	
Architectures ↓	**Micro mobility**	**LEVs (2W/3W)**	**Passenger Vehicles**	**Heavy Commercial vehicles**
Centralized				
Distributed				

7.4.3.2 Scalability

Distributed architecture is preferred when a high level of scalability is required. When battery modules are added or reduced to increase or decrease the battery voltage respectively, with distributed architecture it becomes convenient to add further BMS monitoring units.

7.4.3.3 Fault tolerance

Distributed architecture provides fault tolerance due to multiple BMS monitoring units being distributed throughout the battery system. The faulty battery module can be easily isolated and replaced without impacting the entire system.

7.4.3.4 Communication requirements

In a distributed architecture, all BMS monitoring units need to have a reliable communication channel to exchange the data. This is more complicated than in centralized architecture where there is a single point of communication.

7.4.3.5 Cost considerations

Cost considerations are often dependent on the application. However, in general, the distributed BMS is costlier as it has more hardware components, complex communication infrastructure and software modules.

7.4.3.6 Performance requirements

Distributed architecture has distributed and parallel computation, so it has a faster response compared to centralized architecture where all processing happens centrally.

7.5 BMS functionalities

There are various functionalities of the BMS.

7.5.1 Protections

BMS needs to ensure the protection of the battery and the vehicle user during fault conditions. Types of protection are as given below. All protections are governed by the following conditions.

- Set conditions: This defines the conditions at which the fault needs to be set.
- Reset conditions: This defines the conditions at which the fault needs to be reset.
- Detection delay: This is over all time delay required to detect and set the fault.

All these parameters need to be defined considering the battery boundary condition and vehicle performance requirements.

7.5.1.1 Over and under voltage protection (cell and pack level)

BMS should disconnect the battery pack from load if the cell/pack voltage exceeds or is below the operating range.

7.5.1.2 Over charging and discharging current protection

BMS should disconnect the battery pack if the charging/discharging current exceeds the operating range.

7.5.1.3 Over and under temperature protection during charging/discharging

BMS should disconnect the battery pack if the temperature during charging/discharging exceeds or is below the operating range.

7.5.1.4 Over on-board temperature protection

BMS should disconnect the battery pack if the temperature of BMS circuit boards exceeds the operating range.

7.5.1.5 Pre-charge protection

BMS should protect the battery and load from inrush pre-charge currents.

7.5.1.6 Short circuit protection

BMS should protect the battery from short circuits.

7.5.1.7 Reverse polarity protection

BMS should protect the battery from any reverse polarity connections.

7.5.1.8 Isolation protection

In high-voltage systems, BMS needs to ensure the isolation between high-voltage battery and vehicle chassis.

7.5.1.9 Fault management

As per regulatory requirements (e.g., Indian regulation AIS 156) for any protection violations, BMS needs to record the fault conditions in non-volatile memory. Fault setting and resetting conditions need to be defined based on battery parameters and vehicle performance expectations. In some BMS solution before faults, additional conditions regarding alarms are also defined.

7.5.2 Cell balancing

BMS needs to ensure all their cells are balanced to the same voltages. This helps in the efficient charging and discharging of the battery. If cell balancing is absent, the life of the battery would be defined by the weakest cell among the modules. Figure 7.12 shows the basic concept of cell balancing. BMS ensures the balancing by following approaches.

7.5.2.1 Passive cell balancing

In this method, BMS identifies the weakest cell (cell with the lowest voltage) and dissipates the other cells to bring it to the same voltage level. Sometimes this method is also called dissipative balancing as it dissipates the energy.

As shown in Figure 7.13, the bleeder resistor (R1 ... R4) is used along with a switch (Q1/Q2) to control the balancing. The specification of the bleeder resistor is determined by the balancing current requirement.

Typical balancing current in automotive applications range from 50 to 300 mA.

Balancing current specification has a tradeoff. The higher the current, the faster would be balancing and the better would be efficiency of charging and discharging. However, the higher the current, the more heat is generated and requires thermal management.

Passive balancing currents dissipates the heat and, hence, are inefficient and reduces the range of the vehicle. However, due to simplicity and cost benefits, this method is mostly used in the applications.

Figure 7.12 Basic concept of cell balancing

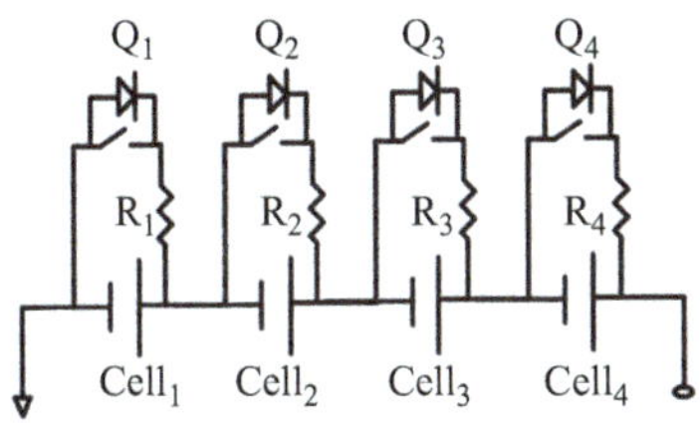

Figure 7.13 Passive cell balancing

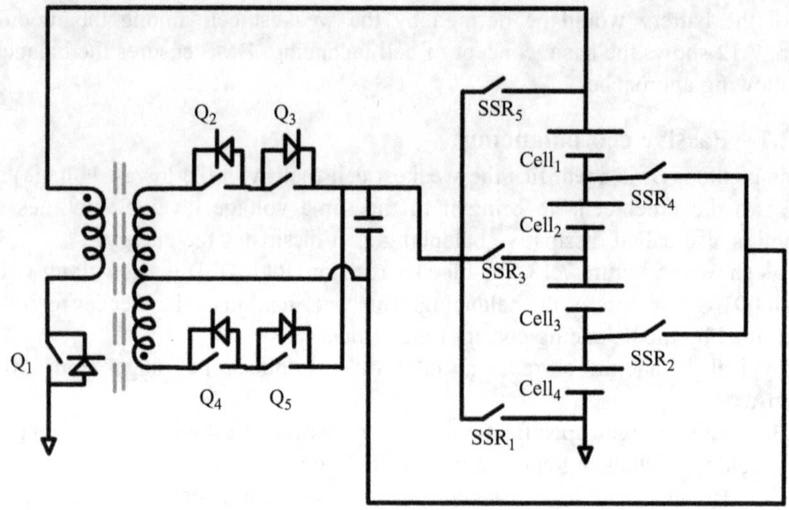

Figure 7.14 Active cell balancing

7.5.2.2 Active cell balancing

In this method, instead of dissipating the charge via the resistor, it is transferred to the other cells in the pack. With this approach the wastage of charge is reduced. However, the transfer of charge requires complex and bulky power electronics to be connected between each cell. Figure 7.14 shows active balancing and it uses bidirectional flyback converter between cells to transfer charge.

7.5.3 Battery-inferred parameter estimation

BMS needs to implement an algorithm for the estimation of SoC, SoH, SoP and SoE [11]. Correct estimation of these parameters and timely communication to BMS algorithm shall result in an effective protection of the battery from any malfunction.

7.6 BMS hardware

7.6.1 BMS hardware architecture

Typical BMS HW architecture shown in Figure 7.15 consists of the following main blocks: analog front end (voltage and temperature sensing, signal conditioning unit, and cell balancing unit), protection, power distribution, and monitoring unit (PDMU), communication unit, and power supply section.

7.6.1.1 Power supply section

The power supply section is responsible for providing power to control and the sensing circuits. It plays a crucial role in ensuring the safe and optimal operation of BMS components. The section comes with a protection mechanism that includes

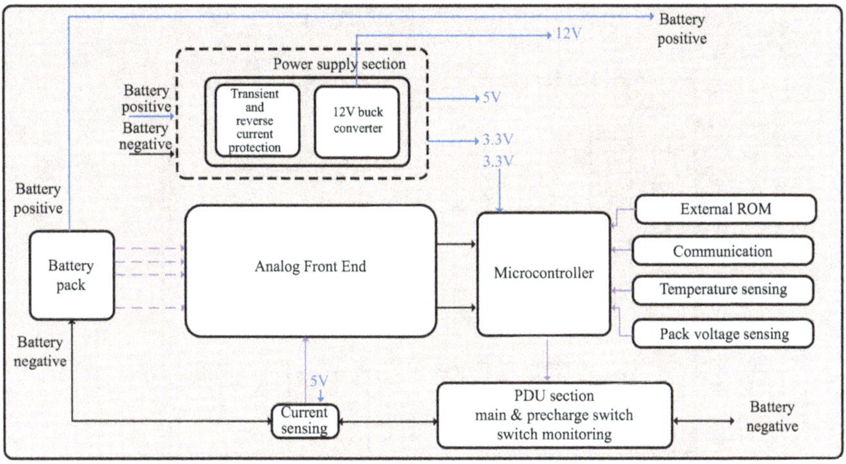

Figure 7.15 BMS hardware architecture

overcurrent, UVLO (under voltage lock out), and the transient and reverse polarity protection. The battery voltage is stepped down to various low voltage levels based on the requirement of the different components (e.g., CAN controller, micro-controllers). For high voltage BMS, a galvanic isolation is incorporated to separate the low-voltage electronics from the high-voltage battery pack.

7.6.1.2 Analog front end (AFE)

The most crucial component in any BMS is the AFE that is controlled by a microcontroller via a serial communication interface. It is responsible for the cell voltage and temperature sensing, and cell balancing. It also senses out of range conditions for cell voltage and temperature. The AFE is provided by many semiconductor companies (like Texas Instruments, Analog Devices, Infinion) and supports multiple cell voltage channels in the battery pack. The cell voltage measurement has redundant architecture with two ADCs (main and auxiliary) to measure the cell voltages. The battery pack temperature measurement is done through the analog input channels.

AFE consists of multiple comparators with programmable thresholds. These comparators are used to detect out of range conditions for cell voltages and pack temperatures.

For cell balancing, AFE has channels controlled by internal balancing switches. In the case of high-end AFEs, these channels are separated from sensing channel while for low-end AFEs, they are multiplexed with sensing channel. AFE is a slave device controlled by microcontroller.

A case study of the ASIC BQ79716 is shown for the understanding of AFE as shown in Figure 7.16, where the number of channels are 16 and for a greater number of channels two AFEs are used and connected through a daisy chain communicating with the master AFE BQ79600 and to the micro-controller. The specifications of BQ-79716 are summarized in Table 7.3.

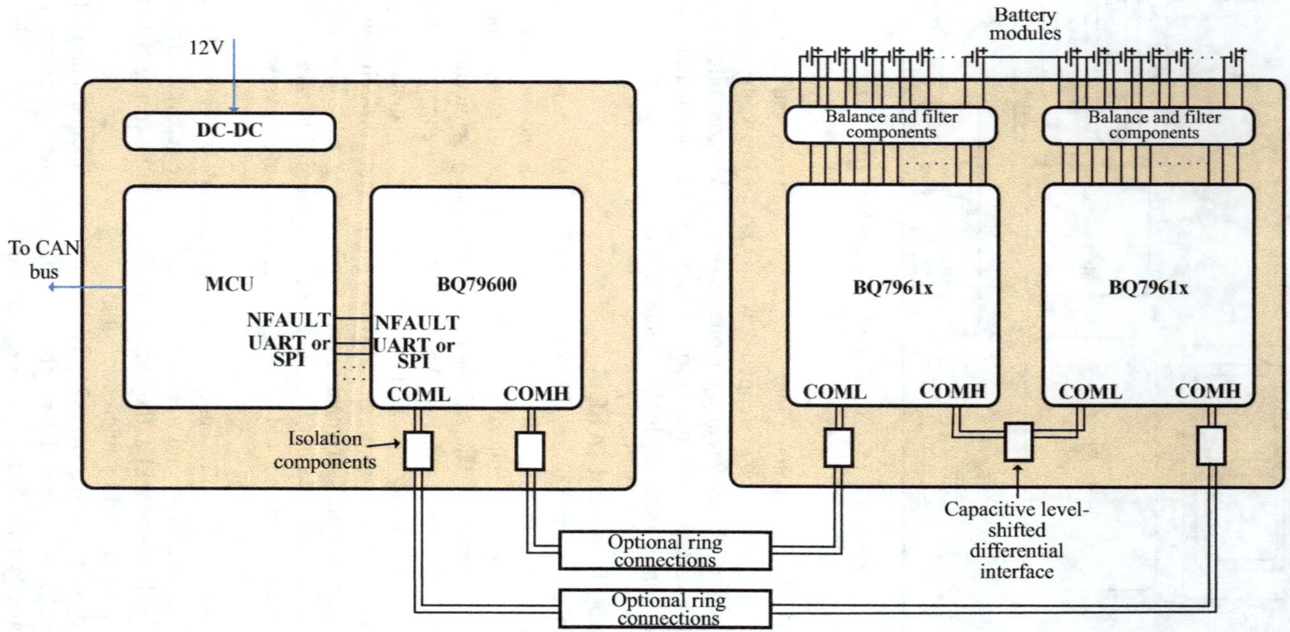

Figure 7.16 BMS AFE hardware architecture for case study

Table 7.3 Specifications of BQ-79716

Description	Supports from 6S to 16S battery modules
AEC-Q100 standard	Qualified
ASIL	D
ADC accuracy	1.5 mV
ADC resolution	Main 16 bits Aux 14 bits
Cell balancing	Supports internal cell balancing with a balancing current of 240 mA
Communication	UART/SPI
ADC conversion time	8 μs for single channel (8 * no. of channel) in round robin manner

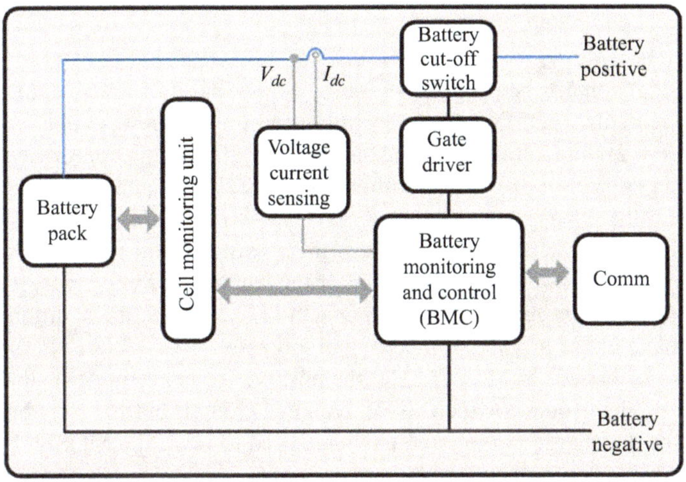

Figure 7.17 Battery cutoff switch placed on the positive terminal of battery (high side cutoff)

7.6.1.3 Power distribution unit

It consists of three subunits, a battery cut-off switch, a pre-charge circuit, and a switch monitoring unit.

Battery cut-off (main) switch

It is also called the power delivery unit. It engages or disengages the battery to the load and charger. Based on the voltage levels of application and size constraints either solid-state switches like MOSFETs or electromechanical switches like contactors are used for the purpose. Figures 7.17 and 7.18 show battery pack with high side cut-off and low side cut-off, respectively.

In the case of MOSFETs, conduction losses are kept in consideration for selection. Response time and life of the MOSFETs are better than contactors. On

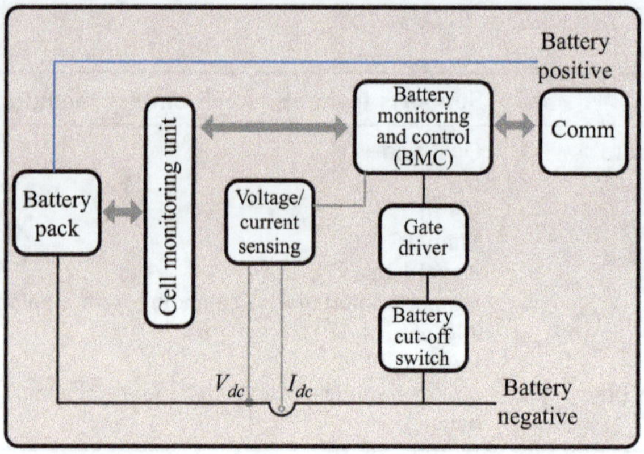

Figure 7.18 Battery cutoff switch placed on negative terminal of the battery (low side cutoff)

the other side, contactors are better in terms of thermal performance and replacement requirements. MOSFETs are generally used for low-power applications.

In a few architectures, charging and discharging paths for the battery are separate and, hence, in this case, additional switches are required to be placed and controlled in the other path.

Based on architecture considerations, these switches are placed either on the positive terminal or the negative terminal of the battery. In some cases, it is placed at both ends to ensure redundancy. As per AIS 156 standard, the additional fuse is added to make entire protection mechanism fail safe.

The major benefit of MOSFETs at the positive battery terminal is that there is no bypass of the ground or there is no hanging ground, but the drawback is the bootstrap, or the charge pump circuitry is required to drive the MOSFETs.

Pre-charge circuit

When initially connecting a battery to a load with capacitive input, there is an inrush of current as the load capacitance is charged up to the battery voltage. With large batteries (with a low source resistance) and powerful loads (with large capacitors across the input), the inrush current can easily peak 1,000 A. A pre-charge circuit limits that inrush current, without limiting the operating current.

The pre-charge circuit consists at the minimum of:

- A pre-charge resistor, to limit the inrush current (R1)
- A contactor (high power relay) across the pre-charge resistor (R1) to bypass the resistor during normal operation.

Additionally, the pre-charge circuit may have:

- A pre-charge relay (K1), to keep the load from being powered through the pre-charge resistor when the system is off.
- A contactor (MOSFET or electromechanical relay) in line with the other end of the battery (K3) to isolate the load when the system is off.

In the typical pre-charge circuit as shown in Figure 7.19, the pre-charge resistor is on the positive terminal of the battery, though it could just as easily be on the negative terminal. Typical charging time and its representation in terms of voltage and current are shown in Figure 7.20.

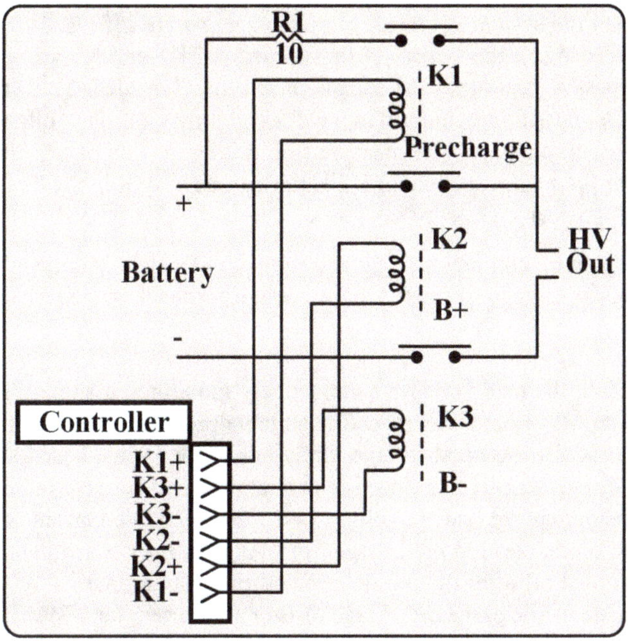

Figure 7.19 Battery pre-charge circuit

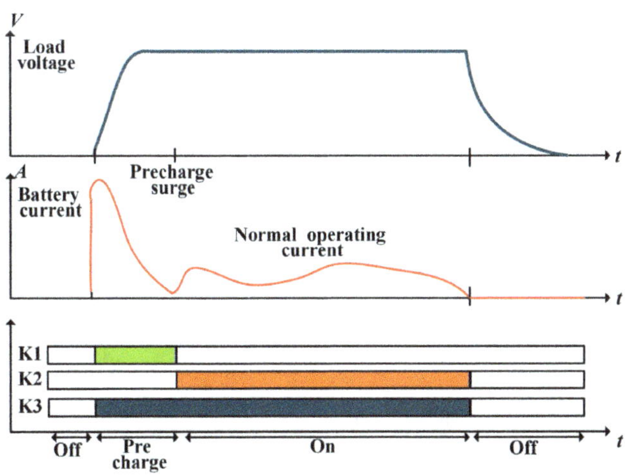

Figure 7.20 Pre-charging sequence

R1 is the pre-charge resistor.
K1 is the pre-charge contactor at the positive terminal.
K2 is the normal contactor.
K3 is the Pre-charge contactor at the negative terminal.
In the most basic form, the pre-charge circuit is operated as follows:

- Off: When the system is off, all relays/contactors are off.
- Pre-charge: When the system is first turned on, K1 and K3 are turned on, to pre-charge the load, until the inrush current has subsided. This can be either done time based (preconfigured switch on time) or the voltage feedback (capacitance voltage can be measured) to decide the end of pre-charge.
- On: After pre-charge, contactor K2 is turned on (relay K1 may be turned off to save coil power).

The decision to stop precharging and move to ON state is done either in open loop: time-based strategy or closed loop: voltage feedback strategy.

Time-based strategy
In this approach, the load voltage is not actively monitored assuming that the battery and capacitors of load will reach the suitable pre-charge level within the specified time. It is also essential to note that an open-loop, time-based strategy may not be as precise or adaptive as closed-loop strategies that actively monitor the load voltage. Considering the battery voltage and the pre-charge current which is calculated based on the desired pre-charge time. As it is timer-based control, the pre-charging is turned off after the pre-determined pre-charge time.

Sample calculation of the pre-charge resistor and current for 48 V battery pack to charge a load to 98% of the battery voltage in 2 ms. The load capacitance is 5 mF.

Battery voltage = 48 V
Pre-charge time = 2 ms
Load capacitance = 50 μF
Time required to charge the load to 98% = $4RC$
where R is the pre-charge resistor and C is the load capacitance:

$$2 * 10^{-3} = 4 * R * 50 * 10^{-6}$$

$$R = \frac{4 * 50 * 10^{-6}}{2 * 10^{-3}}$$

$R = 10\ \Omega$
Initial pre-charge current = $\frac{V}{R}$

$$= \frac{48}{10} = 4.8\ \text{A}$$

Closed loop – voltage feedback strategy
In voltage feedback strategy, the load voltage is monitored and given as feedback to BMS controller. When the load voltage reaches to the battery voltage, the controller switches off the pre-charge contactor and switches on the main contactor to

engage the load. In some cases, this feedback is also used to control the current using a current controller circuit instead of a constant precharge resistance.

Switch monitoring unit
To ensure that switches (MOSFETs or contactors) are operating well, additional circuitry which includes voltage sensing is used to conduct the diagnostic tests. This becomes very relevant and critical as the operations of this switch are very important specially during fault conditions.

7.6.1.4 Pack voltage sensing and current sensing

Pack voltage sensing is used to monitor the entire pack voltage. It can alternatively be done via summing up individual cell voltages; however, as redundancy and efficient pre-charging, it is recommended to add this circuit.

Current sensing is realized via shunt or hall.

In some architectures, due to precision requirements, multiple shunts are used to measure the currents in different ranges.

7.6.1.5 Onboard temperature sensing

BMS has a lot of heat-generating components like pre-charge resistors, cell balancing resistors, and MOSFETs. It is important to monitor the onboard temperature to ensure that BMS does not heat up above design considerations. Additionally, this heating should not impact battery temperature. Hence it is recommended to add onboard temperature sensing and take appropriate actions if temperatures cross the threshold limits.

7.6.1.6 Short-circuit protection

Usually, all the protections are provided with SW intelligence; however, for short-circuit protection to ensure response times within microseconds, an additional circuit is added to protect the battery from short circuit. A low side shunt current sensing is used as a protective measure. The Op-Amps used are chosen with high a common mode rejection ratio (CMRR) so that the small differential voltage across the shunt element is sensed. The digital IOs to the microcontroller have pull-up or pull-down circuitry if the logic is of active low and active high, respectively. The microcontroller pin inputs are clamped to 5 V by using Zener diodes to protect the pins against the short to battery voltages.

7.6.1.7 Micro controller

All the intelligence lies in this section of the BMS HW. The microcontroller of the BMS can be programmed with all the features and protection and estimation of battery internal states is ensured.

The selection of micro controller can be on the following aspects.

- BMS functional safety – ASIL level
- Algorithms memory and computation requirements
- Communication requirements
- Input and output pin requirements

- Sleep current requirements
- Cost
- Memory size

7.6.1.8 Communications

BMS communicates all its state information and safety information to multiple controllers that are inside the vehicle network. CAN being standard communication in automotive, BMS needs to support this protocol.

An isolated or non-isolated CAN section is used for the data transfer from the BMS to the external ECUs. The CAN transceivers used are selected for maximum battery voltage such that the CANH and CANL pins do not get affected due to the high-voltage stress. Additional back-to-back diode protections are given to the transmitter and receptor pins.

7.6.2 HW design considerations BMS performance parameters

To ensure optimal performance and safety of the batteries, BMS is designed to monitor and control its operation. Some key performance parameters are typically considered when evaluating the effectiveness of the BMS which includes SoC, SoH, cell voltage monitoring, current monitoring, temperature monitoring, and more.

7.6.2.1 Leakage currents

The term leakage current refers to the small amount of the current that flows through the BMS when the battery is not in use or during the standby period. The consideration of leakage current is essential in the BMS design because it can lead to battery drainage if left unattended. There are various factors that are responsible for the leakage currents which include circuitry imperfections, component tolerances, parasitic capacitance, or residual charges.

7.6.2.2 Accuracies of measurement

For monitoring the battery and managing its performance accurately, the measurement of the physical parameters which includes the cell voltage, temperature, and battery pack current should also be done with high accuracy. Thus, it starts with the sensors used for measurements, their offset values, and ruggedness. These values play a major role in the estimation of the other parameters like SoC, SoH, and SoP of the batteries.

7.6.2.3 Thermal performance

The heat-dissipating components in the BMS include the MOSFET switches, the balancing resistor, shunt element (for current sensing), and a considerable amount of the heat generated from the power supply section also. The heat generated from these components has to be handled and dissipated wisely otherwise the overall battery ambient temperature may increase. Thus, MOSFETs selection is made considering the temperature rise during the operation because of the huge current flow. MOSFETs with top-side cooling are chosen as they have the superior thermal performance than the normal FETs.

7.6.2.4 Response times

The response time defines the time taken for the cut-off switches to readily act for any fault and disconnect the battery pack during charging and discharging conditions. The more critical of all faults is the short-circuit fault that requires the cut-off switches to disconnect in micro-seconds. The entire response time includes the response time of the current sensing circuitry, over-current protection circuit, the gate driver circuits and the turn-off time of the MOSFETs. Apart from the short circuit protection, all other faults are handled by the software which has pre-defined time to set and reset the faults and trigger the battery switches accordingly.

7.6.3 *BMS software*

Figure 7.21 shows typical software architecture for BMS applications.

7.6.3.1 BMS SW architecture

BMS SW architecture consists of two sections:

1. Application software: It includes software implementation of the functions of BMS.
2. Base software: This contains driver configuration and handler implementations accessed by the applications. Usually, this is made according to standard AUTOSAR implementations. In a few cases, this can be customized to vehicle needs

BMS state manager

The BMS has different types of states and the power consumption in each state differs. The common states in the BMS are normal, sleep and fault. In the normal

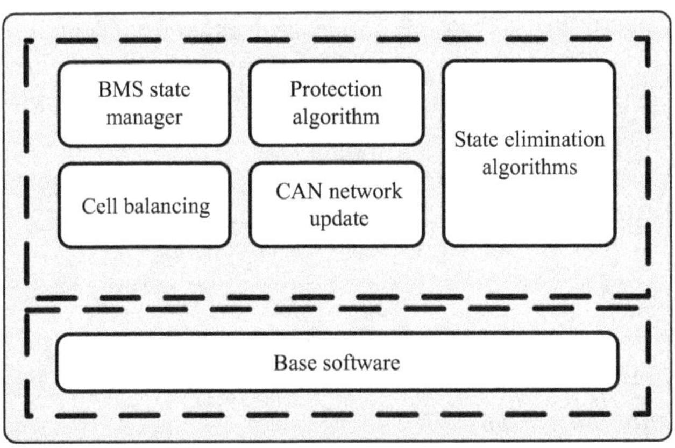

Figure 7.21 Typical software architecture for BMS

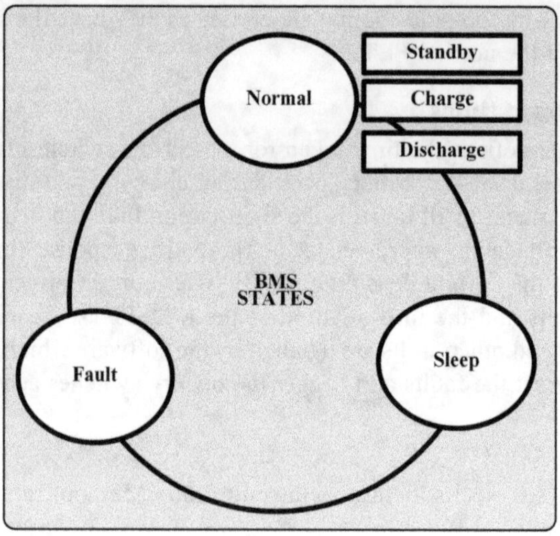

Figure 7.22 BMS states

state, the BMS can be in charge, discharge, or standby mode. To conserve power, the BMS enters to sleep state which can be based on a timer mechanism. There should also be some wakeup mechanism to bring the BMS back to normal operation. These wakeup mechanisms include a wakeup from user via Ignition switch or CAN wakeup from other ECUs of the vehicle. There are also some functionalities being carried out in the sleep mode considering the safety critical compliances. Thus, sleep mode reduces the power consumption of the battery in turn saving the energy in the battery. Another common state in BMS is fault state. Here, BMS enters based on the faults detected. BMS comes out from this state back to normal either by user intervention or via intelligent strategies to reset the fault. Considering safety criticality, many vehicles prefer to reset the fault via user intervention. Figure 7.22 shows a typical software architecture for BMS.

Whenever BMS transits towards normal state either from fault or from sleep, it is important to go through a pre-charge phase. As the load capacitances that are going to be connected during normal state, it can demand lot of inrush currents.

Protection algorithms
This application layer incorporates all the checks of battery operating conditions. Any fault detected would lead the BMS to enter the fault state. Recovery from the fault is also based on BMS reset or release conditions.

Cell balancing
BMS balances the series-connected cells to maintain same level of charge. This function is usually done during charging. In some cases, it is also allowed during standby.

State estimation algorithms
This SW application is responsible for estimating all the battery's internal states including SoC and SoH [12].

CAN network update

BMS updates all the information over CAN to communicate wit the other controllers.

7.6.3.2 BMS algorithms

State estimation algorithms

SoC and SoH are two most important battery states that need to be monitored by BMS and shared with other systems in the EV [12]. This information is then used for updating driver with battery conditions, Motor controllers and charger controllers (Figure 7.23).

SoC and SoH

SoC – state of charge is ratio of remaining capacity to actual capacity of the battery.

SoH – it measures the capacity degradation or Internal resistance increase due to battery ageing.

(a) Look up table method for SoC

In this method, SoC versus open circuit voltage (OCV) characteristic is obtained from offline measurements using the battery. This characteristic is then used by BMS as a look-up table to infer SoC from the OCV measurements.

Figure 7.24 shows battery OCV versus SoC for a lithium-ion polymer battery (LiPB). Thus we can infer battery SoC by looking-up at the table between OCV and SoC. This is the most commonly used method in BMS.

It can be efficiently used for correcting offset and providing initial value in other SoC estimation techniques like Coulomb counting and model based estimation. However, it is hard to measure the precise OCV in real-time. This type of SoC estimation method is more suitable for being applied to the laboratory environment.

(b) Coulomb counting method for SoC

Figure 7.25 presents the Coulomb counting method for SoC estimation. In this SoC estimation technique, electric charge passing through the battery (during charging and discharging) is measured.

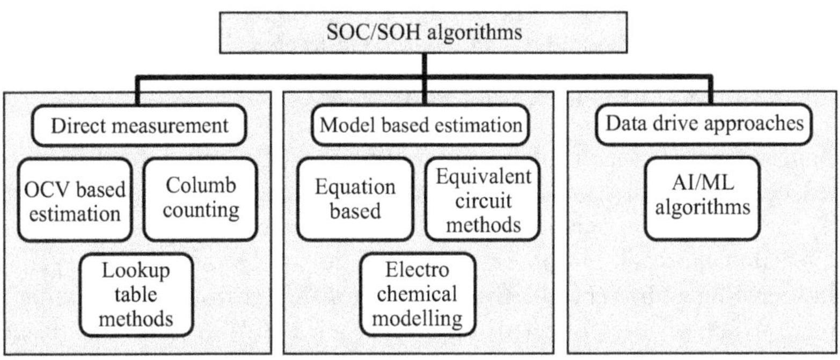

Figure 7.23 Battery algorithm strategies

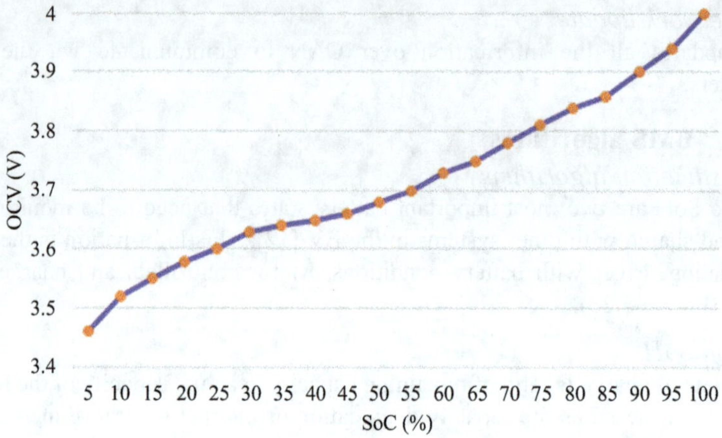

Figure 7.24 OCV versus SoC for a lithium-ion polymer battery

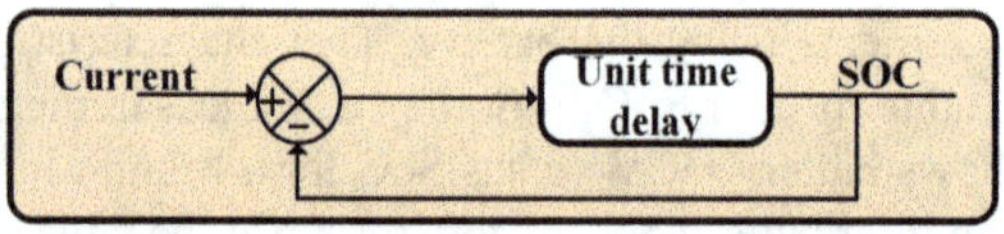

Figure 7.25 Coulomb counting methods to obtain SoC of the battery

$$z_k = z_{k-1} - \frac{deltaT * i_k}{Q}$$

z_k is the SoC of the cell at the *k*th instant; Q is the discharge capacity of cell; i_k input current; and $deltaT$ is the sampling time.

Although this is a simple estimation method, it has certain limitations.

The initial state of SoC is unknown and hence the estimations could be wrong.

The current sensor noises can lead to integration errors.

Battery degradation over time due to factors like aging, temperature, etc. are not factored in this methodology and would lead to error in estimation.

(c) Empirical model (data driven) for SoH

This approach utilizes ageing data collected from battery/cell under different predefined operating conditions. This data is used to develop look-up tables to specify the capacity degradation and IR increase over specific intervals of cycles. However, this approach is very basic and can go wrong as the actual cyclic and calendric ageing of batteries on the vehicle is different from the operating conditions at which data is collected. This approach is further optimized to incorporate the operating conditions to be as close as the actual vehicle conditions (e.g., operating conditions are defined by driving cycles).

This data can also used in development of AI based and machine learning approaches to formulate the ageing effect. This approach required huge amount of data beforehand for learning and testing purposes. Accuracies of these models highly depend on the training data. Additionally, the learning process can be reinforced to adapt the unexpected changes.

Coulomb counting in combination with OCV-based compensation (for initial SoC information and minimize the integration errors) is the most used SoC algorithms for NMC batteries. Additionally, the look-up table approach to define the capacity degradation is used to estimate SoH parameter. The capacity fade information from SoH is used to calibrate the actual capacity of the battery. This incorporates the ageing effect in SoC estimation as well.

(d) Model-based estimation using Kalman filter for SoC and SoH

Figure 7.26 presents a state-of-the-art method based on Kalman filter for SoC and SoH estimation. In this method, the dynamic battery model is developed to emulate the battery under different operating conditions. It takes into consideration of different battery-related parameters like current, temperature, voltage, ageing, self-discharge rates, etc. to describe time-based estimation of SoC/SoH. This estimated SoC/SoH is dynamically corrected using the Kalman filter [9].

Battery models change as per the battery chemistry. There are three most widely used techniques for battery model development: equation-based modeling, electrochemical model (EM), and equivalent circuit model (ECM). In the model-based SoC/SoH estimation methods, battery models are expressed as state equations.

Kalman filter acts as a state observer for SoC/SoH. There are several other possible nonlinear state estimation algorithms and adaptive filters that can be employed to estimate SoC/SoH. The typical algorithms are Luenberger observer, PI (proportion integration) observer, H∞ observer, and sliding-mode observer.

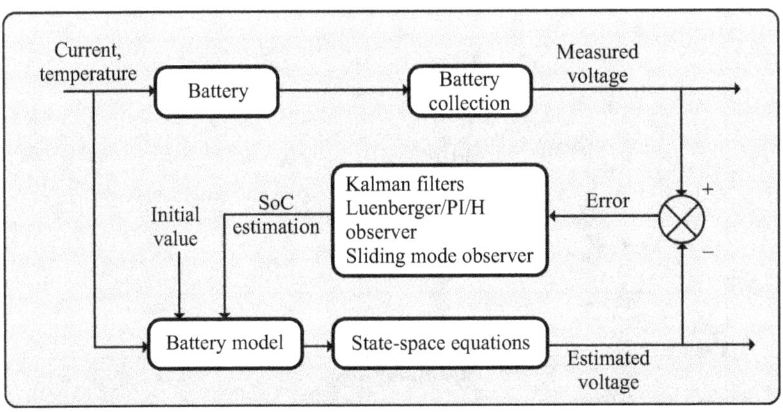

Figure 7.26 State of art methods to obtain SoC/SoH of the battery

7.6.3.3 Battery modeling techniques

Lithium ion battery is a non-linear system that is a combination of electrical and chemical phenomenon. The aim of modeling the battery is to replicate the battery behavior as close to the real state in the software environment. This model then can be used in the development of closed-loop algorithms for battery-state estimation (SoC and SoH) and also for BMS SW validation. Battery is a safety critical system, performing all testing activities using the real battery is not always possible. In such scenarios, the battery model can be safely used.

Battery as a model consists of the following subsystem models:

1. Electrical model
2. Thermal model
3. Chemical model (only in electro chemical modeling approaches)
4. Ageing model

Each of these sub-systems (battery behaviors) can be modeled using any of the following techniques:

1. Equation-based model
2. Equivalent circuit modeling (nRC models)
3. Electrochemical modeling
4. Data-driven modeling using ML

Figure 7.27 shows a comparison of different battery models with respect to CPU time and predictability. Also, a typical battery model architecture is represented in Figure 7.28.

Equation-based modeling

The equation-based approach uses polynomial equations to estimate the electrical, ageing, and thermal behavior of the cell [13]. These equations are obtained using

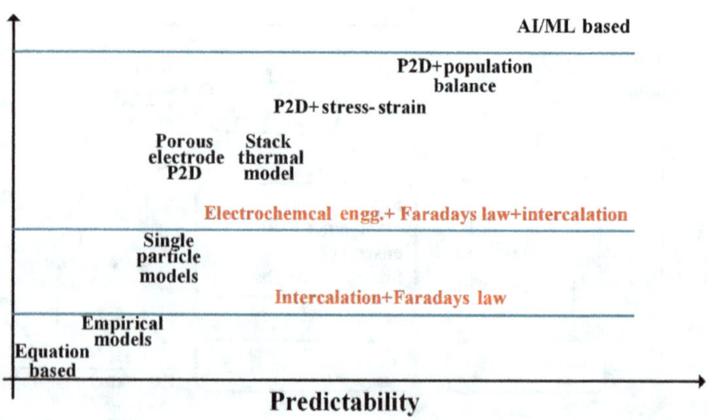

Figure 7.27 Comparison of different battery models with respect to CPU time and predictability

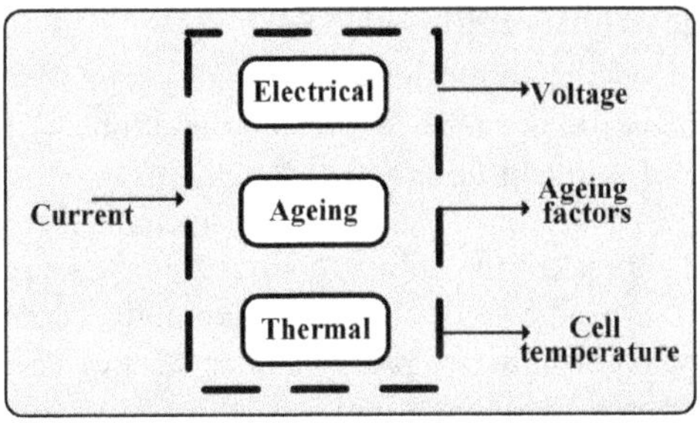

Figure 7.28 Typical battery model architecture

data collected from multiple characterization tests on the cells. These tests include the following:

1. Hybrid pulse power characterization test (HPPC test): The output is the function of the open circuit voltage and cell capacity.
2. Capacity tests: The output is the overall capacity of the cell.

Equivalent circuit modeling

Here, the cell characteristics are represented via resistance capacitance-lumped circuits. The similar tests are used to obtain the RC coefficients for the lumped circuits. The RC circuits can represent the dynamic behavior (transients) of the battery. This is a "state-of-the-art" modeling technique that used batteries in the automotive applications.

Electro chemical modeling

The electrochemical modeling approach uses cell chemical information to model the battery behavior. This approach is cross-functional development of the cell chemical phenomenon's leading to the electrical outputs. Additional inputs include the chemical information of the cell.

For example, anode characteristics, cathode characteristics, and electrolyte characteristics.

This approach is governed by defined sets of assumptions on the cell. Accordingly, they are classified into the following:

1. Single particle models (SPMs)
2. P2D models (porous electrode) models

As we add more details, the complexity of these models increases.

Machine learning approaches

Machine learning (ML)-based approaches work on data and develop suitable models against it. This approach requires the cell life data and expected output characteristics. These characteristics are modeled using ML approaches and integrated to give desired output parameters. The accuracies of these models highly depend on the training data. Additionally, the learning process can be reinforced to adapt the unexpected changes.

7.6.3.4 BMS algorithm validation

Algorithms need to be validated for their functionality and accuracy. SoC and SoH have a direct impact on the distance to empty estimations. Any deviations in the estimations would lead to misinformation to the end user.

Since SoC and SoH are only estimated parameters, there is no direct measurement possible. This makes the validation a complex process. A standard validation strategy for these algorithms, for the automotive application, is not yet available. As of now, algorithms are proven under a set of defined operating conditions based on the application needs and regulatory considerations.

Typical operating conditions for a 48 V BMS using NMC batteries for 2 W application in Indian conditions are summarized in Table 7.4.

The most commonly used test setup for BMS includes a battery cycler and a climate/temperature test chamber. Battery cycler has three main functionalities:

Measurement unit: It is a high precision measurement unit to measure temperature, voltage, and currents of all cells and overall battery pack. Besides this it also includes estimating functions for other parameters like internal resistance, capacity, SoC, SoH, etc.

Programmable load: It consists of a load that can be programmed as per required dynamic or static current profiles.

Data logger: It records all the test results and observations. A typical cell cycler specification is given in Table 7.5.

Table 7.4 Typical operating conditions for a 48 V BMS

Parameters	Values
Operating temperatures	20°C–70°C
Discharge rates	0.2 C, 0.5 C, and 1 C
Charge rates	0.2 C, 0.5 C, and 1 C
Operating cell voltage range	3.0–4.2 V
Operating pack voltage	39–54.6 V
Dynamic profiles	Indian drive cycle (maximum current 1 C)

Table 7.5 Typical battery cycler specification

Parameters	Values
No. of channels	8
Voltage range	0–100 V
Maximum continuous current per channel	100 A
Accuracy voltage measurement	0.01% Full-scale range
Accuracy current measurement	0.02% Full-scale range
Data acquisition rate	10 ms
Profile configuration	Supported

The climate test chamber provides controlled environmental conditions like temperature, humidity, pressure, etc., while the temperature test chamber provides only a controlled temperature environment. Cells or battery pack to be used for BMS validation is kept inside the climate/temperature test chamber for simulating different test conditions.

7.7 Future trends in BMS

BMS plays a very important role in BEVs. Lot of innovations and research are happening in this domain in both industry and academics. New architectures are coming into the vehicles to support additional features like battery swapping, increased redundancy and fault tolerance, reduce wiring complexity, provide cybersecurity, Over-the-air (OTA) updates, support for Artificial Intelligence/ML, switchable architecture (400–800 V for charging), etc.

7.7.1 Wireless BMS

In wireless BMS, each individual cell is wirelessly connected with other cells via wireless communication technology like Bluetooth, ZigBee, Wi-Fi or any other proprietary protocols. This reduces the weight of battery pack, enhances flexibility, increases safety, and makes battery-pack maintenance easy.

7.7.2 Cloud-connected BMS

In cloud-connected architecture, GPRS connectivity is provided to BMS. This enables direct connectivity to Internet and BMS data can be uploaded in the cloud. BMS can then make use of cloud-computing capabilities to perform data analytics and execute advanced battery algorithms which otherwise are difficult to run on the embedded system. It also allows OTA software updates and easy monitoring of the BMS.

7.7.3 Switchable architecture

This architecture allows manual or automatic switching of battery configurations based on different operating conditions. In this architecture, multiple battery packs or modules are configured to achieve specific performance objectives like, for example, different driving modes. Although this provides greater control and adaptability over battery system, ensuring safety of operations becomes critical.

7.7.4 Battery swapping

Battery swapping refers to a mechanism that allows the exchange of discharged batteries with the charged batteries in a vehicle [14]. This de-links the vehicle and battery thereby providing several advantages over the normal charging mechanism.

In this case, Internet of things-based BMS is required and it needs to be an integral part of the battery pack. It needs to support remote monitoring and immobilization capabilities and several other control features to ensure safety, adaptability with different vehicle systems and misuse of the battery pack.

7.8 Summary

The chapter aims to present various aspects of BMS required for a typical electric vehicle. The industrial perspective with prevailing architecture and algorithms has been presented for the understanding of the readers. The graphical and pictorial representations of the available industrial systems are also included to give the reader a feel of real-time systems. The future trends has also been included to encourage the coming development and the research in the prospective areas.

Symbols

V	voltage
I	current
$C_{remaining}$	available capacity that can be extracted from battery at time t
C_a	actual capacity
C_r	rated capacity
R_a	actual internal resistances
R_r	rated internal resistances
$P_{max}(t)$	peak power
$P_{nominal}(t)$	nominal power
E_N	nominal energy amount
$p(\tau)$	power
τ	time
R1/R2	bleeder resistor
Q1/Q2	switch
K1	pre-charge contactor at positive terminal
K2	normal contactor
K3	pre-charge contactor at negative terminal
C	load capacitance
R	pre-charge resistor
z_k	SoC of cell at kth instant
Q	discharge capacity of cell
i_k	input current
deltaT	sampling time

Glossary

AFE	analog front end
Ah/Wh	Ampere-hours (Ah)/Watt-hours (Wh)

AIS	Automotive Industry Standard
AUTOSAR	automotive open system architecture
BEV	battery electric vehicle
BMS	battery management system
CAN	controller area network
DEC	diethyl carbonate
DMC	dimethyl carbonate
DoD	depth of discharge
DTE	distance to empty
EC	ethylene carbonate
EFC	equivalent full cycle
EM	electrochemical model
EMC	ethyl methyl carbonate
ECM	equivalent circuit model
EIM	electrochemical impedance model
IBS	intelligent battery sensor
ICE	internal combustion engine
LEV	light electric vehicle
LFP	lithium iron phosphate
MOSFET	metal-oxide-semiconductor field-effect transistor
NiMH	nickel–metal hydride battery
NMC	nickel manganese cobalt
NTC	negative temperature coefficient
OEMs	original equipment manufactures
PCM	parallel-cell-module
PDMU	protection, power distribution, and monitoring unit
PDU	power distribution unit
SCM	series-cell-module
SoC	state of charge
SoH	state of health
SoP	state of power
OCV	open circuit voltage
UVLO	under voltage lockout

References

[1] https://www.hella.com/hella-in/assets/media_global/HELLA_Group_Overview. pdf

[2] https://www.bosch-engineering.com/portfolio/engineering-services/electrified-mobility.

[3] https://conti-engineering.com/components/battery-management.

[4] B-M. Claudio, M. E. Orchard, M. Kazerani, R. Cárdenas, and D. Sáez. "Particle-filtering-based estimation of maximum available power state in lithium ion batteries", *Applied Energy*, 2016.

[5] D. Roosevelt, "Battery management system for li-ion batteries for electric vehicle application", M.Tech. Thesis, MANIT Bhopal, India, April 2023.

[6] S. Mateen, M. Amir, A. Haque, and F. I. Bakhsh, "Ultra-fast charging of electric vehicles: a review of power electronics converter, grid stability and optimal battery consideration in multi-energy systems", *Sustainable Energy, Grids and Networks*, vol. 35, 2023, article no.101112.

[7] R. Xiong, J. Cao, Q. Yu, H. He, and F. Sun. "Critical review on the battery state of charge estimation methods for electric vehicles", *IEEE Access*, vol. 6, pp. 1832–1843, 2018.

[8] R. Xiong. "*Battery Management Algorithm for Electric Vehicles*", Springer Science and Business Media LLC, Berlin, 2020.

[9] Y. Ye, Z. Li, J. Lin, and X. Wang, "State-of-charge estimation with adaptive extended Kalman filter and extended stochastic gradient algorithm for lithium-ion batteries", *Journal of Energy Storage*, vol. 47, article no. 103611, 2022, doi: 10.1016/j.est.2021.103611.

[10] H-A. Pan, O. Ghodbane, Y-T. Weng, *et al.*, "Investigating mechanisms underlying elevated temperature-induced capacity fading of aqueous MnO polymorph supercapacitors: cryptomelane and birnessite", *Journal of the Electrochemical Society*, vol. 47, no. 5, article no. A5106, 2015, doi: 10.1149/2.0171505jes.

[11] T. Wang, L. Pei, R. Lu, C. Zhu, and G. Wu, "Online parameter identification for lithium-ion cell in battery management system", *2014 IEEE Vehicle Power and Propulsion Conference (VPPC)*, 2014.

[12] M-H. Hung, C-H. Lin, L-C. Lee, and C-M. Wang, "State-of charge and state-of-health estimation for lithium-ion batteries based on dynamic impedance technique", *Journal of Power Sources*, vol. 268, pp. 861–873, 2014, doi: 10.1016/j.jpowsour.2014.06.083.

[13] N. Bhushan, S. Mekhilef, K. S. Tey, M. Shaaban, M. Seyedmahmoudian, and A. Stojcevski, "Overview of model- and non-model-based online battery management systems for electric vehicle applications: a comprehensive review of experimental and simulation studies", *Sustainability*, vol. 14, no. 23, article no. 15912, 2022, doi: 10.3390/su142315912.

[14] R. Dwivedi, S. Singh, and B. Singh, "A single phase modified bridgeless single ended primary induction converter based EV battery charger with enhanced power quality", *IEEE Students Conference on Engineering and Systems (SCES)*, 2020, pp. 1–6, doi: 10.1109/SCES50439.2020.9236769.

Chapter 8

Fault–tolerant operation of electric vehicles

Paramjeet Singh Jamwal[1], Vinay Kumar[2] and Sanjeev Singh[3]

This chapter discusses various probable faults of power electronic switches in an electric vehicle and preferable solutions for a fault–tolerant operation of electric vehicles.

8.1 Introduction

In presently available electric vehicles (EVs), a high number of power electronic components and their drivers are being used in the motor control unit (MCU) and DC–DC converters. It increases the probability of faults in power electronic components and their drivers.

Any unexpected variation in the switching device parameters (specifically voltage and current through it), from its desired specification is defined as a "fault." Further, the fault may be categorized as a malfunction of an individual device or the complete system.

The faults in the devices used in power electronic converters (PEC) decrease its reliability, lower its efficiency, increase its production losses, increase user hazards, and increase its repair costs. Therefore, several techniques are reported for fault detection, diagnosis, and restoration in PEC [1–4]. In an induction motor (IM) drive system, 38% fault occurs in the PEC. Among these faults, 34% of faults are occurring in the power semiconductor devices. The occurrence of faults in PEC and their devices arises the need to make it fault-tolerant (FT) [5].

The primary reason for failure in PEC is an open-circuit (OC) fault in the switches used in it [6]. The probability of the OC fault increases with the increase in the number of switches. The other reason for failure in PEC is a short-circuit (SC) fault which leads to the flow of current in the absence of a gating signal. This SC fault leads to blowing the switch due to excessive current and overheating of the junction. Therefore, the switch becomes open and shown as the OC fault later.

[1]Department of Electrical Engineering, NIT Hamirpur, India
[2]Department of Electronics and Communication Engineering, GLBITM Greater Noida, India
[3]Department of Electrical Engineering, MANIT Bhopal, India

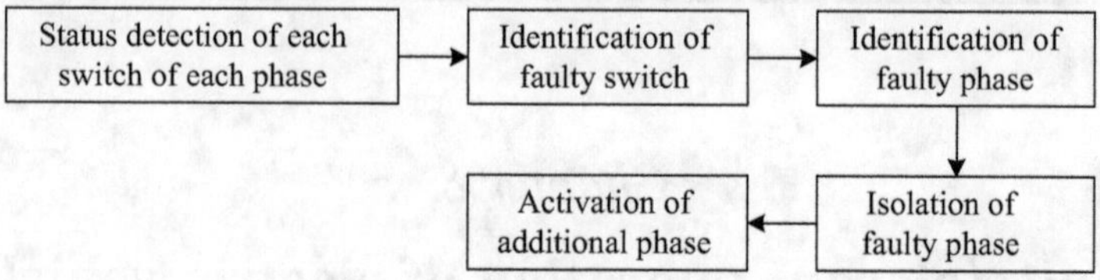

Figure 8.1 Flow chart of fault–tolerant operation

A cascaded H-bridge (CHB) three-level (3L) voltage source converter (VSC) is the first category of CHB multi-level (ML) VSC. There are four most common FT techniques based on switch level, leg level, module level, or system level by which a CHB-ML VSC can be made fault–tolerant [7]. As there is modularity in the structure of CHBML VSC, therefore, FT technique based on module level is mostly preferred.

The fault–tolerant operation of VSC consists of five steps, i.e., status detection of each switch of each phase, identification of faulty switch, identification of faulty phase, isolation of faulty phase, and activation of additional phase. A flow chart of fault–tolerant operation is shown in Figure 8.1.

A technique for the detection of switch status based on phase-voltage error, load current, and gating signal is reported in [8,9]. In this detection technique, a faulty gating signal was considered to create the OC fault. This detection technique took one switching period to detect the fault in two switches of H-bridge while it took around five switching cycles to detect the fault in another two switches. A technique for the detection of switch status based on the half-cycle mean voltage (HCMV) is reported in [10]. This detection technique takes around a half-cycle of output voltage to detect the OC fault. In this chapter, a technique for the detection of switch status is described which took one sample or first gating signal of that switch to detect the fault (OC or SC).

8.2 Types of faults in a VSC and their detection

The voltage source converters (VSCs) can have two types of faults, categorized as hard faults and soft faults. These faults can be further classified as shown in Figure 8.2. Any physical damage to the device is designated as hard fault, such as switch faults, diode faults, gate driver faults, and DC link capacitor faults. The misfiring of switches and errors in the control algorithm are considered as soft faults. The hard faults lead to maloperation or a complete halt of the system, whereas with soft faults, the system continues to operate with altered behavior. The most common types of hard faults in switches and diodes of VSCs are OC and SC faults, known as open switch and short switch faults also.

8.2.1 OC fault in VSC

The OC fault may occur in any switch, either MOSFET or IGBT, of a VSC, due to many reasons such as lifting of soldering track or excessive heat across the junction due to overloading or SC fault across the switch or physical damage of the switch.

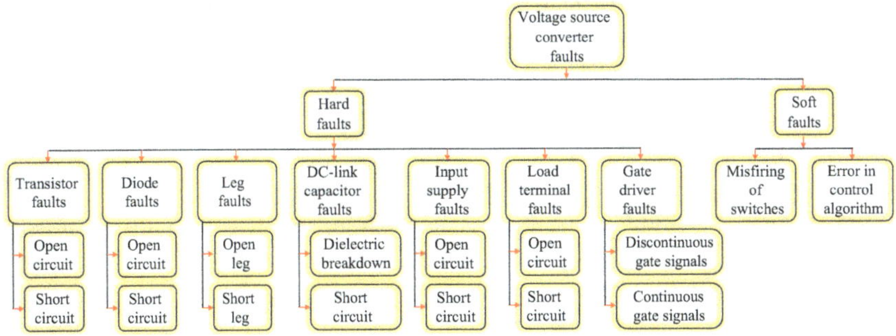

Figure 8.2 Types of faults in a VSC

Table 8.1 The detection logic for OC fault in S_{1a} switch of VSC

Switch condition	Sensed signal			Signal status			OCF logic status
	G_{S1ad}	v_{S1a}	i_{S1a}	S_{vS1a}	S_{iS1a}	$\overline{S_{iS1a}}$	$S_{OCF} = G_{S1ad} \times S_{vS1a} \times \overline{S_{iS1a}}$
Healthy	1	V_{SW}	I_a	0	1	0	0
	0	V_{dc}	0	1	0	1	0
OC fault	1	V_{dc}	0	1	0	1	1
	0	V_{dc}	0	1	0	1	0

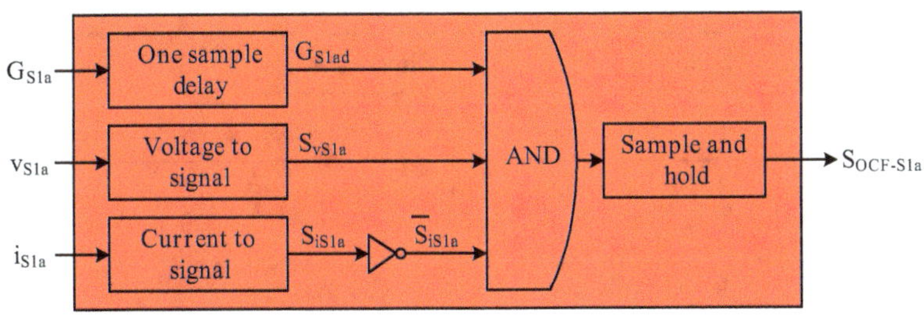

Figure 8.3 Block diagram of OC fault detection scheme [14]

The OC fault can be characterized by rated voltage across the switch and zero current through it when the gate signals are high (ON state). The detection logic of OC fault in S_{1a} switch is presented in Table 8.1 [11]. The logic operation for OC fault in S_{1a} switch is explained under ON condition of the switch as the gating signal (G_{S1a}) and the sensed switch voltage (v_{S1a}) shall be high (rated value). As the switch is open during OC fault, hence, the sensed switch current (i_{S1a}) shall be low (zero value). The schematic block diagram for the detection scheme of OC fault is shown in Figure 8.3. This OC fault detection scheme begins with delaying the gating signal by one sample (G_{S1ad}). Thereafter, sensed switch voltage and current are converted into switch voltage and current signal (S_{vS1a} and S_{iS1a}). While the switch current signal is further inverted ($\overline{S_{iS1a}}$) for the successful detection of OC fault.

8.2.2 *SC fault in VSC*

The SC fault may occur in any switch, either MOSFET or IGBT, of a VSC, mostly due to current overloading or SC across the switch. This fault results in excessive heat generation across the junction leading to blowing the switch and finally open circuits of the switch.

The SC fault can be characterized by zero or very low voltage across the switch (usually equal to ON voltage drop) and high current through it when the gate signals are low (OFF state). The detection logic of SC fault in S_{1a} switch is presented in Table 8.2 [12]. The logic operation for SC fault in S_{1a} switch is explained under OFF condition of the switch as the gating signal (G_{S1a}) shall be low (zero value). As the switch is short during SC fault, hence, the switch current (i_{S1a}) shall be high (rated or more value). The block diagram of SC fault detection scheme is shown in Figure 8.4. This SC fault detection scheme also begins with delaying the gating signal by one sample (G_{S1ad}). Thereafter, the delayed gating signal is further inverted ($\overline{G_{S1ad}}$) while the sensed switch current is converted into switch current signal (S_{iS1a}) for the successful detection of SC fault.

Table 8.2 The detection logic for SC fault in S_{1a} switch of VSC

Switch condition	Sensed signal		Signal status		SCF logic status
	G_{S1ad}	i_{S1a}	$\overline{G_{S1ad}}$	S_{iS1a}	$S_{SCF} = \overline{G_{S1ad}} \times S_{iS1a}$
Healthy	1	I_a	0	1	0
	0	0	1	0	0
SC Fault	1	I_a	0	1	0
	0	I_a	1	1	1

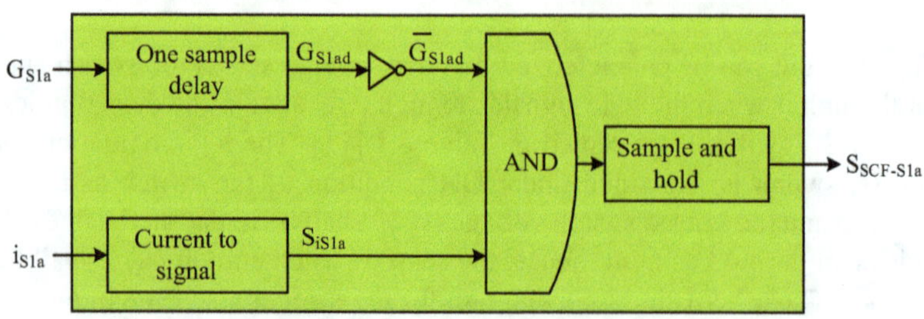

Figure 8.4 Block diagram of SC fault detection scheme [12]

8.3 Identification of faulty phase in VSCs

The detection logic results in the type of fault on any switch, individually, but the switch and the respective phase must be identified by the logic to enable the fault removal process. A cumulative logic to identify the switch status of the VSC is presented in Table 8.3. The cumulative logic status results in "1" (high value) fault status, under both the OC and SC faults. The respective phase of the faulty switch is detected based on the logic status received for each switch in the phase from Table 8.3, as shown in Figure 8.5. In a similar way, the status of other phases is obtained.

Table 8.3 Cumulative detection logic for OC and SC faults in VSC switch

Switch condition	Logic status		Switch status
	$S_{OCF\text{-}S1a}$	$S_{SCF\text{-}S1a}$	S_{S1a}
OC fault	1	0	1
SC fault	0	1	1

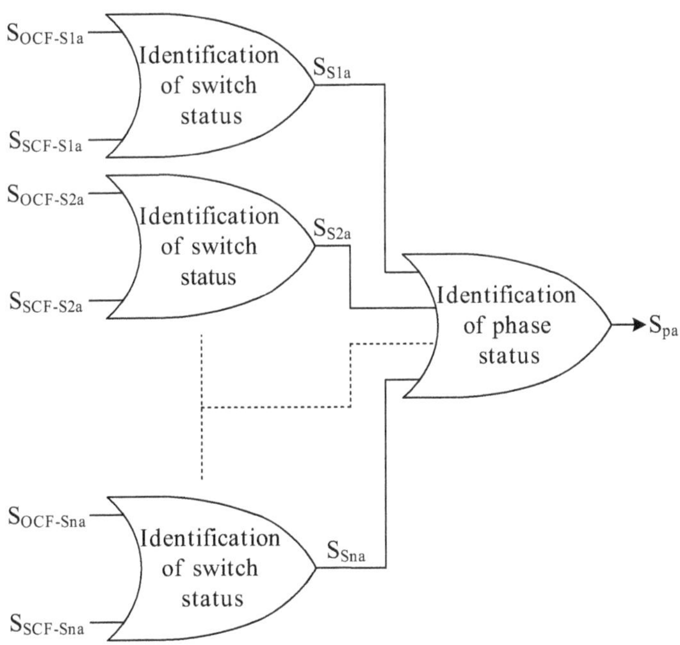

Figure 8.5 Schematic diagram of faulty phase identification scheme

8.4 Removal of fault in VSCs

A process in which the switches of the faulty phase are turned-off and switches of an additional phase are turned-on is termed as VSC fault removal scheme. This fault removal scheme consists of two steps as faulty phase switches isolation (removal) and additional phase switches activation in the same phase.

8.4.1 Isolation of faulty phase switches of VSCs

The schematic diagram for the isolation of faulty phase switches is shown in Figure 8.6. The logic is designed in such a way to turn-off all the switches of faulty phase with gating signals ($G_{S1ai\text{-}Snai}$). For this, the status of that phase (S_a) is complemented with the NOT gate. After that, the complemented status of that phase is multiplied by each gating signal ($G_{S1a\text{-}Sna}$) of the switches of that phase. During healthy condition, the status of that phase is "low" hence each signal of that phase is multiplied with high value and switches of that phase remains in operation. During fault in any switch of that phase, the status of that phase becomes "high," hence each signal of that phase is multiplied by low value and switches of that phase are turned-off. In this way, the switches of the faulty phase are isolated from the VSC.

8.4.2 Activation of additional phase switches of VSCs

The schematic diagram for the activation of an additional phase switches is shown in Figure 8.7. The logic is designed in such a way to turn-on all the switches of

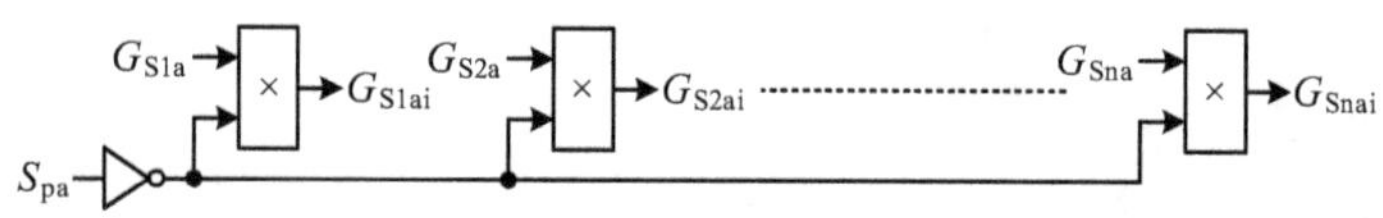

Figure 8.6 Schematic diagram for isolation of faulty phase switches

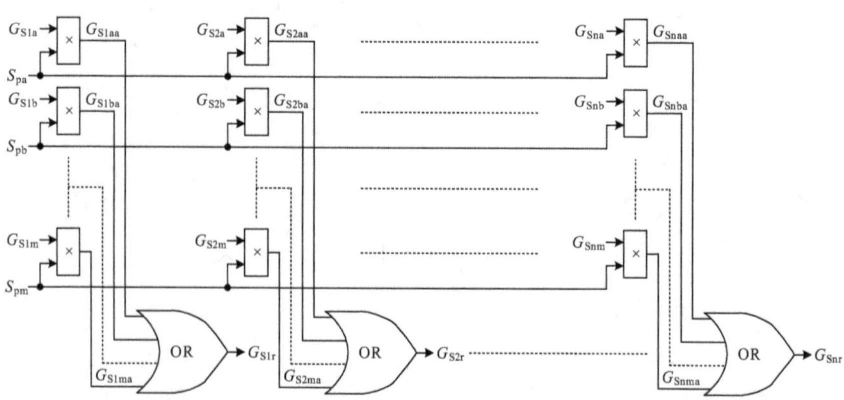

Figure 8.7 Schematic diagram for the activation of additional phase switches

additional phase with the gating signal of the faulty phase. For this, the status of each phase (S_{pa}, S_{pb}, ..., S_{pm}) is multiplied by the gating signals of the switches of that phase ($G_{S1a\text{-}Sna}$, $G_{S1b\text{-}Snb}$, ..., $G_{S1m\text{-}Snm}$) to obtain the activation signals ($G_{S1aa\text{-}Snaa}$, $G_{S1ba\text{-}Snba}$, ..., $G_{S1ma\text{-}Snma}$). The activation signal of the first switch of each phase is fed to OR gate to obtain the gating signal for the first switch of additional phase (G_{S1r}). In a similar way, gating signals for all the switches of an additional phase are obtained (G_{S2r}, ..., G_{Snr}).

During the healthy condition, the status of each phase is "low," hence, each signal of that phase is multiplied with a low value and switches of the additional phase remain inactive. During fault in any switch of any phase, the status of that phase becomes "high," hence, each signal of that phase is multiplied with a high value and switches of additional phase are turned-on with the gating signals of that phase. In this way, the switches of an additional phase of the VSC are activated.

8.5 Fault–tolerant VSC topologies for EVs

VSC topologies are used in the EV to convert DC power into AC power and vice versa. Any fault (OC or SC fault) in the VSC topology of EV disturbs its operation. Therefore, it arises the need to make VSC topology fault–tolerant. The fault–tolerant two-level and three-level VSC topologies are discussed in the following section.

8.5.1 Two-level VSC topologies

The two-level VSC topologies are most preferred topology because it uses the lowest number of components. The schematic diagram of fault–tolerant two-level VSC topology-fed IM is shown in Figure 8.8. During the healthy condition, the fault–tolerant two-level VSC behaves as a normal VSC and the additional phase remains inactive. When any switch of any phase becomes faulty, the status of that phase becomes high and activates the respective bidirectional switch (S_a, S_b, S_c). In this way, an additional phase replaces the faulty phase and maintains the smooth operation of IM.

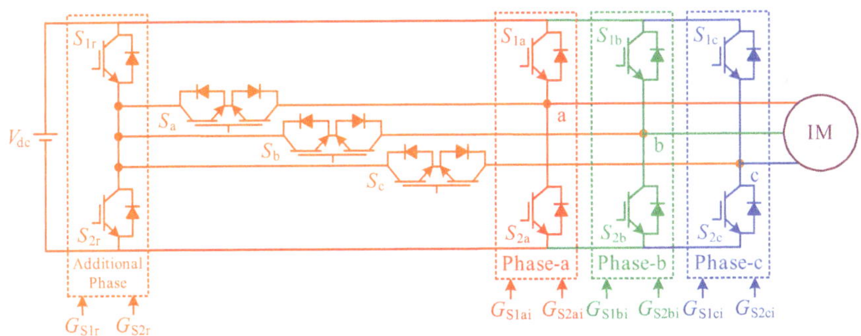

Figure 8.8 Fault–tolerant two-level VSC-fed IM for EVs [13]

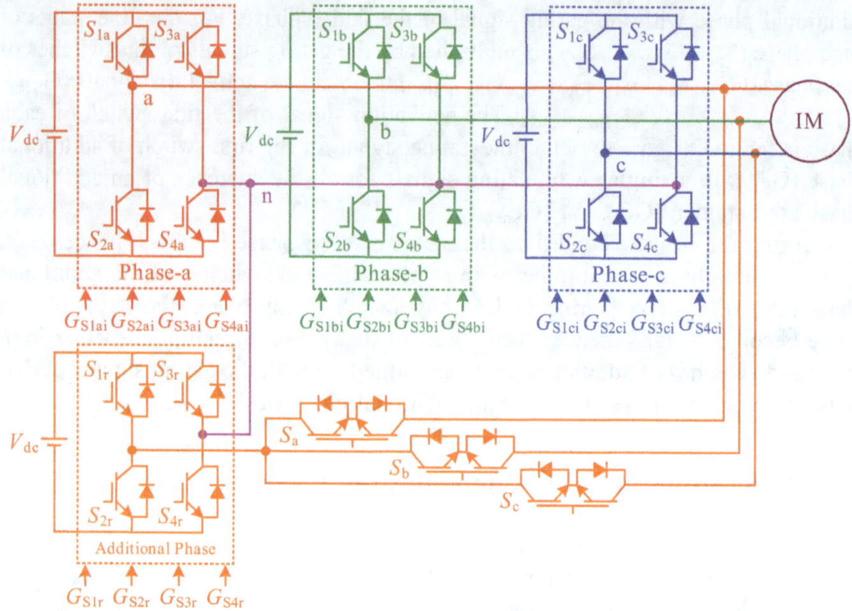

Figure 8.9 Fault–tolerant CHB three-level VSC fed IM for EVs

8.5.2 Three-level VSC topologies

Several topologies of three-level VSC are available in the literature. Among them, cascaded H-bridge (CHB) three-level VSC is popular where isolated DC sources are available. The schematic diagram of fault–tolerant CHB three-level VSC-fed IM is shown in Figure 8.9. During the healthy condition, the fault–tolerant CHB three-level VSC behaves as a normal VSC and an additional phase remains inactive. When any switch of any phase becomes faulty, the status of that phase becomes high and activates the respective bidirectional switch (S_a, S_b, S_c). In this way, an additional phase replaces the faulty phase and maintains the smooth operation of IM.

8.6 Results and discussion

The performance of fault–tolerant VSC is analyzed after observing the waveforms of gating signal, phase status, phase currents, voltage across motor terminals and speed, and current through motor terminals and torque as shown in Figures 8.10–8.14, respectively. The gating signals of a switch of each phase are shown in Figure 8.10. The gating signal of the switch of each phase is transferred to the switch of an additional phase during the fault in the respective phase. The status of each phase is shown in Figure 8.11. The status of each phase becomes high during the fault in the respective phase. The currents of each phase are shown in

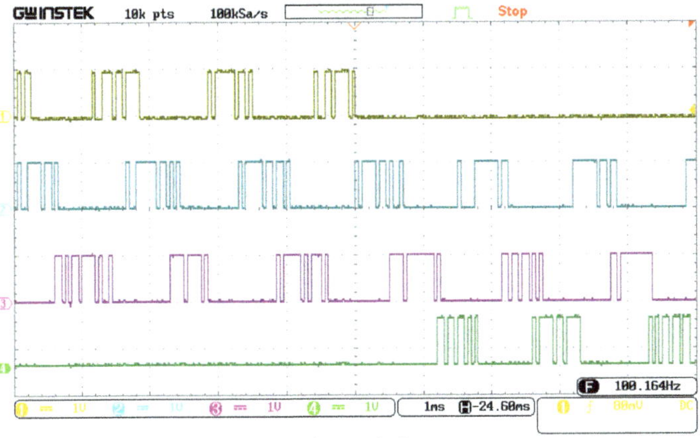

(a) Phase-a fault

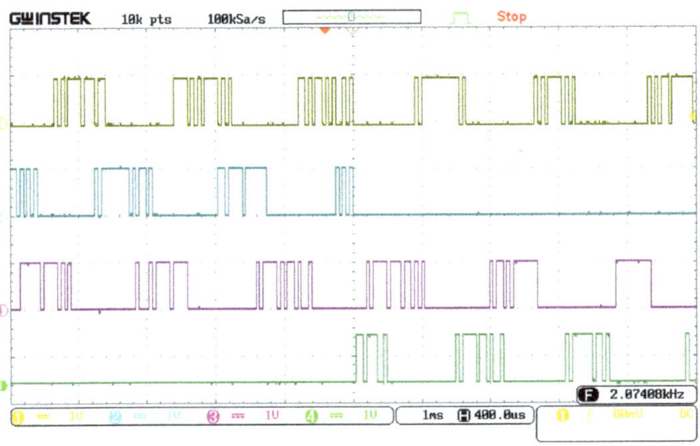

(b) Phase-b fault

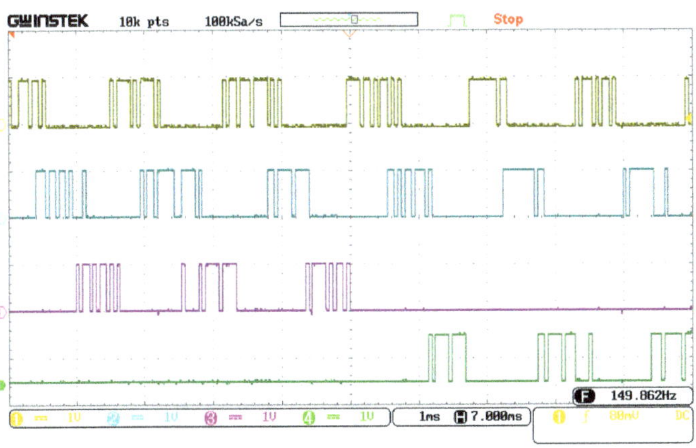

(c) Phase-c fault

Figure 8.10 Gating signal waveforms

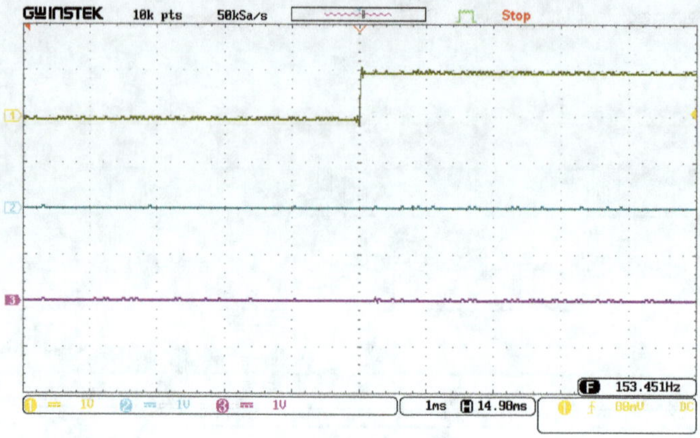

(a) Phase-a fault

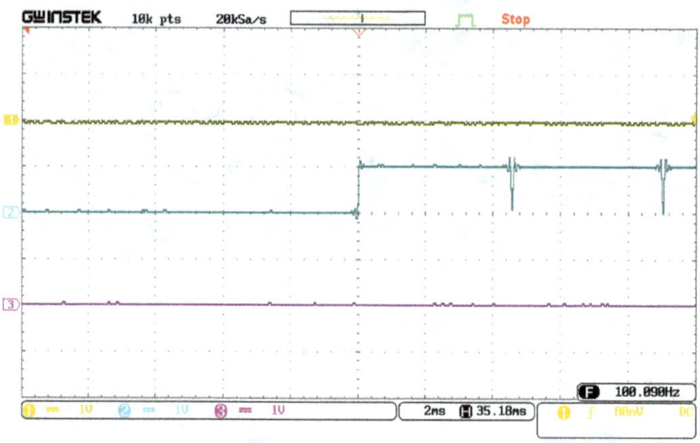

(b) Phase-b fault

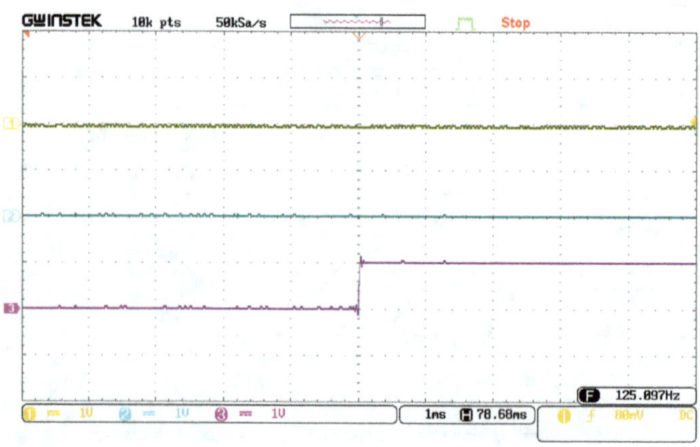

(c) Phase-c fault

Figure 8.11 Phase status waveforms

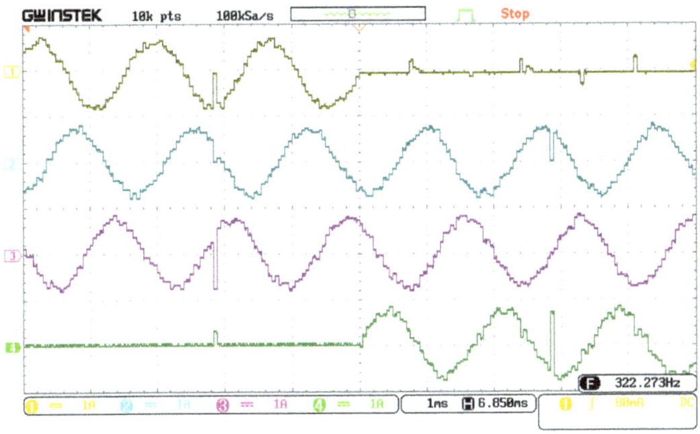

(a) Phase-a fault

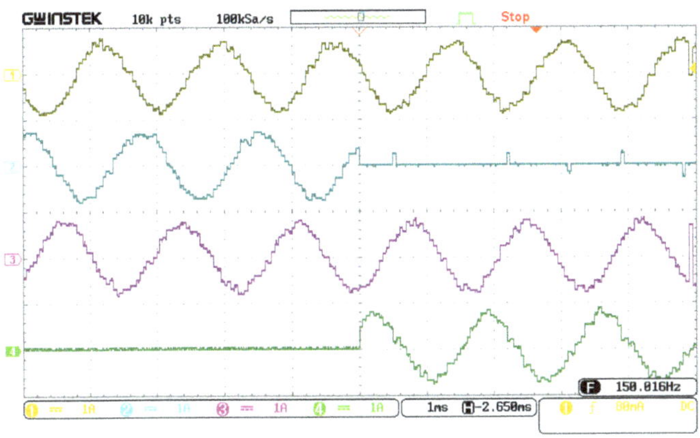

(b) Phase-b fault

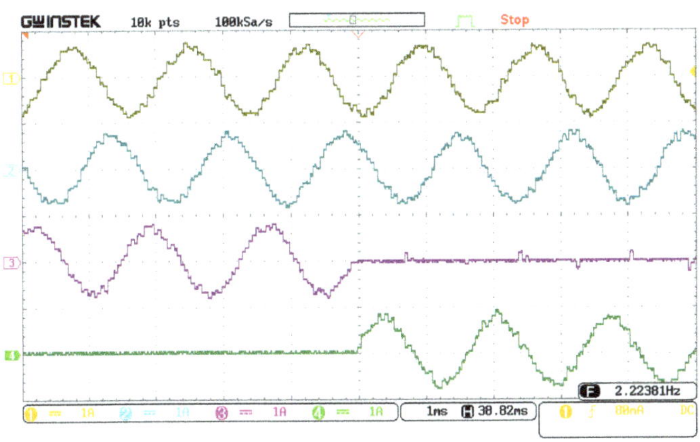

(c) Phase-c fault

Figure 8.12 Phase current waveforms

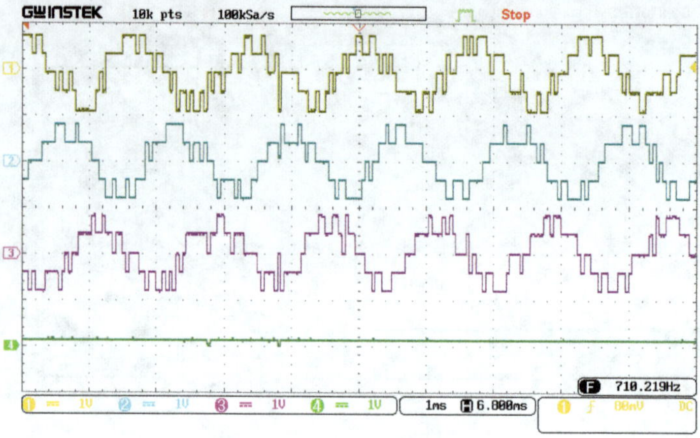

(a) Phase-a fault

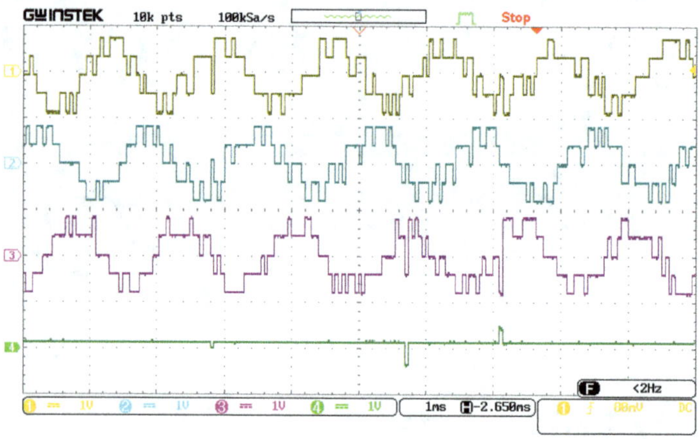

(b) Phase-b fault

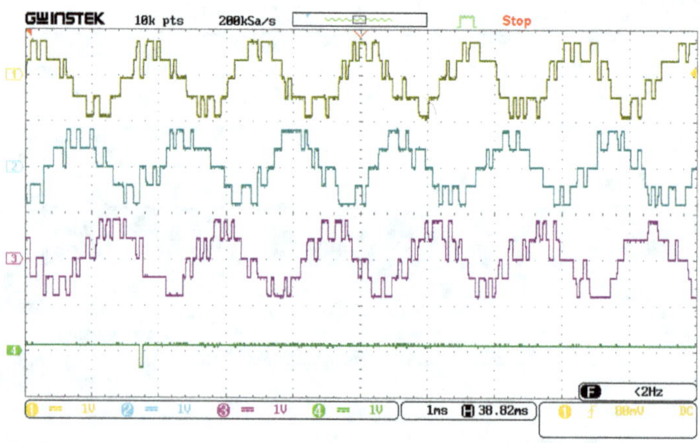

(c) Phase-c fault

Figure 8.13 Voltage across motor terminals and speed waveforms

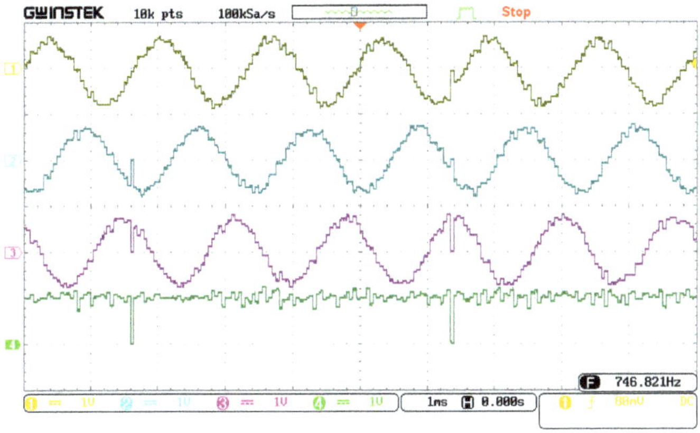

(a) Phase-a fault

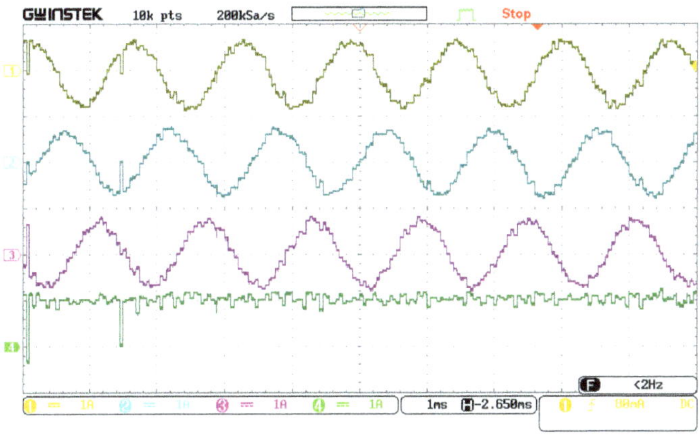

(b) Phase-b fault

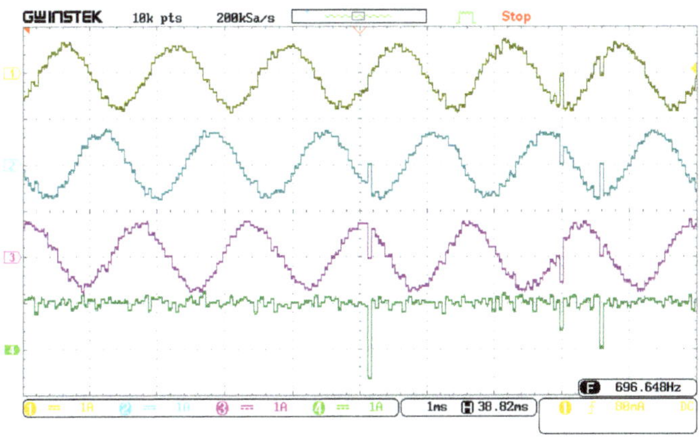

(c) Phase-c fault

Figure 8.14 Current through motor terminals and torque waveforms

Figure 8.12. The current of each phase is supplied by an additional phase during the fault in the respective phase. The voltage across motor terminals is shown in Figure 8.13. The voltage across motor terminals is applied by an additional phase during the fault in the respective phase. The current through motor terminals is shown in Figure 8.14. The current through motor terminals is supplied by an additional phase during the fault in the respective phase.

8.7 Summary

In this chapter, the fault–tolerant operation of various converters used in an electric vehicle has been discussed with an overview of the types of possible faults in the EV system. The fault detection, identification, and removal schemes used on a fault–tolerant VSC for EV traction motor have been discussed in detail with a generalized concept. The results of a fault–tolerant system have also been presented to showcase the practicability of such systems. This results in an uninterrupted operation of the EV under OC and SC faults in VSC. There is a wide scope for fault–tolerant operation of converters for EV system and expected that this chapter has given thought-provoking ideas in this direction.

Symbols

G_{S1a}	gating signal for S_{1a} switch
G_{S1ai}	gating signal for Isolation of S_{1a} switch
G_{S1b}	gating signal for S_{1b} switch
G_{S1m}	gating signal for S_{1m} switch
G_{S1r}	gating signal for S_{1r} switch
G_{S2a}	gating signal for S_{2a} switch
G_{S2ai}	gating signal for Isolation of S_{2a} switch
G_{S2b}	gating signal for S_{2b} switch
G_{S2m}	gating signal for S_{2m} switch
G_{S2r}	gating signal for S_{2r} switch
G_{Sna}	gating signal for S_{na} switch
G_{Snai}	gating signal for Isolation of S_{na} switch
G_{Snb}	gating signal for S_{nb} switch
G_{Snm}	gating signal for S_{nm} switch
G_{Snr}	gating signal for S_{nr} switch
G_{S1ad}	one sample delay in G_{S1a}
i_{S1a}	current through S_{1a} switch
S_{1a}	first switch of phase-a
S_{1b}	first switch of phase-b

S_{1c}	first switch of phase-c
S_{1r}	first switch of additional phase
S_{2a}	second switch of phase-a
S_{2b}	second switch of phase-b
S_{2c}	second switch of phase-c
S_{2r}	second switch of additional phase
S_{3a}	third switch of phase-a
S_{3b}	third switch of phase-b
S_{3c}	third switch of phase-c
S_{3r}	third switch of additional phase
S_{4a}	fourth switch of phase-a
S_{4b}	fourth switch of phase-b
S_{4c}	fourth switch of phase-c
S_{4r}	fourth switch of additional phase
S_a	bidirectional switch connected between additional phase and phase-a
S_b	bidirectional switch connected between additional phase and phase-b
S_c	bidirectional switch connected between additional phase and phase-c
S_{iS1a}	switch current signal of current through S_{1a} switch
$S_{OCF\text{-}S1a}$	status of OC fault in S_{1a} switch
$S_{OCF\text{-}S2a}$	status of OC fault in S_{2a} switch
$S_{OCF\text{-}Sna}$	status of OC fault in S_{na} switch
S_{pa}	status of phase-a
S_{pb}	status of phase-b
S_{pm}	status of phase-m
S_{S1a}	status of S_{1a} switch
S_{S2a}	status of S_{2a} switch
S_{Sna}	status of S_{na} switch
$S_{SCF\text{-}S1a}$	status of SC fault in S_{1a} switch
$S_{SCF\text{-}S2a}$	status of SC fault in S_{2a} switch
$S_{SCF\text{-}Sna}$	status of SC fault in S_{na} switch
S_{vS1a}	switch voltage signal of voltage across S_{1a} switch
v_{S1a}	voltage across S_{1a} switch

Glossary

CHB	cascaded H-bridge
DC	direct current
EV	electric vehicle

FT fault–tolerant
IM induction motor
MCU motor control unit
OC open circuit
OCF open circuit fault
SC short circuit
SCF short circuit fault
VSC voltage source converter
2L two-level
3L three-level
5L five-level

References

[1] J. N. Apruzzese, S. B. Monge, J. Bordonau, S. Alepuz, and A. C. Prado, "Analysis of the fault-tolerance capacity of the multilevel active-clamped converter," *IEEE Trans. Ind. Electron.*, vol. 60, no. 11, pp. 4773–4783, 2013.

[2] X. Kou, K. A. Corzine, and Y. L. Familiant, "A unique fault-tolerant design for flying capacitor multilevel inverter," *IEEE Trans. Power Electron.*, vol. 19, no. 4, pp. 979–987, 2004.

[3] E. Esfandiari, N. Mariun, M. H. Marhaban, and A. Zakaria, "Switch-ladder: reliable and efficient multilevel inverter," *Electron. Lett.*, vol. 46, no. 9, p. 646647, 2010.

[4] K. Ambusaidi, V. Pickert, and B. Zahawi, "New circuit topology for fault tolerant H-bridge DC–DC converter," *IEEE Trans. Power Electron.*, vol. 25, no. 6, pp. 1509–1516, 2010.

[5] M. Ali, Z. Din, E. Solomin, K. M. Cheema, A. H. Milyani, and Z. Che, "Open switch fault diagnosis of cascade H-bridge multi-level inverter in distributed power generators by machine learning algorithms," *Energy Rep.*, vol. 7, pp. 8929–8942, 2021.

[6] V. Kumar, S. Singh, and S. Jain, "A reduced switch count symmetric T-type multilevel inverter with single and multiple switch open circuit fault tolerant capabilities," *IETE J. Res.*, pp. 1–23, 2022.

[7] W. Zhang, D. Xu, P. N. Enjeti, H. Li, J. T. Hawke, and H. S. Krishnamoorthy, "Survey on fault-tolerant techniques for power electronic converters," *IEEE Trans. Power Electron.*, vol. 29, no. 12, pp. 6319–6331, 2014.

[8] M. Kumar, "Open circuit fault detection and switch identification for LS-PWM H-bridge inverter," *IEEE Trans. Circuits Syst.—II: Express Briefs*, vol. 68, no. 4, pp. 1363–1367, 2021.

[9] G. Zhang and J. Yu, "Open-circuit fault diagnosis for cascaded H-bridge multilevel inverter based on LS PWM technique," *CPSS Trans. Power Electron. Appl.*, vol. 6, no. 3, pp. 201–208, 2021.

[10] A. Anand, A. V. B. N. Raj, J. G and S. George, "A generalized switch fault diagnosis for cascaded H bridge multilevel inverters using mean voltage prediction," *IEEE Trans. Ind. Appl.*, vol. 56, no. 2, pp. 1563–1574, 2020.

[11] P. S. Jamwal, S. Singh, and S. Jain, "Fault-tolerant operation of the cascaded H-bridge three-level inverter for electric vehicle application," *Int. J. Ambient Energy*, vol. 44, no. 1, pp. 1649–1662, 2023.

[12] P. S. Jamwal, *Multilevel Inverter fed Induction Motor Drive for Battery Electric Vehicle*, PhD Thesis, EIE Department, SLIET Longowal, India, July 2023.

[13] R. L. d. A. Ribeiro, C. B. Jacobina, E. R. C. d. Silva, and A. M. N. Lima, "Fault-tolerant voltage-fed PWM inverter AC motor drive systems," *IEEE Trans. Ind. Electron.*, vol. 51, no. 2, pp. 439–446, 2004.

[14] V. Kumar, *Investigations on Fault Tolerant Operation of Multilevel Inverter*, PhD Thesis, EIE Department, SLIET, Longowal, India, 2022.

Chapter 9

Design, simulation, and control of battery charger for electric vehicle

Anjanee Kumar Mishra[1] *and Ankit Kumar Singh*[1]

This chapter is intended to provide insight into the design and development of single-stage battery charging systems for on-board applications of plug-in electric vehicles (PEVs), their classification and issues related to them, different configurations, size, and compactness improvement, and possibility and technique to incorporate solar energy into charging system.

9.1 Introduction

Petroleum-based goods are mostly used in land transportation. To fulfill this expectation, numerous steps were taken. On the one hand, rapid trains, the metro, and tramways all now run on electric trains. On the other hand, adopting lighter ICEs boosted the efficiency of personal transportation by reducing vehicle weight. Internal combustion engine advancements, however, were insufficient to lower fuel usage. To accomplish this, scientists and automakers have investigated a number of alternatives, such as biodiesel, ethanol, compressed natural gas (CNG), liquefied natural gas (LNG), hydrogen, compressed air, electric vehicles (EVs), and others.

New advances have directed the entry of electric cars as an alternative to cars using internal combustion engines. Good policies in both developed and developing countries and the development needed to prevent urban pollution to create the most favorable environment for these developments. Some countries have established emission standards to limit the environmental pollution of cars to be sold in these countries. Additionally, continued research on battery technology leads to improved and cheaper batteries for future EVs; therefore, the overall cost of EVs is falling rapidly. The EV sales are expected to continue to rise due to two key factors: market scenario and other is innovation in battery technology. In terms of the market economies, various manufacturers have developed their own storage systems to cut its costs. Also, batteries are used for power distribution, drone aviation, etc. It should be noted that it is widely used in areas and its prices will decrease rapidly.

[1]Department of Electrical Engineering, Netaji Subhas University of Technology (NSUT), India

The majority of EVs are charged by coupling the car straight to the low-voltage grid. The PEVs can be broadly categorized as fuel cell plug-in hybrid EVs (FC-PHEV), plug-in hybrid EVs (PHEV), and battery EVs (BEVs). On the other hand, hybrid EVs (HEVs) are not the same as PEVs. These vehicles do not require a mains charge because their batteries may be charged by ICE-made electricity that is produced inside. PEVs, which produce clean energy for automobiles using non-polluting batteries, are now a potential approach to reducing pollution. PEV is a combination of an additional charger, battery, and inverter drive system. Batteries play an important factor in the development of EVs. The battery charger for the car should be compact and lightweight as it is located inside the car. So, to achieve these objectives, scientist and researchers are focusing on the development and production of electronic components for pure EVs.

9.2 Classifications of chargers

PEVs use two types of battery chargers; standalone (non-built-in) chargers and OBCs. Use a built-in charger when more charging power is needed. The size and the weight of the charger are easier to carry with a built-in charger, and the vehicle can be charged anywhere [1,2]. The desired properties of OBCs are lightless, high-energy density, and high performance [3–7]. An OBC may be configured as a single-stage converter (Figure 9.1) or a two-stage converter (Figure 9.2). Since there are more components in a two-stage charger [3,8–12], a single-phase converter charger [13–17], it is more attractive for car use, but the problem associated with its the presence of low-frequency ripple at the DC link.

OBC adds heaviness and costs to the EV; thus, it is generally designed for lower power levels, usually less than 3.5 kW [18,19]. However, using the transformer as a separate two-phase converter is a convention in OBC design. Also, multiple output voltages and high-frequency transformers for galvanic isolation can affect performance and power density [20–22]. There is no electrical limit to the need to isolate the battery from the AC input because the battery floor usually floats with the vehicle's body floor. However, for safety reasons, a protective relay can be

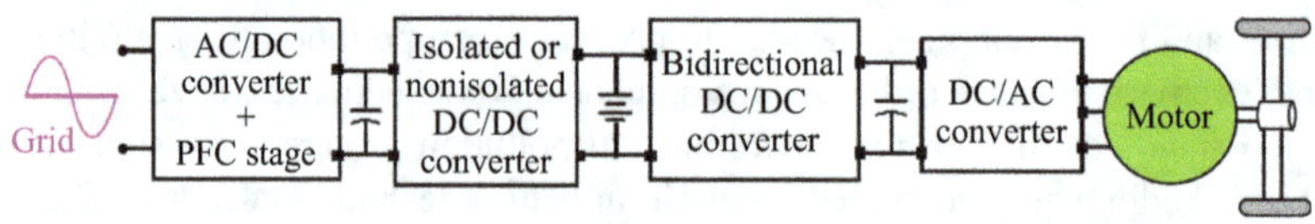

Figure 9.1 Single-stage EV battery charging scheme

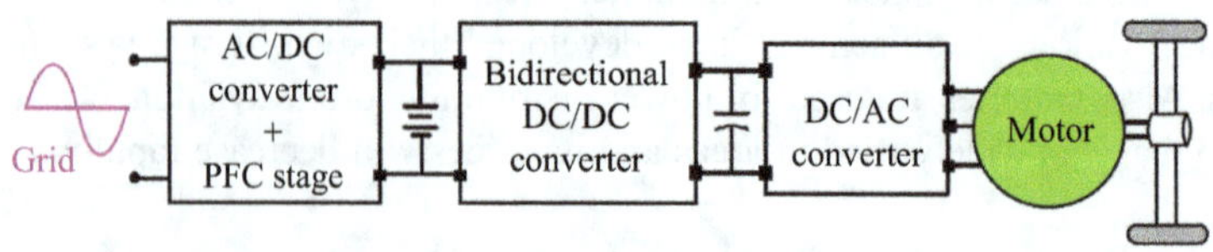

Figure 9.2 Two-stage EV battery charging scheme

added to the battery discharge. The power capacity and the charging time of different types of chargers can also be categorized. The distribution was chosen based on the current power levels in the country [20], as shown in Table 9.1. In addition, the charger is also divided into unidirectional and bidirectional chargers.

A bidirectional charger allows for charging from the grid and the ability to inject battery energy back into the grid (V2G operation mode). Figure 9.3 displays a typical block schematic of one such charger. The bidirectional charger comprises two stages: a bidirectional DC/DC converter to control battery current and an active bidirectional AC/DC converter connected to the grid that enforces the power factor. However, there are drawbacks to bidirectional power flow, including battery deterioration from repeated cycling, the high cost of a charger with this capability, metering problems, and the requirement for distribution system upgrades. Additionally, using a bidirectional charger will call for rigorous safety precautions.

A single-stage OBC may be more desirable than an off-board charger in areas where there has been considerable growth in PEVs since it prevents the dispersal of fewer off-board charging stations and helps to manage traffic at charging spots. To decrease the price, heaviness, and dimension of the on-board charger, scientists have designed integrated converter-based chargers. These can be classified as integrated chargers utilizing machine windings and traction converter into the charging circuit and integrated chargers incorporating bidirectional DC/DC converter into the charging circuit. In the subsequent sections, single-stage-based chargers (conventional and integrated) are reviewed.

Table 9.1 Class of chargers according to power level

Level	Placement of charger	Utilizing location	Power capacity
Level-1			
120 V_{ac} (USA)	OBC, single-phase	Residential charging	1.4 kW (12 A)
230 V_{ac} (EU)			1.9 kW (20 A)
Level-2			
240 V_{ac} (US)	OBC, single-or three-phase	Charging at private or public outlets	4 kW (17 A)
400 V_{ac} (EU)			8 kW (32 A)
Level-3			
208–600 V_{ac} or V_{dc}	Off-board, three-phase	Commercial	Commercial

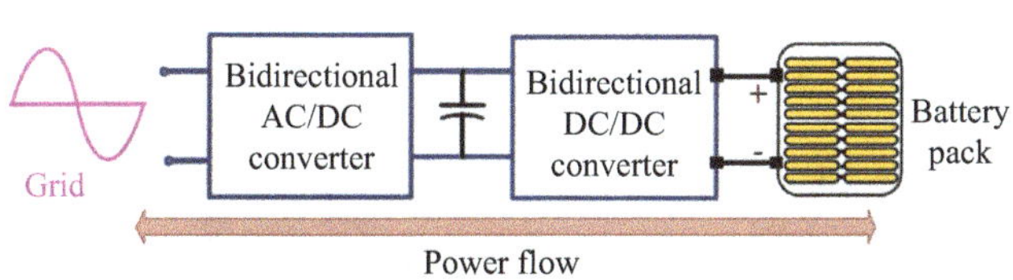

Figure 9.3 Block diagram of bidirectional charger

9.3 Integrated charging system

The block diagram of a typical single-phase charger is shown in Figure 9.4. The EV battery is connected to the DC/DC converter via a bidirectional DC–AC converter [18], which only needs one converter to support a full working mode, power factor adjustment stage [4,23], and other features. The entire system is less than a single charge, as various indices, combining semiconductor devices and some passive components, are distributed as a pattern. Many of these chargers are listed in [17], and all have been checked.

In [24], a combination with each type of throwing boost function is proposed for each vehicle type, as shown in Figure 9.5. The regenerative braking power of the converter's buck/boost function provides better control and performance. In this class of converter, a total of two to four semiconductor components charge the battery at a voltage of 200–450 V using the global grid. In the current operation, the transmission loss is exacerbated, and the power efficiency is reduced.

The topology discussed in [25], shown in Figure 9.6, involves a total of nine switches to achieve the desired operation of the vehicle. Due to the presence of detachment in PIC mode, the converter evades thermal supervision problems and avoids interference with bridge rectifiers. Nevertheless, the large number of semiconductor devices makes on-board charging less preferable. In addition, many control strategies in this switch are required to integrate IGBT switches.

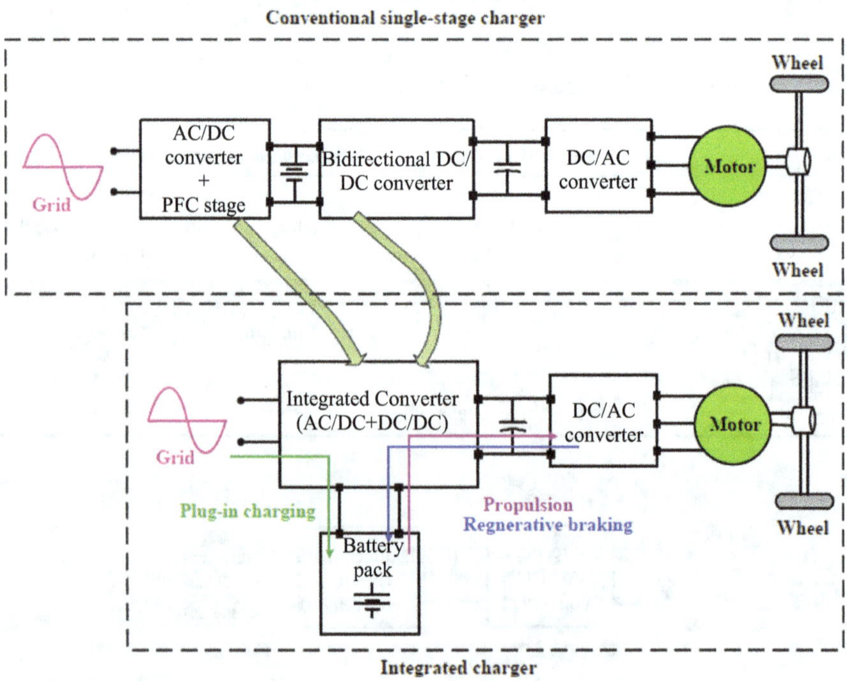

Figure 9.4 Scheme of integrated charging arrangement for EV

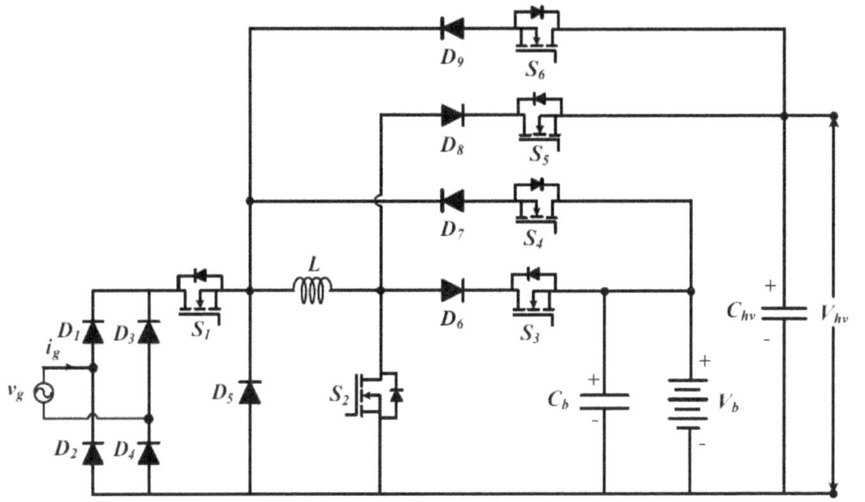

Figure 9.5 Integrated charger [24]

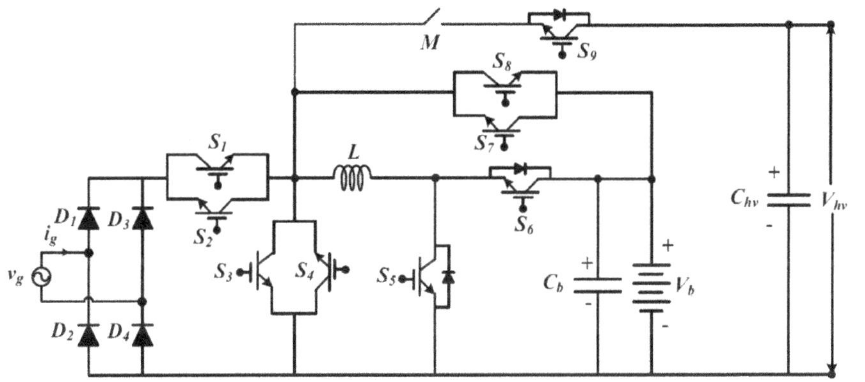

Figure 9.6 Integrated charging scheme [25]

This study's [26] integrated charger approach features one power drive switch and one inductor and performs both charging and discharging functions. The buck–boost converter also performs recharging functions. The following modes are the plug-in, regenerative, and propulsion modes for the inverting buck/boost and buck and boost operational converters, respectively. Also, the result of this converter is better than the converter [24], depending on the number and quality of each type of product. Therefore, it provides inexpensive translation for developers. However, the main disadvantage of this converter is increased voltage and current stress on electrical components and current stress on magnetic components. Additionally, the dual-stage converter having buck–boost ability in PP and RB mode has greater transmission loss as compared to the existing structure associated with mechanical switches Figure 9.7).

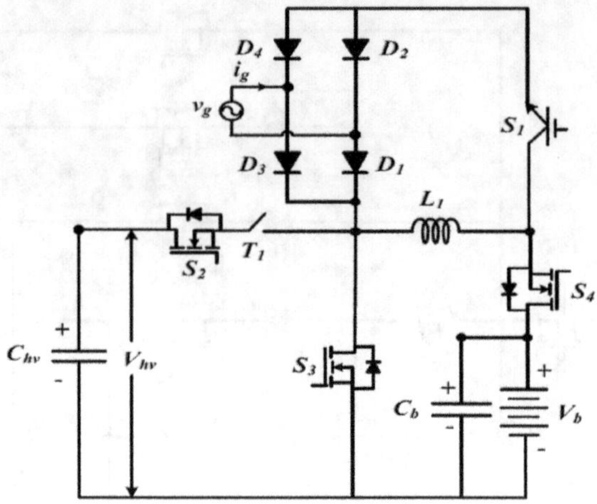

Figure 9.7 Integrated charger [26]

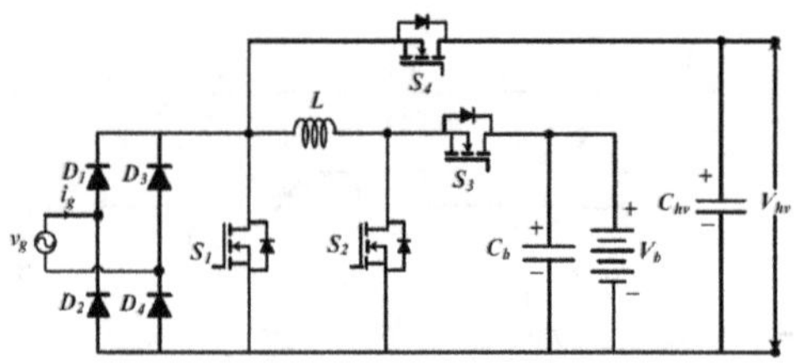

Figure 9.8 Integrated charger [27]

Dusmez and Khalid [27] proposed an integrated charging circuit using four switches and an inductor in the biased channel with various semiconductor devices (depending on the model), as shown in Figure 9.8. Also, the flexibility and the performance loss of individual technologies are compared to their traditional equivalents (persistent/supportive switching with one or two switches). Also, since the voltage/current in this converter is very low, there is less loss and less heat gain, which ultimately makes the circuit thinner. Furthermore, by measuring the current through the inductor, only one sensor measurement is needed for one type, reducing the overall design and reducing feedback. Inactive during the pay period; it cannot control the power of the battery when the voltage is higher than the battery voltage.

Researchers in [28] proposed another expansion of the bridgeless integrated converter with carrier base control, as illustrated in Figure 9.9. Plug-in charging techniques are used with non-linear charger controllers and are PFC compatible in permanent connection. Therefore, the charger becomes more compact with less input circuitry. The automatic bridge design of the converter is designed to upsurge the efficiency of the converter in PIC mode by minimizing the count of components utilized in the current path. However, this converter follows the standard inverting

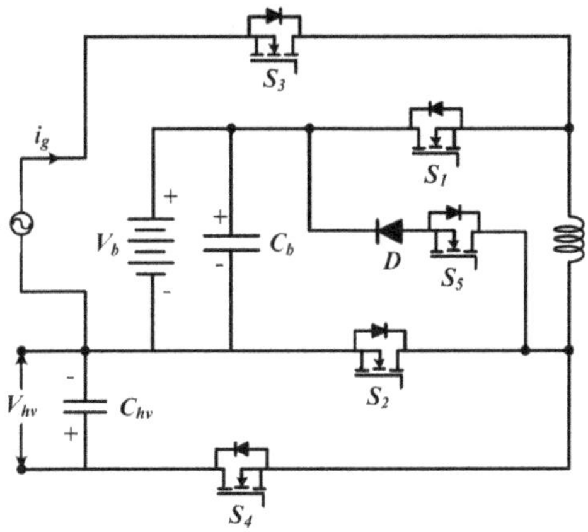

Figure 9.9 Integrated charger [28]

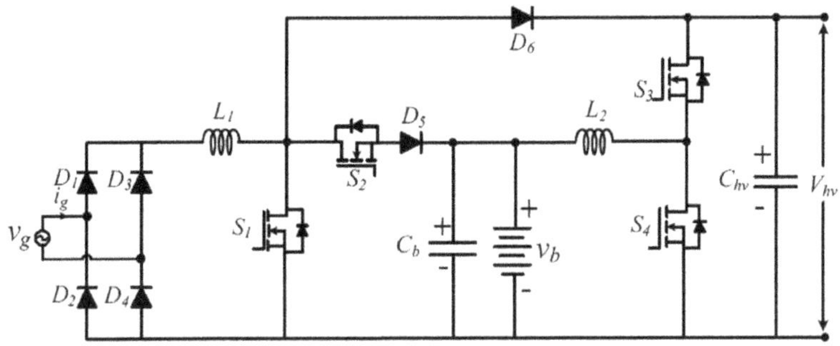

Figure 9.10 Integrated charger [29]

converter having the buck–boost capability. During the second half duration of the mains supply, great stress is created, which causes losses and characteristics (behavior and change). As a result, the benefits of disconnected features are somewhat sacrificed on the negative half of the grid. Additionally, the battery's voltage range is limited because there is no single mode that has both buck and boost capabilities. Since there is no single operating mode with both buck and boost operation, the voltage range of the battery is limited [29].

For various battery voltages, semi-two-phase chargers have been developed as illustrated in Figure 9.10. When battery voltage $v_b >$ peak mains voltage, $v_{g:max}$, the input converter behaves like a typical step-up converter (single-stage operation); when v_b, $v_{g:max}$, the designed converter works as a semi-two-stage converter. Due to the additional conversion, the variation of this conversion in plug-in charging mode is slightly larger than that of the standard powered converter with PFC capability. However, this circuit has lower semiconductor losses due to the uniqueness of its three-level output. Additionally, in PP and RB modes, the claimed circuit operates in boost/buck modes.

9.4 Assessment of existing integrated charging circuits

In this section, the integrated chargers are compared according to the voltage/current stress of the semiconductor device, the buck/boost operation of each type of converter, and the components. Table 9.6 is a list of each semiconductor device's stress.2. "Less" denotes an input or output value [(v_{out} or v_{in}) and (i_{out} or i_{in})]. and "More" denotes the sum of input and output values [$v_{out} + v_{in}$ and $i_{out} + i_{in}$]. From Tables 9.2 and 9.3, the integrated charger [24] has buck/boost functionality and low pressure in all modes at the expense of a high number of semiconductor components. By lowering the number of components in the current path, the converter [25] improves the efficiency of the converter [24], but the charger's price rises as a result of the numerous switches. The chargers [26] were better than the chargers [24,25] in terms of component count. The charger [27] has fewer components and stress than any prevailing converters and has smaller current/voltage stress in all modes. The charger in [29] has buck/boost operation in plug-in charge mode, but a tedious control approach is required in this mode. Additionally, the converter's two-stage operation in buck mode loses its advantages.

Table 9.2 Stresses in terms of voltage/current on semiconductors

Integrated chargers	PIC		PP		RB	
	Voltage	**Current**	**Voltage**	**Current**	**Voltage**	**Current**
Figure 9.5	Less	Less	Less	Less	Less	Less
Figure 9.6	Less	Less	Less	Less	Less	Less
Figure 9.7	More	More	Less	Less	Less	Less
Figure 9.8	Less	Less	Less	Less	Less	Less
Figure 9.9	More	More	Less	Less	Less	Less
Figure 9.10	Less	Less	Less	Less	Less	Less

PIC, plug-in charging; PP, propulsion; RB, regenerative braking.

Table 9.3 Comparison between the existing charging schemes

Chargers Type	Operating modes			Components count		
	PIC	**PP**	**RB**	**Diode**	**Inductor**	**Capacitor**
Figure 9.5	Bu/Bo	Bu/Bo	Bu/Bo	9	1	2
Figure 9.6	Bu/Bo	Bu/Bo	Bu/Bo	4	1	2
Figure 9.7	Bu/Bo	Boost	Buck	5	1	2
Figure 9.8	Boost	Bu/Bo	Bu/Bo	4	1	2
Figure 9.9	Bu/Bo	Boost	Buck	1	1	2
Figure 9.10	Boost	Bu/Bo	Bu/Bo	6	2	2

Bu/Bo, buck–boost.

As can be seen from the literature cited above, there is a possibility for additional upgrading in the combination by creating converters having the functionality of buck/boost in every mode, reducing the number of devices, and improving PP and RB mode performance. Further, reducing the complication of the feedback offers less in the converter dimension and increases its trustworthiness.

9.5 Modified zeta-based integrated converter for battery charging

This part discusses the design, operation, and control of an integrated converter for battery charging for PEVs that is also capable of operating in two other modes, i.e., propulsion and regenerative modes. The diagram of the charging circuit with the designed converter is illustrated in Figure 9.11. The proposed scheme is derived from an existing Zeta converter followed by a buck/boost operation in propulsion and RB mode and a traditional Zeta power factor correction converter in PIC mode, respectively. Since Zeta works in the PIC mode, it can charge the battery from any electrical input supply. In addition, due to the capability to buck/boost, the energy stored in the devices may be completely composed during RB. Moreover, this converter has fewer components than other conventional converters. The control logic for this converter is formed simply by utilizing one switch throughout the operation. The voltage/current stresses and loss studies of the converter were also carried out to select the semiconductor devices and to demonstrate the feasibility of the designed converter.

9.6 Working of integrated converter

The integrated converter illustrated in Figure 9.11, consists of three switches, four diodes, two capacitors (apart from filter capacitors), and two inductors (excluding filter inductors). Table 9.4 tabulates the role of switches and diodes in all working modes. The following sections discuss the operation of the converter using the operating state and corresponding waveforms.

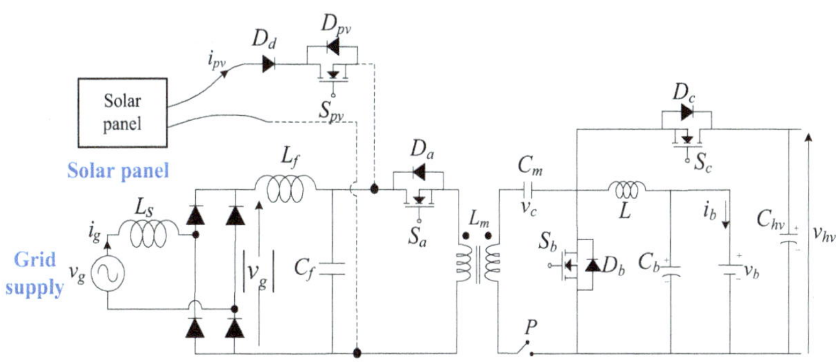

Figure 9.11 Designed battery charging scheme

Table 9.4 Devices state in different operating scenarios

Operating mode	S_a	S_b	S_c	S_{pv}	D_a	D_b	D_c	D_{pv}
PIC	1	0	0	0	0	1	0	0
PV	1	0	0	1	0	1	0	0
PP	0	1	0	0	0	0	1	0
RB	0	0	1	0	0	1	0	0

1: active and 0: in-active.

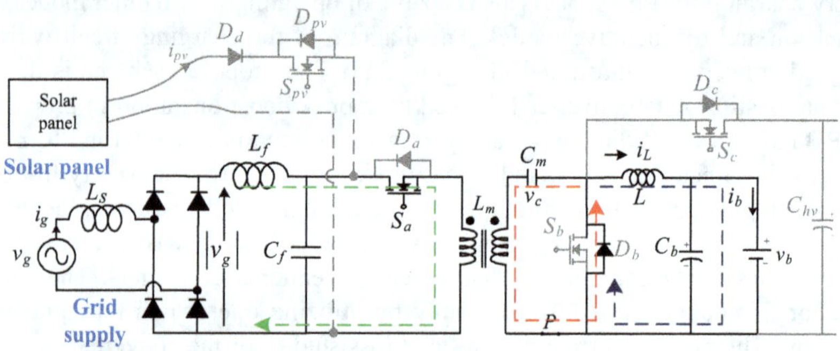

Figure 9.12 Utilizing a grid supply to charge the vehicle battery

9.6.1 PIC mode of operation

This mode only works when the output of the solar panel drops underneath a particular threshold. In this case, the designed converter works as a separate Zeta converter because the "S_{pv}" switch is disabled, and only the mains are used to charge the battery. A high-frequency signal is sent to the switch "S_a" to activate it when the single pole, double-thrown switch "*P*" is in operation. The magnetizing inductor "L_m" retains and charges the battery using a capacitor "C_m" and an inductor "*L*" by following the track indicated by the dotted lines in Figure 9.12. Additionally, switch "S_b" and "S_c" are inactive, and diode "D_b" is used in this mode.

Assuming "d_1" has a duty ratio, the voltage-sec. balancing technique for inductors L_m /*L* for a single switching period, T_s can be expressed as,

$$V_{g\,\max}|\sin(\omega t)| \times d_1(t) = V_b \times \{1 - d_1(t)\} \times T_s \tag{9.1}$$

Utilizing (9.1), the voltage gain of the converter "M_1" can be written as,

$$M_1 = \frac{V_b}{V_{g\,\max}|\sin \omega t|} = \frac{d_1(t)}{1 - d_1(t)} \tag{9.2}$$

9.6.2 Charging through solar power

This mode is activated when the PV power is higher than a critical value. In this case, both the switches "*P*" and "S_a" are active, but the "S_b" and "S_c" switches are not used. In this situation, the converter improves the behavior of the photovoltaic panel in all environmental conditions by using MPPT control [30–36] to produce the PWM signal for "S_a." Figure 9.13 illustrates the equivalent circuit for such an operation. The voltage gain for the converter in this mode is the same as before, as this mode also has the same ratio.

9.6.3 Driving mode of vehicle

This mode initiates when the battery starts providing power to the motor drive. In this mode, switches other than switch "*P*" are inactive, while switch "S_b" is operated by the PWM signal. As shown in Figure 9.14, when "S_b" switch is closed, the inductor "*L*" is charged via "V_b–L_1–S_b–V_b" (red dotted line) by another inductor through diode "D_c" when "S_b" is not active (blue dotted line).

The reconfiguration of converter components in this mode forms the boost converter configuration. Assuming "d_2" is the duty ratio, the magnitude of the

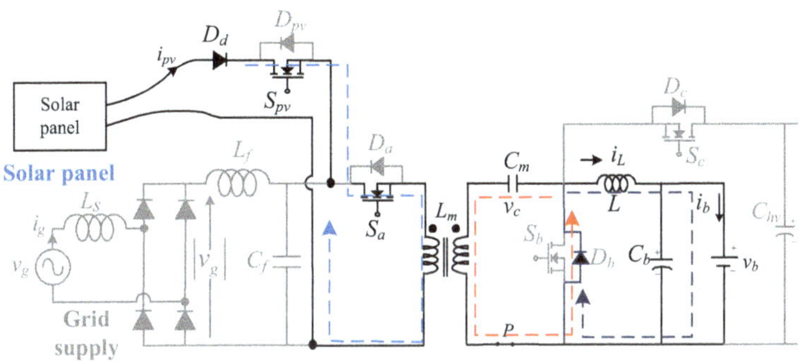

Figure 9.13 Vehicle battery charging utilizing PV power

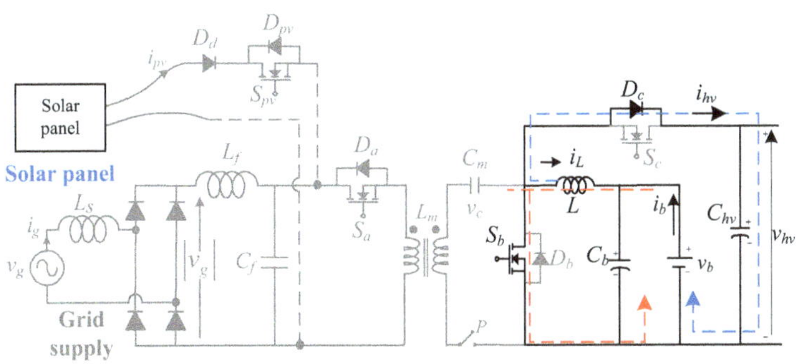

Figure 9.14 Functioning of the integrated converter in driving mode

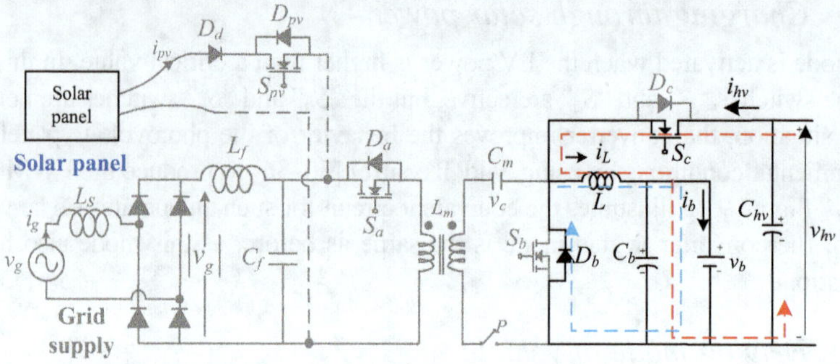

Figure 9.15 Operation in the regenerative mode

average voltage across the inductor, "L" over a single switching duration, "T_s" is null, and henceforth,

$$(V_b - V_{hv})(1 - d_2) \times T_s = V_b \times d_2 \times T_s \tag{9.3}$$

Following (9.3), the voltage gain "M_2" can be written as,

$$M_2 = \frac{V_{hv}}{V_b} = \frac{1}{1 - d_2(t)} \tag{9.4}$$

9.6.4 Regenerative braking mode

This mode is very critical for vehicle operation and contributes a major part to extending the vehicle range corresponding to per-hour charging. The battery gets recharged in this mode by the mechanical energy gained by the motor during propulsion mode. In this case, switch "P" is remaining to be inactive, and switch "S_c" is in operating condition. On the other side, the switches labeled "S_a" and "S_b" are both inactive. The suggested converter's performance in this mode is shown in Figure 9.15. Based on the average inductor voltage for one switching time, which serves as the foundation for (9.5), the voltage transfer function M_3 can be expressed as (9.6):

$$(V_{hv} - V_b)d_3 \times T_s = V_b \times (1 - d_3) \times T_s \tag{9.5}$$

$$M_3 = \frac{V_b}{V_{hv}} = d_3(t) \tag{9.6}$$

9.7 Design of the battery-charging converter

In general, the component ratings of the converter in each mode are entirely different. Thus, the selection of devices is governed by the particular operating mode. Therefore, the ratings of the devices are chosen based on the stresses, mainly in terms of voltage/current stress.

9.7.1 Design of switching devices

A few diodes and switches are unglued into AC/DC and DC/DC phase groups as a result of the bidirectional operation in DC/DC stages. The voltage and the current flowing through the switches and diodes will vary due to the variations in power and capacity in various modes. As a result, the choice of voltage/current values depends on the switch's peak voltage/current. The voltage/current stresses on semiconductor switches in each mode are shown in Table 9.5. The switch "S_a" has a peak current rating of $i_g(t) = \pi/2 + I_b$, while switches "S_b" and "S_c" have maximum voltage ratings of max. (V_{hv}, V_b). The rated peak voltage of "S_a" is stated as: $S_a = [v_g(\omega t)_{\omega t=\pi/2} + V_b]$.

9.7.2 Aspects of passive component selection

The inductor "L_2" contributes to all operating modes of the converter. So, by choosing the inductor value with allowable current ripple for each mode, the CCM nature of the converter can be maintained in each mode. The maximum value of these parameters is selected according to the value of the inductor "L_2."

The size of "L_2" for maintaining the CCM working in PIC mode is expressed as

$$L_2 > \left(\frac{R_L d_1'(t) T_s}{2}\right) > \frac{R_{L,\max}}{2} \frac{|V_{g,\max}|}{|V_{g,\max}| + V_b f_s} \frac{1}{} \tag{9.7}$$

Corresponding the parameters shown in Table 9.1 value of "L_2" in the PIC mode is estimated as

$$L_{2plug-in} > \frac{90}{2} \frac{311}{311 + 300} \frac{1}{20,000} = 1.14 \text{ mH} \tag{9.8}$$

The size of "L_2" for CCM, (9.7)–(9.8) define working in PP, and RB modes are

$$L_{2prop} = \frac{V_b d_2}{\Delta i_b f_s} \tag{9.9}$$

$$L_{2reg} = \frac{V_b(1 - d_3)}{\Delta i_b f_s} \tag{9.10}$$

where "Δi_b" signifies the ripple current at the battery side. Corresponding to parameters tabulated in Table 9.6 and selecting the ripple in battery current, $\Delta i_b =$ 20 % of I_b, the size of "L_{2prop}" and "L_{2reg}" is estimated as

$$L_{2reg} = L_{2reg} = \frac{300 \times 0.25}{1.33 \times 20,000} = 2.83 \text{ mH}$$

The following is chosen as "L_2" final value:

$$L_2 = \max\left(L_{2plug-in}, L_{2pop}, L_{2reg}\right) \tag{9.11}$$

Table 9.5 Ratings of peak voltage and current for semiconductor devices

Operations	S_a (D_a)		S_b (D_b)		S_c (D_c)
	Voltage	Current	Voltage	Current	Voltage/current
PGC and PV	$\left[v_g(\omega t)_{\omega t=\pi/2} + V_b\right]_{S_a}$	$\left[i_g(\omega t)_{\omega t=\pi/2} + I_b\right]_{S_a}$	$\left[v_g(\omega t)_{\omega t=\pi/2} + V_b\right]_{D_b}$	$\left[i_g(\omega t)_{\omega t=\pi/2} + I_b\right]_{D_b}$	$\left[v_g(\omega t)_{\omega t=\pi/2} + V_b\right]_{D_c}$ /NO
PR	$[\text{Zero}]_{S_a(D_a)}$	NO	$[V_b]_{S_b}$	$\left[\langle i_{Lprop}\rangle_{T_s} + \frac{\Delta i_{Lprop}}{2}\right]_{S_b}$	$[V_{hv}]_{D_c}/\left[\langle i_{Lprop}\rangle_{T_s} + \frac{\Delta i_{Lprop}}{2}\right]_{D_c}$
RB	$[\text{Zero}]_{S_a(D_a)}$	NO	$[V_b]_{D_b}$	$\left[\langle i_{Lrege}\rangle_{T_s} + \frac{\Delta i_{Lrege}}{2}\right]_{D_b}$	$[V_{hv}]_{S_c}/\left[\langle i_{Lrege}\rangle_{T_s} + \frac{\Delta i_{Lrege}}{2}\right]_{S_c}$

Table 9.6 Various parameters of the designed system

Variables	Values
DC link voltage (V_{hv})	400 V
Grid supply voltage ($V_{g,min}-V_{g,max}$)	90–220 V
Frequency (f_g)	50 Hz
L_2	3 mH
$C_b/C_1/C_{hv}$	1,500/10/550 μF
Operating frequency of converter (f_{sw})	20 kHz
Charging power (P_b)	1 kW
Battery voltage, (V_b)	320 V

9.7.3 Size of capacitor (C_m)

The link capacitor "C_1" plays a vital role in the suggested converter due to its significant impact on establishing the quality input current. To prevent input applications from being too diverse at each half-line cycle and to maintain a constant voltage during such a switching period, the resonance frequencies of "L_2" and "C_m" in CCM operation must be greater than the line frequency "f_L" and smaller than the switching frequency "f_s." As a result, the design of this capacitor will be subject to the following limitations:

$$f_L < f_r < f_s \tag{9.12}$$

where

$$f_r = \frac{1}{2\pi\sqrt{(L_1 + L_2)C}} \tag{9.13}$$

The switching frequency "f_s" in this study is set to 20 kHz, and "f_r" is set to 1 kHz. Both in simulation and in hardware, the capacitor "C_1" is set at 10 F. Because this capacitor's voltage characteristics are chosen in accordance with the battery voltage range and during all operations, the voltage in the coupling capacitor is selected according to the "v_b."

9.7.4 Size of capacitor (C_b)

Higher-order switching harmonics are presumptively ignored because a capacitor, "C_b" is directly connected to the vehicle battery. However, the double-frequency voltage ripple is demonstrated and has a major impact on the battery pack's lifespan.

The low-frequency voltage ripple in "V_b" is written as

$$\Delta v_b \geq \frac{P_b}{4f_g C_{b,\,\min} V_b} \tag{9.14}$$

where "f_g" denotes the grid supply frequency and, "Δv_b" denotes the voltage ripple across the capacitor, "C_b." Table 9.6 tabulates the converter's chosen values for usage in the system.

9.8 Control strategy

A schematic of the controller employed by the recommended converter is shown in Figure 9.16. The designed converter undergoes three operating scenarios, and different control concepts are utilized in each operating mode. The mode selection logic primarily chooses the operating mode, and then it begins the baseline parameter for that mode. Since the input reference is dependent on the operating modes, it may be any one of the parameters like voltage at the DC bus, braking torque, power for charging, etc. Corresponding to this operating mode, the relevant controller will be activated and will guarantee the desired operation. The overall control strategy of the operation can be divided into two main classes. The first part controls the PIC and solar PV charging modes, and the second section controls the PR and RB modes.

The input selector logic's PGC and solar PV modes are included in this control parameter. The logic developed for this mode is illustrated in Figure 9.17. In the PIC mode, a 2-loop controller is used to control power on the grid side and charge the LPEV. The controller "$G_{ib}(z)$" calculates the expected battery current and the

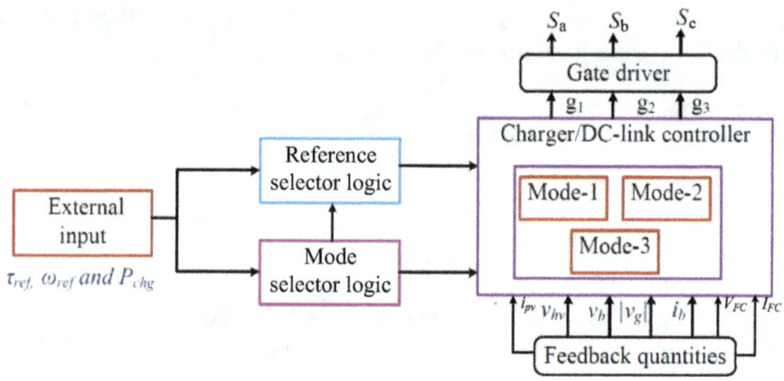

Figure 9.16 Charging mode control technique

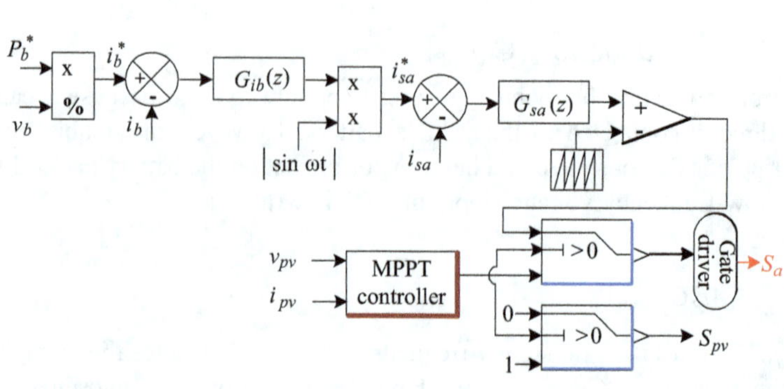

Figure 9.17 Battery charging with solar energy

actual battery current and outputs the difference as

$$G_{ib}(Z) = K_p + \frac{K_i T_s}{z - 1} \tag{9.15}$$

Combining the "$G_{ib}(z)$" output with a sinusoidal signal unit template, the reference current for "G_{iL}" is produced. This reference is created by multiplying the "$G_{ib}(z)$" output by a sinusoidal signal unit template. After power factor adjustment, it produces the high-frequency gate control signal for switch "S_a" at the utility grid. The term "G_{iL}" can be expressed as

$$G_{iL}(Z) = K_{pc} + \frac{K_{ic} T_s}{z - 1} \tag{9.16}$$

To keep the controller bandwidth at 1/6th and 1/10th of the operating switching frequency, the gains (K_{pc} and K_{ic}) should be changed. Similar to this, the "G_{ib} (z)" controller's bandwidth is adjusted to less than 120 Hz to reduce the impact of the current reference loop's second harmonic. In PV charging mode, the "S_{pv}" switch is passed through a switch selector with 0 and 1 logic, while the drift-independent MPPT P&O control method is used to produce high-frequency signals for switch "S_a." Therefore, the loss associated "S_{pv}" is very low.

9.8.1 *Control for the modes of propulsion and regenerative braking*

The average current mode logic is used for both propulsion and regenerative modes of operation. Depending on the operating mode, appropriate switches are operated utilizing a mode picker. Combining logic gates with stop command and changeable input voltage, the mode picker logic system is formed. In the propulsion mode, the VSC voltage input is controlled by the controller $G_{hv}(z)$ and the inner loop controller $G_{ibc}(z)$.

In the regenerative mode, the switch "S_c" plays a significant part in returning energy from the motor to the battery through the motor inverter switches. Figure 9.18 shows the high-frequency signals for switch "S_c" produced by the

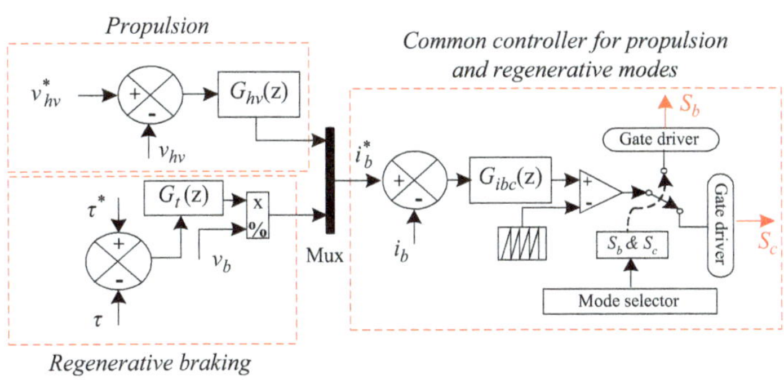

Figure 9.18 Control logic for PP and RB modes

logic. The braking torque is the command utilized in this mode. As shown in Figure 9.18, the command then generates a reference for employing a $G_t(z)$ PI controller that produces a current reference for the vehicle battery.

9.9 Result and analysis

The converter characteristics shown in Table 9.6 are used to model a battery charger that uses a modified Zeta-based converter. Plotting and individual examination of the system's behavior in the three modes have been studied in this section. The circuit of the proposed system has been modeled using the 2013a MATLAB® version of the Sim Power-Systems toolbox. To finish the power generation, the simulation time is maintained at 1e−6s.

The waveforms of the grid current and the voltage corresponding to the PIC mode of operation are illustrated in Figure 9.19. It can be observed that both the quantities are exactly in phase and authenticate the unity power factor operation of the system. The Fourier technique is utilized to estimate the total harmonic distortion (THD) and power factor of the system, which are obtained at 3.78% and 0.99, respectively. While charging the battery, the battery's base voltage is set to 300 V, including an initial SoC of 20%, a grid voltage of 220 V, and 1 kW charging power. Figure 9.20 shows simulated graphs for battery voltage (v_b) and battery current (i_b).

From Figure 9.20, it can be seen that there is a presence of lower order frequency (100 Hz) oscillation in the battery current in a single-phase system. The presence of a series inductor attached to the battery is responsible for the oscillating

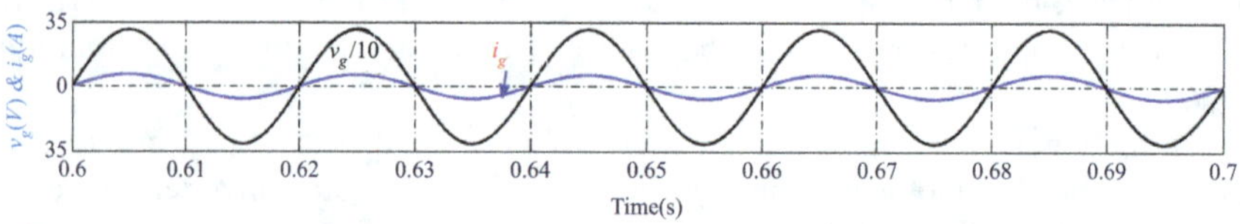

Figure 9.19 Grid current and voltage waveforms in the PIC mode of operation

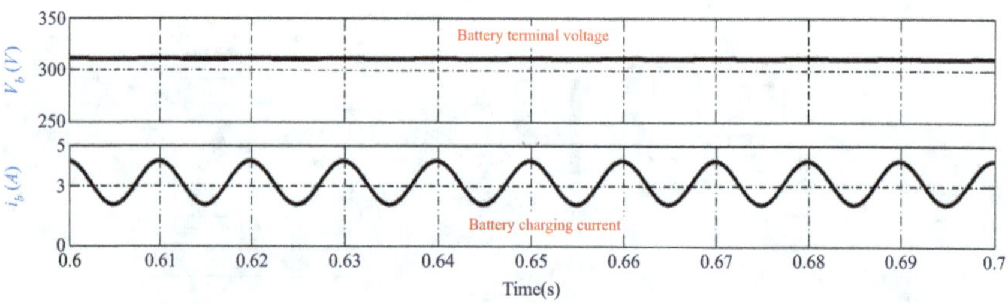

Figure 9.20 Battery voltage (V_b) and current (i_b) waveforms during PIC mode of operation

nature of the battery current. This is a trade-off situation between the charging circuit size and the overall service life of the battery.

The voltage across the filter capacitor and current through the filter inductor at the grid side is illustrated in Figure 9.21. The nature of capacitor voltage is realized to be constant and pulsating in this mode. The current through the filter inductor has also been seen to be of a similar nature but of dissimilar magnitude. The estimated maximum current through "L_f" is nearly 9.7 A, while the peak voltage across "C_f" is approximated to be 311 V.

The plot showing the THD in grid current in PIC mode is demonstrated in Figure 9.22, which is noted to be 1.57%. This value of THD clearly verifies the UPF operation of the integrated converter. This type of operation helps to minimize the overall power consumption and also diminishes the reactive power loss.

The nature of the voltage across the middle capacitor "C_m" is shown in Figure 9.23(a). The highest voltage across "C_m" is exactly 212 V, which is also the maximum supply voltage. According to Figure 9.23(b), the voltage across switch "S_a" is determined by adding the supply and battery voltages.

As the designed converter has the capability to act as both PFC as well as MPPT converter, thus, the examination of the behavior of PV and battery indices is also very critical. Figure 9.24 shows the PV voltage and current waveforms under the solar charging mode of operation. In this case, the voltage across the battery is comparably smoother as compared to the PIC charging. It is due to the elimination of the rectifying operation of the grid voltage to supply the DC voltage at the battery side after the converter processing.

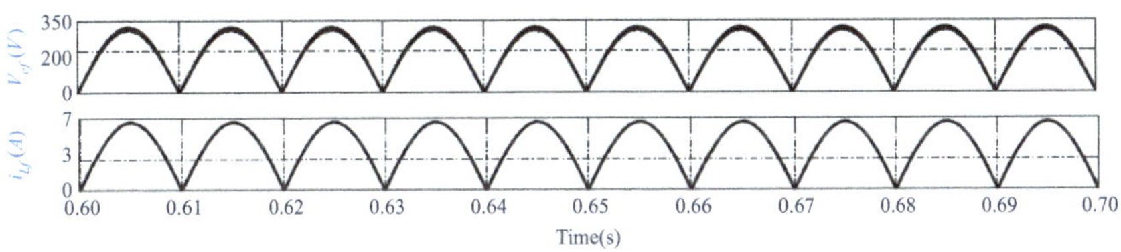

Figure 9.21 Voltage and current waveforms across filter capacitor inductor in PIC mode of operation

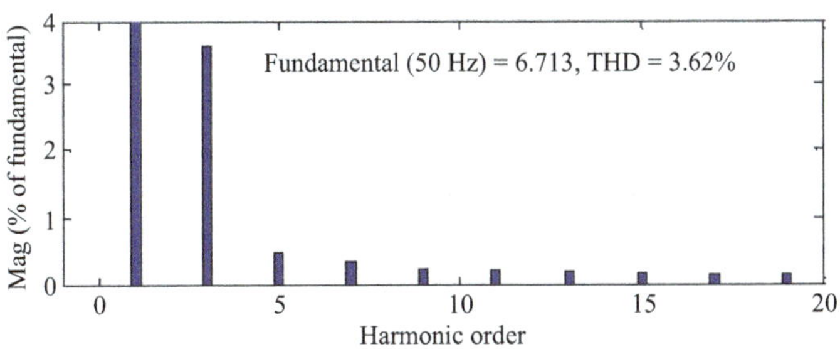

Figure 9.22 Total harmonic distortion of grid current in the PIC mode of operation

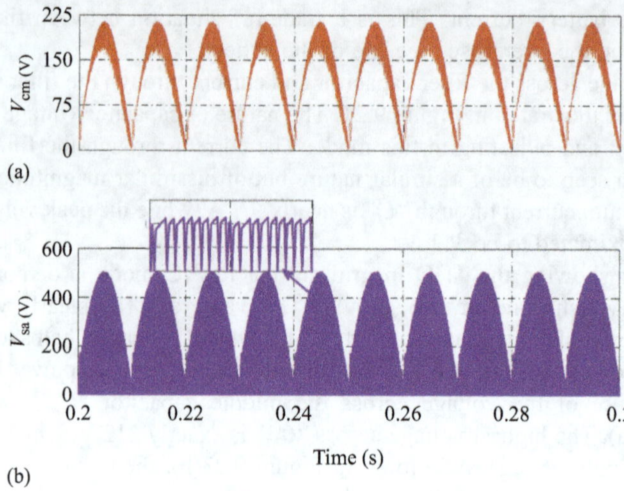

Figure 9.23 Voltage across C$_m$ *and voltage stress across switch "*S$_a$*" in PIC mode*

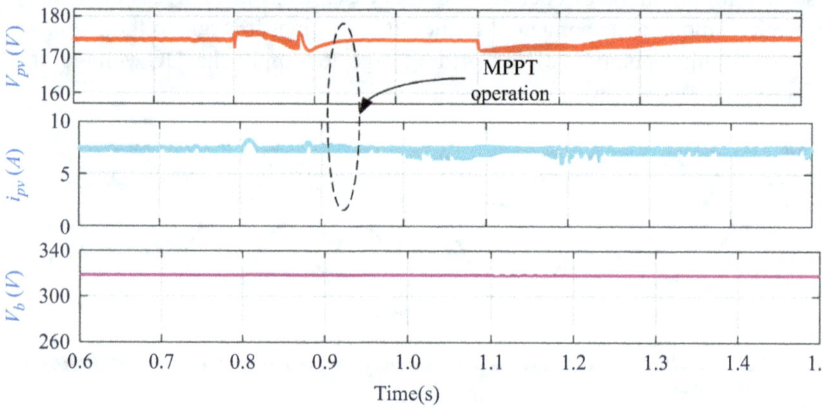

Figure 9.24 Parameters of solar PV and battery solar charging mode of operation

The waveforms of various system parameters corresponding to the propulsion mode of operation are illustrated in Figures 9.25 and 9.26. This mode is realized in the system by utilizing step load variation. The change in loading power is done from 1 kW to 2 kW at t = 1.2 s and from 2 kW to 1 kW at t = 1.8 s. The controller design for this mode has the basic objective of maintaining the DC link voltage at the drive side. The effective working of the designed controller is well established by observing Figure 9.25 that the DC link regulated successfully at the desired voltage level.

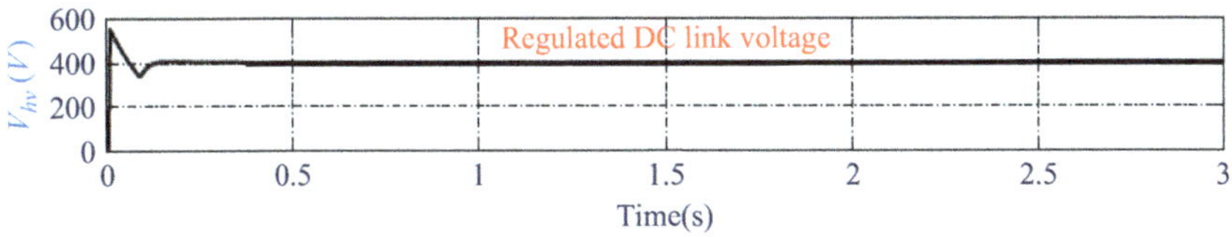

Figure 9.25 DC link voltage nature in the propulsion mode

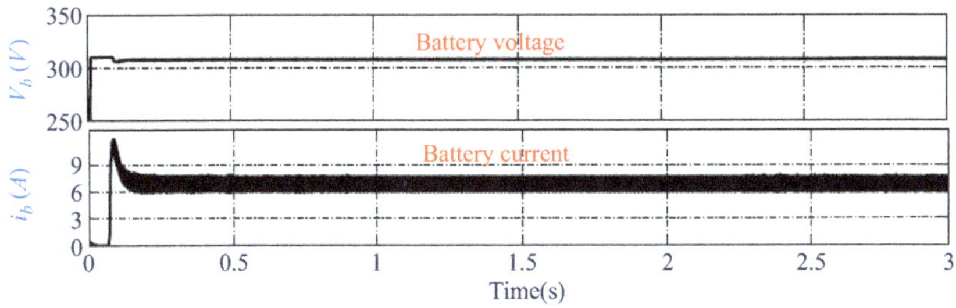

Figure 9.26 Nature of battery voltage and current under dynamic scenario

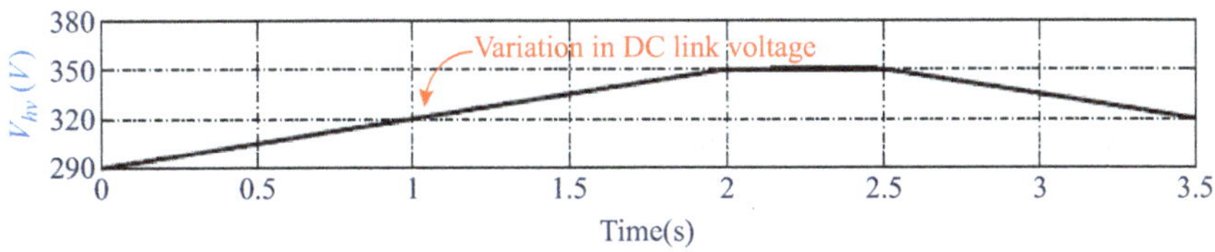

Figure 9.27 DC link voltage nature under regenerative braking operation

The nature of battery indices in the propulsion mode of operation is demonstrated in Figure 9.26. The magnitude of current through the battery is observed to increase from 3.45 to 9.9 A at the time, $t = 1.2$ s. In the same way, it decreases to 3.45 A from 9.9 A at time $t = 1.8$ s by changing the load value from 2 kW to 1 kW. As can be seen that the nature of battery voltage is almost constant under all these dynamic scenarios.

In the last, the behavior of some designed system parameters in the RB mode is illustrated in Figures 9.27 and 9.28. This mode plays a very crucial role in enhancing the vehicle's performance, particularly in terms of range/km. Therefore, the performance of the system under this mode is examined and analyzed very carefully. The variation in DC link voltage at the inverter side is shown in Figure 9.27. The variation is very smooth, which validates the effectiveness of the designed controller. The magnitude changes from 290 to 350 V. On the other side, the variation in battery voltage and current in this mode is shown in Figure 9.28. Irrespective of the change in DC link voltage, the magnitude of the battery current has a constant nature and is maintained at 3.5 A.

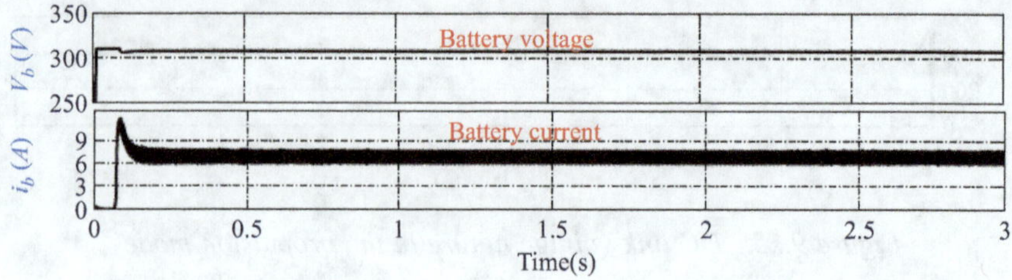

Figure 9.28 Battery voltage behavior under regenerative braking operation

9.10 Summary

This chapter presented the evolution of power converter topologies for on-board applications of PEVs, which were essentially based on single-stage conversion. These converters are capable of multi-mode operation to achieve battery charging, vehicle propulsion, and regenerative braking. This feature of multi-mode operation is reflected in calling them integrated converters. Priority is devoted to classifying chargers and reviewing conventional single-stage and integrated chargers. The non-isolated converters are found more suitable for on-board EV applications. The review of battery chargers has been divided into two groups: (a) integrated chargers using a diode rectifier in the AC/DC stage and (b) bidirectional DC/DC converter incorporated into the charging circuit. A review of each topology is focused on its merits and demerits. After evaluating the existing converter-based battery charging typologies, the design, analysis, and control of one such new battery charging system are discussed. To validate the effectiveness of this new charging scheme, MATLAB-based simulation results are examined and provided in this chapter. For plug-in battery charging and regenerative braking modes, the simulated integrated converter-based battery charging system operates as a ZETA converter; for propulsion mode, it operates as a buck-mode converter. Therefore, the designed converter has efficiently achieved all the desired working performances under every operating scenario.

Symbols

P_g, P_b, P_{pv}	Power output from the grid, battery power, and solar panel power, respectively
V_g, V_b, V_{PV}, V_{hv}	RMS voltage of the grid, battery, solar panel @1000W/m^2, and DC-bus voltages, respectively
v_g, v_b, v_{pv}	Instantaneous voltages of the grid, battery, and solar panel, respectively
$V_{g,max}$	Peak voltage output from the utility grid
V_{ac}	Grid supply voltage
v_{out}, v_{in}	Output and input voltage of the integrated converter

i_{out}, i_{in}	Output and input current of the integrated converter
S_{pv}	Solar PV switch of charging system
P	Single pole double thrown switch of the converter
L_m, L	Magnetizing and output inductor
C_m	Intermediate energy transfer capacitor
D_a, D_b, D_c	Diodes of the integrated converter
T_s	Switching period of the converter
M_1, M_2, M_3	Voltage gains in charging, propulsion, and regen mode, respectively
v_{cm}	Voltage across the middle capacitor of the integrated converter
i_g, i_b, i_{PV}	Instantaneous currents of the grid, battery, and solar panel, respectively
$i_{g,max}$	Peak current output from the utility grid
f_l, f_s	Supply frequency and operating switching frequency of the integrated converter
i_{Lprop}, i_{Lrege}	Magnitudes of current in driving and regen modes through output inductor, L.
d_1, d_2, d_3	Duty ratio in charging, propulsion, and regen mode, respectively
D_{pv}	Duty ratio of solar PV switch
R_L	Equivalent load resistance at the output of the converter
$G_{hv}(z)$, $G_{ibc}(Z)$	Voltage and current controllers
τ, τ^*	Reference and actual braking torque in regen operation
S_a, S_b, S_c	Switches of the integrated converter

Glossary

CCM	continuous conduction mode
DCM	discontinuous conduction mode
EMC	electromagnetic compatibility
EMF	electromagnetic force
EMI	electromagnetic interference
EVs	electric vehicles
FPGA	field programming gate array
ICEs	internal combustion engines
NLCC	nonlinear carrier control
OBC	on-board charger
PCB	printed circuit board
PEVs	plug-in electric vehicles

PF	power factor
PFC	power factor correction
PI	proportional integral
PLL	phase-locked loop
PV	photovoltaic
PWM	pulse width modulation
RMS	root mean square
SAE	Society of Automotive Engineers
SOC	state of charge
THD	total harmonic distortion
TSBuB	two-switch buck/boost
UPF	unity power factor
V2G	vehicle-to-grid

References

[1] C. C. Chan and K. T. Chau, "An overview of power electronics in electric vehicles," *IEEE Transactions on Industrial Electronics*, vol. 44, no. 1, pp. 3–13, 1997.

[2] A. Emadi, Y. J. Lee, and K. Rajashekara, "Power electronics and motor drives in electric, hybrid electric, and plug-in hybrid electric vehicles," *IEEE Transactions on Industrial Electronics*, vol. 55, no. 6, pp. 2237–2245, 2008.

[3] A. K. Singh and M. K. Pathak, "An improved two-stage non-isolated converter for onboard plug-in hybrid EV battery charger," in *1st International Conference on Power Electronics, Intelligent Control and Energy Systems (ICPEICES)*, Jul. 2016, pp. 1–6.

[4] F. Musavi, M. Edington, W. Eberle, and W. G. Dunford, "Evaluation and efficiency comparison of front-end AC–DC plug-in hybrid charger topologies," *IEEE Transactions on Smart Grid*, vol. 3, no. 1, pp. 413–421, 2012.

[5] B. P. McGrath, D. G. Holmes, P. J. McGoldrick, and A. D. McIver, "Design of a soft switched 6-kw battery charger for traction applications," *IEEE Transactions on Power Electronics*, vol. 22, no. 4, pp. 1136–1144, 2007.

[6] C. C. Chan and K. T. Chau, "Power electronics challenges in electric vehicles," Industrial Electronics, Control, and Instrumentation, *Proceedings of the International Conference on Industrial Electronics, IECON*, vol. 2, Nov. 1993, pp. 701–706.

[7] I. A. Khan, "Battery chargers for electric and hybrid vehicles," in *Proceedings of the 1994 IEEE Workshop on Power Electronics in Transportation*, Oct. 1994, pp. 103–112.

[8] J. C. Gomez and M. M. Morcos, "Impact of EV battery chargers on the power quality of distribution systems," *IEEE Transactions on Power Delivery*, vol. 18, no. 3, pp. 975–981, 2003.

[9] F. L. Mapelli, D. Tarsitano, and M. Mauri, "Plug-in hybrid electric vehicle: modeling, prototype realization, and inverter losses reduction analysis," *IEEE Transactions on Industrial Electronics*, vol. 57, no. 2, pp. 598–607, 2010.

[10] M. M. Morcos, N. G. Dillman, and C. R. Mersman, "Battery chargers for electric vehicles," *IEEE Power Engineering Review*, vol. 20, no. 11, pp. 8–11, 2000.

[11] J. Dixon, I. Nakashima, E. F. Arcos, and M. Ortuzar, "Electric vehicle using a combination of ultracapacitors and ZEBRA battery," *IEEE Transactions on Industrial Electronics*, vol. 57, no. 3, pp. 943–949, 2010.

[12] Z. Amjadi and S. S. Williamson, "Power-electronics-based solutions for plug-in hybrid electric vehicle energy storage and management systems," *IEEE Transactions on Industrial Electronics*, vol. 57, no. 2, pp. 608–616, 2010.

[13] S. Bala, T. Tengnr, P. Rosenfeld, and F. Delince, "The effect of low frequency current ripple on the performance of a lithium iron phosphate LFP battery energy storage system," in *IEEE Energy Conversion Congress and Exposition (ECCE)*, 2012, pp. 3485–3492.

[14] H. Z. Z. Beh, G. A. Covic, and J. T. Boys, "Effects of pulse and DC charging on lithium iron phosphate (LiFePO4) batteries," in *2013 IEEE Energy Conversion Congress and Exposition Energy Conversion Congress and Exposition (ECCE)*, 2013, pp. 315–320.

[15] S. Haghbin, S. Lundmark, M. Alakula, and O. Carlson, "Grid-connected integrated battery chargers in vehicle applications: review and new solution," *IEEE Transactions on Industrial Electronics*, vol. 60, no. 2, pp. 459–473, 2013.

[16] K. M. S. Y. Konara and M. L. Kolhe, "Charging management of grid integrated battery for overcoming the intermittency of resources," in *IEEE International Conference on Information and Automation for Sustainability (ICIAFS)*, Dec. 2016, pp. 1–6.

[17] L. Shi, A. Meintz, and M. Ferdowsi, "Single-phase bidirectional AC–DC converters for plug-in hybrid electric vehicle applications," in *IEEE Vehicle Power and Propulsion Conference*, Sep. 2008, pp. 1–5.

[18] SAE Electric Vehicle and Plug-in Hybrid Electric Vehicle Conductive Charge Coupler, SAE Std. J1772, 2010.

[19] M. Grenier, M. G. H. Aghdam, and T. Thiringer, "Design of on-board charger for plugin hybrid electric vehicle," in *5th IET International Conference on Power Electronics, Machines and Drives (PEMD 2010)*, Apr. 2010, pp. 1–6.

[20] M. Yilmaz and P. T. Krein, "Review of battery charger topologies, charging power levels, and infrastructure for plug-in electric and hybrid vehicles," *IEEE Transactions on Power Electronics*, vol. 28, no. 5, pp. 2151–2169, 2013.

[21] O. C. Onar, J. Kobayashi, D. C. Erb, and A. Khaligh, "A bidirectional high-power quality grid interface with a novel bidirectional noninverted buck-boost converter for PHEVs," *IEEE Transactions on Vehicular Technology*, vol. 61, no. 5, pp. 2018–2032, 2012.

[22] J. C. Bendien, G. Fregien, and J. D. van Wyk, "High-efficiency on-board battery charger with transformer isolation, sinusoidal input current and maximum power factor," *IEE Proceedings B (Electric Power Applications)* vol. 133, no. 4, pp. 197–204, 1986.

[23] M. Pahlevaninezhad, P. Das, J. Drobnik, P. K. Jain, and A. Bakhshai, "A new control approach based on the differential flatness theory for an AC/DC converter used in electric vehicles," *IEEE Transactions on Power Electronics*, vol. 27, no. 4, pp. 2085–2103, 2012.

[24] Y. J. Lee, A. Khaligh, and A. Emadi, "Advanced integrated bidirectional AC/DC and DC/DC converter for plug-in hybrid electric vehicles," *IEEE Transactions on Vehicular Technology*, vol. 58, no. 8, pp. 3970–3980, 2009.

[25] H. Chen, X. Wang, and A. Khaligh, "A single stage integrated bidirectional AC/DC and DC/DC converter for plug-in hybrid electric vehicles," in *IEEE Vehicle Power and Propulsion Conference*, Sep. 2011, pp. 1–6.

[26] S. Dusmez and A. Khaligh, "A novel low cost integrated on-board charger topology for electric vehicles and plug-in hybrid electric vehicles," in *Conference Proceedings – IEEE Applied Power Electronics Conference and Exposition* – APEC, 2012, pp. 2611–2616.

[27] S. Dusmez and A. Khaligh, "A compact and integrated multifunctional power electronic interface for plug-in electric vehicles," *IEEE Transactions on Power Electronics*, vol. 28, no. 12, pp. 5690–5701, 2013.

[28] S. Dusmez and A. Khaligh, "A charge-nonlinear-carrier-controlled reduced-part single-stage integrated power electronics interface for automotive applications," *IEEE Transactions on Vehicular Technology*, vol. 63, no. 3, pp. 1091–1103, 2014.

[29] Y. Tang, D. Zhu, C. Jin, P. Wang, and F. Blaabjerg, "A three-level quasi-two-stage single-phase PFC converter with flexible output voltage and improved conversion efficiency," *IEEE Transactions on Power Electronics*, vol. 30, no. 2, pp. 717–726, 2015.

[30] N. Karami, N. Moubayed, and R. Outbib, "General review and classification of different MPPT techniques," *Renewable and Sustainable Energy Reviews*, vol. 68, Part 1, pp. 1–18, 2017.

[31] H. Bounechba, A. Bouzid, H. Snani, and A. Lashab, "Real-time simulation of MPPT algorithms for the PV energy system," *International Journal of Electrical Power & Energy Systems*, vol. 83, pp. 67–78, 2016.

[32] T. Esram and P. L. Chapman, "Comparison of photovoltaic array maximum power point tracking techniques," *IEEE Transactions on Energy Conversion*, vol. 22, no.2, pp. 439–449, 2007.

[33] M. A. Elgendy, B. Zahawi, and D. J. Atkinson, "Operating characteristics of the P&O algorithm at high perturbation frequencies for standalone PV systems," *IEEE Transactions on Energy Conversion*, vol. 30, no. 1, pp. 189–198, 2015.

[34] M. Killi and S. Samanta, "Modified perturb and observe MPPT algorithm for drift avoidance in photovoltaic systems," *IEEE Transactions on Industrial Electronics*, vol. 62, no. 9, pp.5549–5559, 2015.

[35] K. L. Lian, J. H. Jhang, and I. S. Tian, "A maximum power point tracking method based on perturb-and-observe combined with particle swarm optimization," *IEEE Journal of Photovoltaics*, vol. 4, no. 2, pp. 626–633, 2014.

[36] H. A. Sher, A. F. Murtaza, A. Noman, K. E. Addoweesh, K. Al-Haddad, and M. Chiaberge, "A new sensorless hybrid MPPT algorithm based on fractional short-circuit current measurement and P&O MPPT," *IEEE Transactions on Sustainable Energy*, vol. 6, no. 4, pp. 1426–1434, 2015.

Chapter 10

Power quality control of battery charging system

Shailendra Kumar[1], Rheesabh Dwivedi[2], Sanjay Gairola[3] and Miloud Rezkallah[4]

10.1 Introduction

The electric vehicle (EV) chargers can be of different categories (OFF-board or ON-board charger), (AC or DC supply type charger), (integrated or dedicated topology), (conductive or inductive type), and (unidirectional or bidirectional charger) [1,2]. EVs may have different types of chargers due to a variation of battery technology and its voltage level. The battery of any EV requires matching DC voltage with a controlled current to charge it. The most popular method of obtaining DC is to rectify the available AC power from the grid. Either a single-phase or a three-phase diode bridge rectifier (DBR) is used for the rectification from the AC grid, followed by a capacitor as a filter to provide a ripple-free DC voltage [3].

The voltage distribution network from the utility at low voltage inside the charging stations (CS) can be either AC or DC, as shown in Figures 10.1 and 10.2. The AC distribution network is mature in comparison with the DC distribution and so preferred by most of the popular CSs.

It is worth mentioning here that a solid-state transformer can be used in place of the conventional electromagnetic wire wound transformers for better performance when used for fast charging.

Due to the capacitive filter's uncontrolled charging and discharging, the current drawn from the AC grid has non-sinusoidal and peaky behavior as shown in Figure 10.3. This current is analyzed to have poor power factor (PF), high crest factor (CF), high harmonics, and results in several undesirable effects on both sides, i.e., the utility and the consumer, such as line voltage distortion, increased losses and overheating in the transformer, cables, and shunt capacitors [3,4].

[1]Department of Electrical Engineering, IIT – Bhilai, India
[2]Department of EIE, SLIET – Longowal, India
[3]Department of Electrical Engineering, GBPIET – Pauri, India
[4]Department of Electrical Engineering, ETS – Montreal, Canada

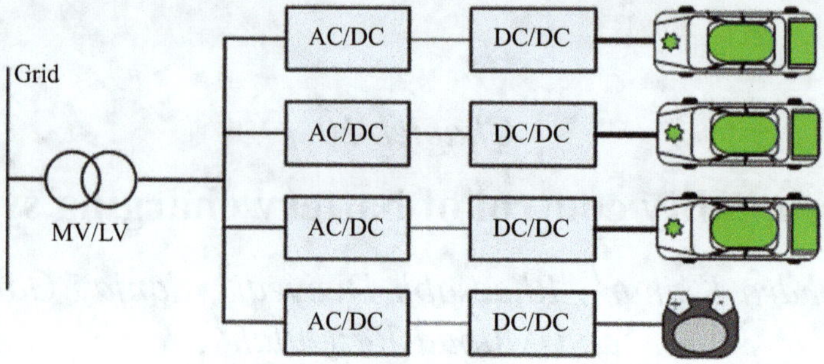

Figure 10.1 Block diagram of battery charging stations with AC distribution

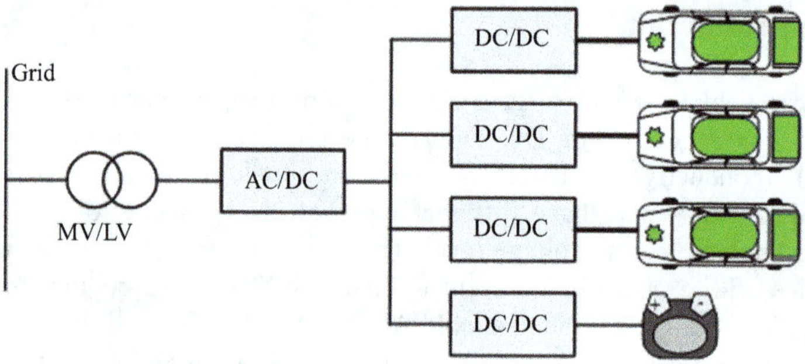

Figure 10.2 Block diagram of battery charging stations with DC distribution

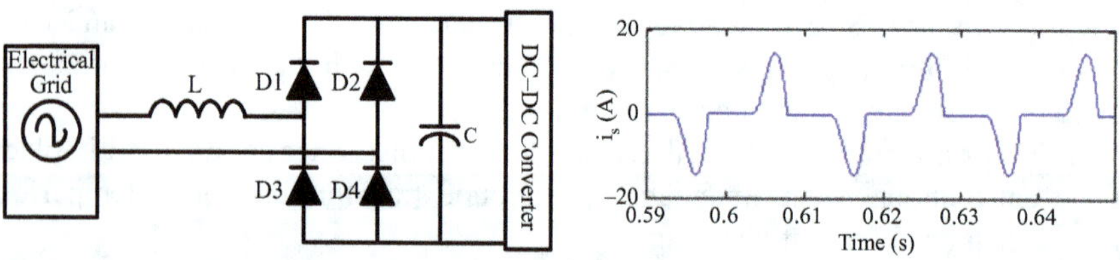

Figure 10.3 Conventional DBR-based battery charger and its current waveform

The major problems in this line frequency-based AC–DC converter are injection of harmonics on AC and DC sides. The capability of tolerating harmonics is affected by the susceptibility of the load (or power source) to them. The most susceptible equipment is one in which the sole objective is heating, as in a heater. In a heater-like application, the energy contained in unwanted harmonics also gets converted into heat and, therefore, becomes useful. The least susceptible is one where the useful output is decided only by the fundamental component.

Therefore, an effective power quality (PQ) control is essential for the battery charging system of EVs with close compliance with international standards. There are various methods reported in the literature to control the PQ problems in battery charging systems. One of the common methods is the use of a DC–DC converter operated with PQ or PF correction (PFC) controller. Another method is to have a controlled rectifier in the place of DBR for the rectification of AC supply with PQ or PFC controller. However, an additional DC–DC converter with galvanic isolation is required as the second stage to provide a well-regulated DC voltage and current for charging the battery in constant voltage (CV) or constant current (CC) mode.

The main question asked by any researcher investigating and analyzing a charging converter is "What factors limit the charging rate of a battery?" The possible answers to this question are given below:

- The converter must provide a very high current for fast charging of the battery which is limited by the capacity of the battery. The size of battery terminals and the chemical reaction rate inside a battery may lead to unwanted heating or deterioration, restricting its life and charging/discharging rate.
- For producing high-charging currents, a high value of source voltage may be needed which is limited by the insulation provided around the battery to be charged.
- As high currents are required in constant-current charging mode, a current source converter may be a preferred choice. Alternately more than one current source converter may be connected in parallel to increase the rating further.
- The minimum number of elements causing voltage drop, between the source and the battery to be charged, should be there. This means only a small number of power electronic switches, inductors, or transformer windings are desired in the path of charging current.
- The battery needs to be investigated if high-voltage pulses are to be permitted for its charging.

Therefore, a PQ controller for the EV battery charger is required to achieve an improved PQ operation at AC mains in a wide input and output voltage range. The other qualities required are the reduction of switching devices in the PFC converters for low-cost system and easy control scheme. Increasing the efficiency of the charger by reducing the conduction and switching losses along with the reduction in charging time are also of prime importance. The power density and the efficiency of the EV charger are adversely affected by the use of high-frequency transformer (HFT) for galvanic isolation in the second stage [5,6]. Isolation requirements in EV battery chargers should be of the least concern, according to SAE J17722 [7].

This chapter discusses these issues and presents various converter topologies and control schemes for PQ control of battery chargers for EVs.

10.2 PQ control for battery charger

The PQ disturbances at AC mains of a charger are represented in terms of various PQ indices such as PF, CF, total harmonic distortion (THD_i) of AC mains current, displacement PF (DPF), and ripples in DC link voltage (ΔV_{dc}). The harmonic voltages/currents in the electrical system produce increased heating by affecting the Core and Cu losses that are frequency dependent. This lowers the efficiency, causes torque pulsations, and produces higher audible noise emissions in machines. Moreover, certain harmonic pairs tend to create mechanical fluctuations in generating station turbines. The other problems are:

- Poor PF.
- Injection of harmonics into the AC mains.
- Voltage distortion due to the voltage drop of harmonic currents flowing through system impedances.
- Capacitor bank overloading due to system resonance.
- Interference on telephone and communication lines due to noise induced from the power conductors.
- Equipment damage from voltage spikes created by high-frequency resonance resulting from notching.
- Secondary effects from the application of large harmonic frequency resonance resulting from notching.

The effect of harmonics on transformers is twofold: current harmonics cause an increase in copper losses and stray flux losses, and voltage harmonics cause an increase in iron losses. The overall effect is an increase in the transformer heating. So the reduction of harmonic currents becomes an important issue. The techniques for harmonic reduction include:

- Use of filters: shunt, series, and hybrid of passive and active.
- Phase multiplication: It is most effective for an installation in which equal-size converters with equal loading and phase retard are used.
- Harmonic compensation or injection: Harmonic currents can be eliminated by inducing harmonic fluxes in the core of transformer with 180° phase shift from the harmonic fluxes induced by the current flowing in the transformer secondary.

Therefore, the PQ control for battery chargers must deal with these indices and observed performance should be as per set national and international standards. There are methods involving passive elements like inductors and capacitors only known as passive methods. The other option is to have active control using fast-switching devices used in various topologies known as active methods.

10.2.1 Passive methods

The passive methods involve the use of tuned filters consisting of inductors and capacitors. The major drawback of these methods is a requirement of bulky inductors and a number of filters for different harmonic frequencies. The losses in the passive methods are also more and, in transient conditions, they may not be effective.

The multi-pulse AC–DC converters (MPCs) are the best choice as the passive method, which employs suitable transformer connections along with converters [8]. Normally the power processor consists of 6-pulse diode bridge rectifiers as the

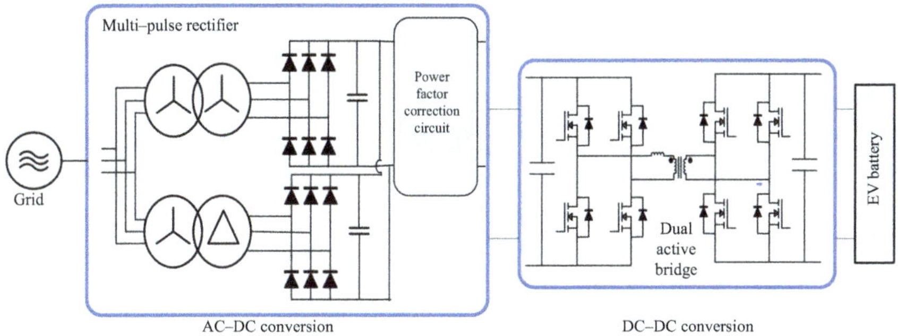

Figure 10.4 MPC topology for PQ improvement in battery charger

front end. In large power ratings, up to several thousand horse powers, controlled bridges are used that too not essentially with isolation transformers. The bidirectional power flow through a rectifier is possible by using a bridge rectifier having thyristors while the diode bridges shall enable unidirectional power flow only. The transformers employed at the front end essentially have multiple secondary windings which produce phase-shifted AC voltages feeding one or more six-pulse AC–DC converter. These transformers help shape the current on the utility side and additionally provide isolation of source to load. A number of multi-pulse topologies have been presented in the literature [8]. One such topology is shown in Figure 10.4 for information of the readers.

For high-power applications, 3-phase bridge rectifiers are placed in most industrial applications for AC–DC conversion. These rectifiers are also applicable at the input of off-board EV battery charging stations which work at high voltages for fast charging. The charging stations nowadays have provision for power flow from either grid-to-vehicle (G2V) or vehicle-to-grid (V2G), thereby demanding bidirectional power flow capability. At higher power ratings, the SCR/GTO-based AC–DC converters are employed at the input for bidirectional power flow, while for unidirectional power flow, diode-based AC–DC converters are used. The multipulse AC–DC conversion stage may additionally have transformers and interphase reactors (IPRs) besides the bridge converters.

10.2.2 Active methods

The active methods of the PQ improvement involve the power electronic controller with switching schemes to control the current or the voltage with a desired shape of the waveforms. There are various topologies reported in the literature to have PQ improvement at AC mains when connected to battery charging applications. A few typical topologies are presented in Figure 10.5.

The suggested topologies are made to operate in a continuous conduction mode with current multiplier control or discontinuous conduction mode (DCM) with voltage follower control to efficiently and simply attain a PF that is close to unity. Additional benefits of DCM operation include zero-current power switch turn-on, zero-current output diode turn-off, and a reduction in the complexity of the control circuitry.

10.2.2.1 Current multiplier control

The first control scheme shown in Figure 10.5 uses current multiplier control, in which the input voltage template is used for the generation of current reference for the DC–DC converter. This is an accurate method and requires more number of sensors for the implementation of the control scheme.

10.2.2.2 Voltage follower control

The control scheme shown in Figure 10.5 as the second part uses voltage follower control, in which the discontinuous conduction mode of operation is ensured for the DC–DC converter. In this process, the current drawn from the input naturally follows the input voltage. This is an approximate method and requires less number of sensors for the implementation of the control scheme.

10.2.3 PQ standards

Research reports state that with the increased use of power electronic equipment, the PQ aspect needs immediate attention. To prevent the ill effects of harmonics on the utility lines, an IEEE Standard IEEE-519 [9] was established in 1981 as the "Recommended Practices and requirements for Harmonics Control in Electrical Power System" giving limits on current and voltage distortion and revised in 1992. The limit specified on current distortion for general distribution systems in voltage range of 120–69,000 V is applicable for the fast-charging stations of higher power ratings, and the THD is desired to be less than 5% at the input.

The electrical equipment is categorized into four groups, classes A, B, C, and D, for the purpose of harmonic current limitation in the IEC 61000 [10]. The balanced three-phase equipment falls under class-A category, and the engineers must follow the limits (in Amperes) on odd harmonic currents specified by it.

10.3 Topologies for PQ control of battery charger

The topologies for PQ control of battery chargers may include unidirectional or bidirectional power flow. To facilitate power flow from V2G with voltage and current regulation during peak load hours of the grid, bidirectional converters are employed. Bidirectional converters provide the function of transfer of power between grid and battery sources in either direction. It plays a crucial role in facilitating the interaction between the vehicle and the grid as an energy interface. Therefore, the bidirectional converter must be reliable, cost-effective, and efficient.

10.3.1 Uncontrolled rectifier topologies

The main power sources of EVs are rechargeable batteries, which confer power to the vehicle. These rechargeable batteries (nickel–cadmium, nickel–metal hydride, lead–acid, and lithium-ion battery) are usually recharged through a conventional

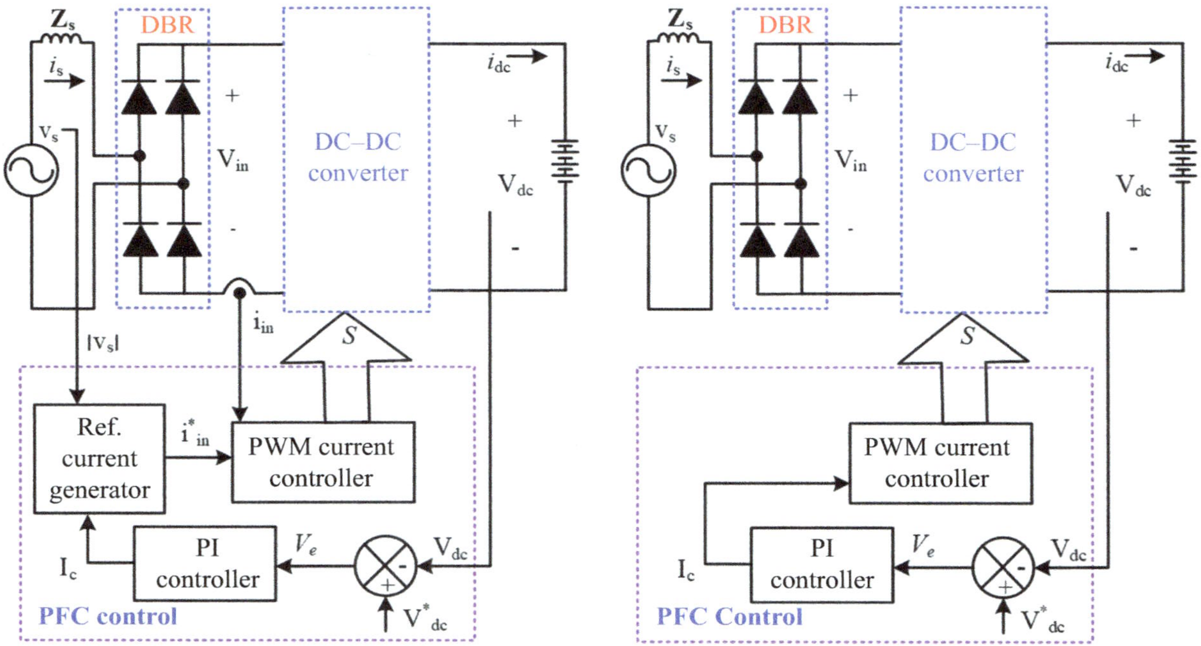

Figure 10.5 Topologies for PQ improvement in battery chargers

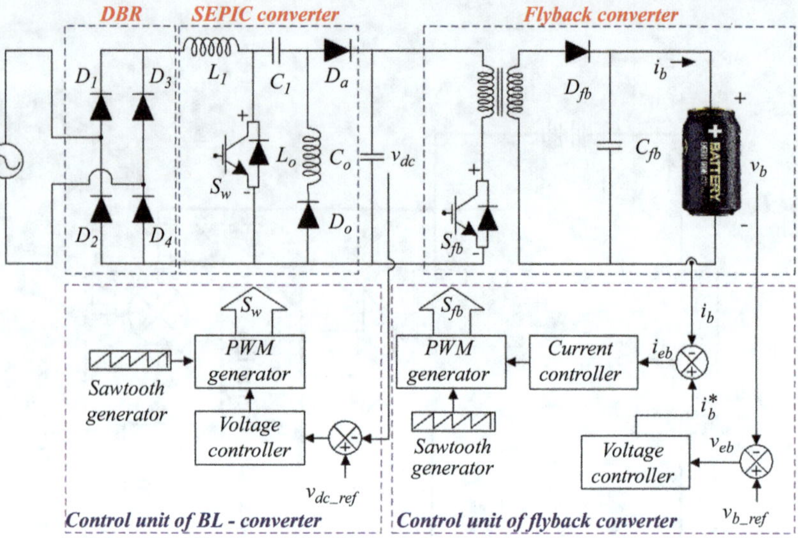

Figure 10.6 DBR-based battery charger topology with BL PQ converter

battery charger, which consists of a DBR to alter the AC mains into DC voltage [11,12] as represented in Figure 10.3. But due to the nonlinear behavior of diodes, severe harmonics appear at the input mains, raising PQ issues. These PQ indices do not adhere to the IEC 61000-3-2 guidelines [13], which affects the performance of other equipment. For PQ control at the input, many PQ improvement techniques for single or double-stage converters are discussed in [14,15], but none have concentrated on efficiency improvement. The uncontrolled rectifier topologies may have a similar approach as shown in Figure 10.5. other than these, there may be options for bridgeless (BL) topologies as shown in Figure 10.6. The BL converter reduces the conduction losses and desired results are obtained with improved efficiency.

10.3.2 Controlled rectifier topologies

The controlled rectifier topology may have single-stage as well as double-stage converters. A single-stage converter topology is shown in Figure 10.7. It has a small range of control as it has to control the voltage at DC link as well as the PQ at the input mains.

10.3.3 Bidirectional converter topologies

The bidirectional converter for EV charger uses two-stage topologies as shown in Figure 10.8. The architecture is capable of operating in G2V and V2G modes. The system has two conversion stages: AC–DC and DC–DC. An AC–DC converter is interconnected with the grid by boost inductors ($L_r = L_y = L_b = L_s$), while the DC–DC converter is coupled with DC-link capacitor (C_{dc}). In each of these stages, a variety of active and passive components like capacitors, inductors, and

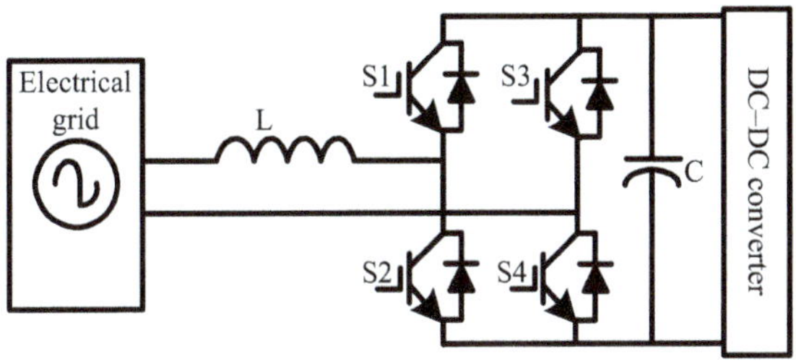

Figure 10.7 Single-phase EV charger with controlled rectifier

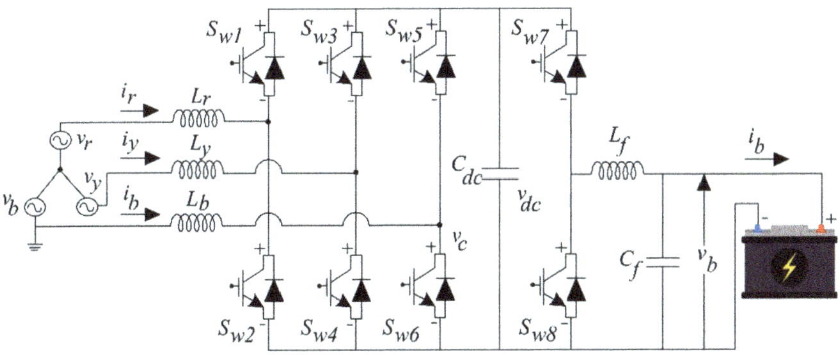

Figure 10.8 Circuit configuration of three-phase EV charger

semiconductors are used. Inductors used on the grid side minimize ripples in the grid currents and help to achieve sinusoidal current. The second-stage functions as a bidirectional buck–boost converter, acting as a buck converter while charging and a boost converter during discharging (i.e., G2V conversion). Active–reactive power command scenarios are handled by the charger architecture. The THD of AC drawn by the battery charger must not exceed 5% irrespective of the battery's state of charge (SoC). Inductor design and grid-side current controller design play an important role in achieving it.

Bidirectional chargers are capable of transferring active power in both directions, so they can work both as G2V and V2G chargers. It allows both active and reactive powers to be controlled. In V2G technology, EV customers are able to transfer the battery energy to the distribution grid and to other vehicles during peak hours, a process also known as V2V sharing. EV chargers that are bidirectional are designed to exchange both active and reactive powers between the vehicle and the grid. As a result, the EV can supply active power to the grid whenever demand increases suddenly and absorb surplus power if surplus power is available.

The control scheme for the bidirectional EV charger is shown in Figure 10.9. The controller has two main functions. The first one is that it charges the EV battery by utilizing active power from the grid (G2V), and the second function is it

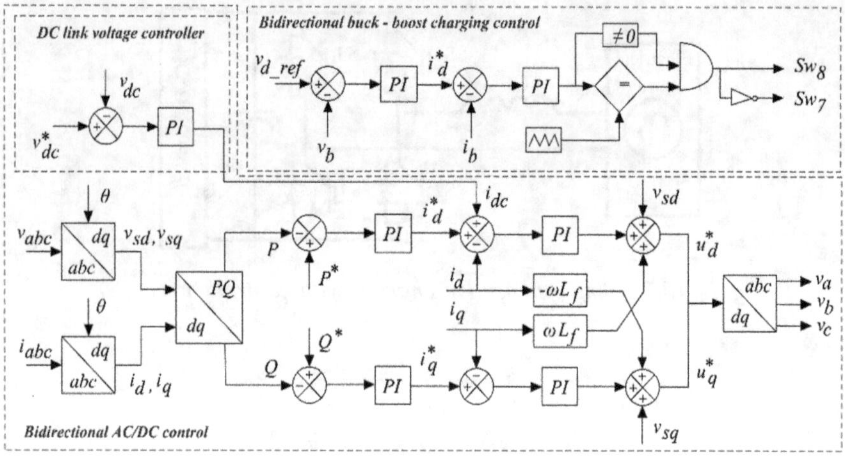

Figure 10.9 Control topology of bidirectional EV charger

sends active power back to the grid when needed or performs a V2G operation. The control function follows the charging power command P* and maintains UPF at its input. In this operating mode, the reactive power command Q* = 0. During charging, the charger operates as a buck converter by controlling the switching of S_{W7} to control the battery charging current.

Second, it supplies the reactive power to the grid when requested from the utility grid. During this mode, the AC/DC bidirectional converter is controlled to maintain 180° phase shift between EV current and grid voltage by setting the reference power command P* to be negative valued and Q* = 0.

10.3.3.1 Performance during G2V mode of operation

The performance evaluation of the bidirectional EV charger during the G2V mode of operation is discussed here. The simulation results of grid side currents $(i_r, i_y, \text{and } i_b)$, DC link voltage (v_{dc}), battery voltage (v_b), battery current (i_b), measured active power (P), and reactive power (Q) are studied during all operating modes. In the G2V mode of operation, the charger draws the required active power from the grid for battery charging. Figure 10.10(a) and (b) shows the simulated dynamic performance of the proposed bidirectional EV charger, which is obtained during the step change in the input voltage at rated condition. Grid currents are sinusoidal and in phase with grid voltages. As shown in these figures, a limited overshoot or undershoot in DC link is obtained during a step change in grid voltage, and the DC link voltage is maintained at the desired value. As the input voltage changes, the output battery charging voltage and current remain constant, demonstrating the G2V mode's satisfactory operation.

10.3.3.2 Performance during V2G mode of operation

EV chargers operate in the V2G mode, delivering stored energy from the battery to the utility grid. When delivering power to the grid, the injected current is

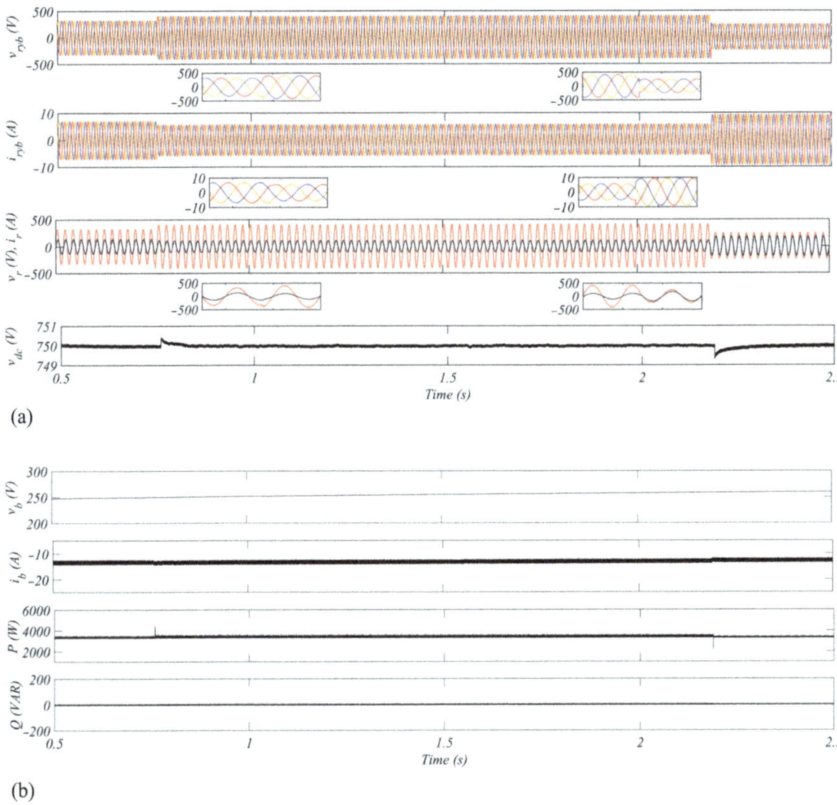

Figure 10.10 Dynamic performance of PFC bidirectional EV charger in G2V mode

in the reverse direction of the grid voltage, which can be seen from 180° phase difference. Each phase current is exactly out of phase with a respective phase in the case of discharging. Figure 10.11(a) and (b) shows the simulated dynamic performance of the proposed bidirectional EV charger, which is obtained during sudden load variation from 3.3 to 2.65 kW at time 1 s, 2.65 to 2 kW at 1.5 s, and sudden increase in load from 2 to 3.3 kW. During each case, a limited overshoot or undershoot in DC link is obtained and the output voltage is almost constant.

As shown in Figure 10.12(a) and (b), a unity PF and very low THD of supply current are observed at the AC mains. The PQ indices thus achieved are within the acceptable limits of IEC 61000-3-2 [13].

10.3.4 BL converter topologies

To improve this, BL PFC converter topology based on dual boost configurations is deliberated as in Figure 10.13. The conduction loss reductions in the switches are

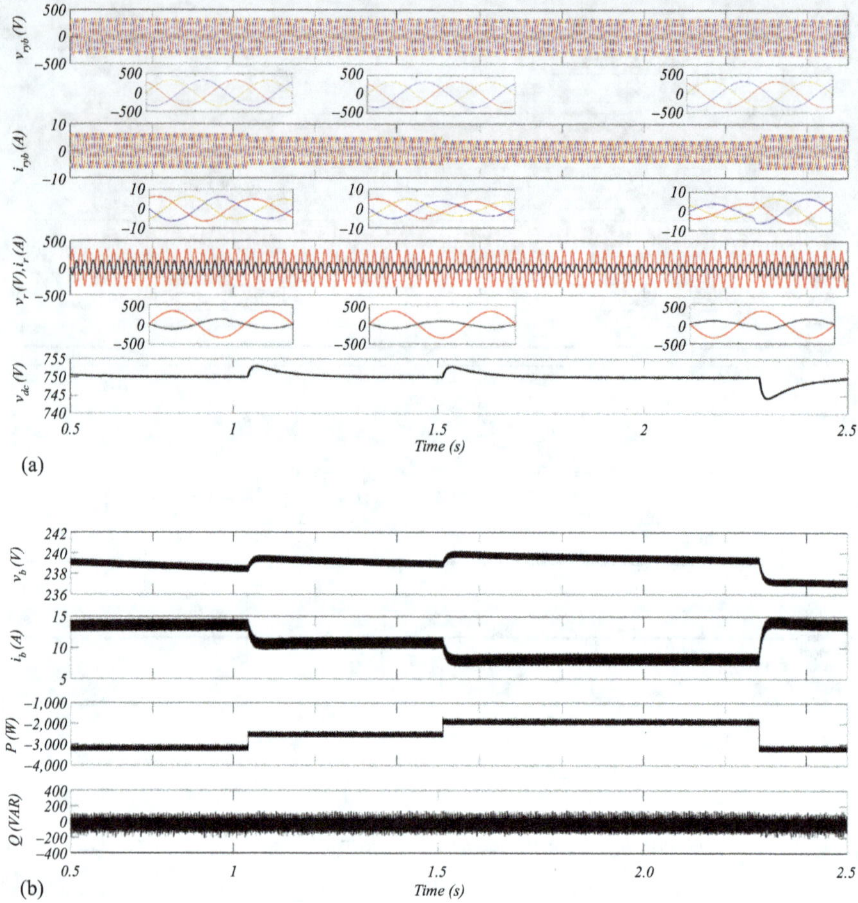

Figure 10.11 Dynamic performance of PFC bidirectional EV charger in the V2G mode

presented by the elimination of DBR with ripple-free input current. The BL-PFC topology reduces the conduction losses and the results are high efficiency as well as power density.

It mainly consists of BL front-end, AC/DC power stage converter based on SEPIC configuration and employing flyback converter to balance the charging current at the time of constant. The front-end BL AC–DC converter is designed in DCM to instate unity PF in an ingenious and emphatic mode. A DCM operation consequence zero current turn-on switching, output diode soft turn-off as well as, helps to degrade the complicacy of the control system due to single-voltage sensor and the flyback converter is operated in DCM of switching. The control loops of both converters are isolated from each other.

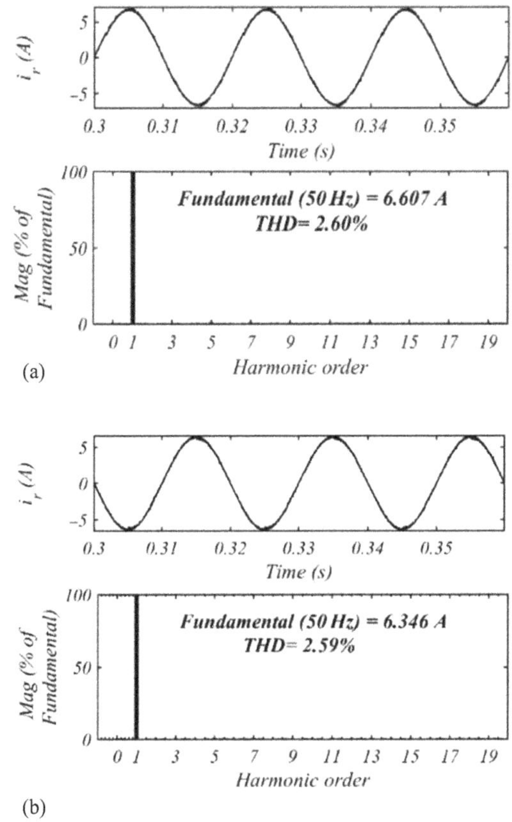

Figure 10.12 Harmonic spectrum of grid current at AC mains for a bidirectional EV charger operating at rated condition. (a) For G2V and (b) V2G mode of operation.

10.3.4.1 Design of BL converter

The design of BL converter includes the detailed analysis of design equations for calculating the values of different components used in their circuit configurations.

The introduced BL-SEPIC has a discontinuous current in output inductors (L_{p2} and L_{n2}). The input AC voltage v_{in} is conferred as

$$v_{ac}(t) = v_{ac\_pk}\ \mathrm{Sin}(\omega_L t) = 220\sqrt{2}\ \mathrm{Sin}(314t) \tag{10.1}$$

where v_{ac_pk} is the peak input voltage and f_L is the line frequency of input mains, i.e., 50 Hz.

Average voltage of the supply mains

$$v_{ac\_avg} = \frac{2\sqrt{2}v_{in}}{\pi} = \frac{2 \times 311.14}{\pi} = 198.06\ V \tag{10.2}$$

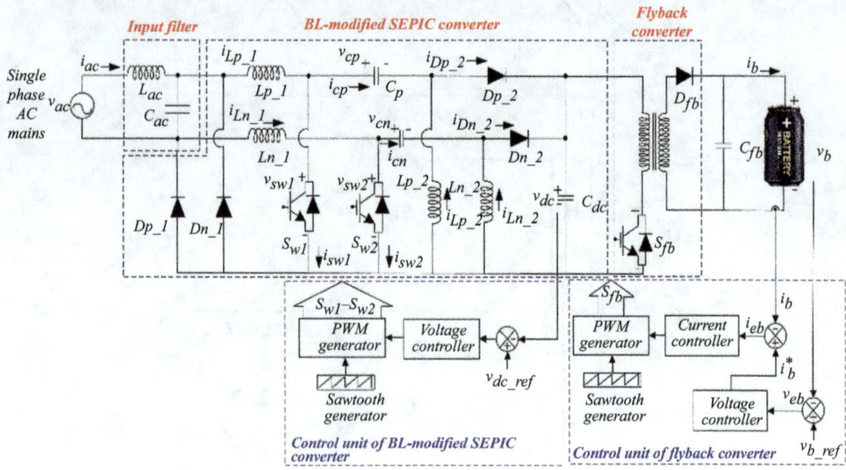

Figure 10.13 BL converter-fed EV charger configuration and control

Therefore, the estimation of duty ratio, D, is articulated as

$$D = \frac{v_{dc}}{v_{ac\_avg} + v_{dc}} = 0.6688 \tag{10.3}$$

Estimated value of input inductors $\left(L_{p1}, L_{n1}\right)$ for the allowed ripple current of ξ as 30% of I_m is calculated at DC link voltage $v_{dc} = 400$ V and expressed as

$$L_{p\_1}, L_{n\_1} = \left(\frac{v_{ac\_\min}{}^2}{P_{in}}\right)\frac{1}{0.2 f_s}\left(\frac{v_{dc}}{v_{in\_\min} + v_{dc}}\right) = 1.979 \text{ mH} \tag{10.4}$$

Designed value of output inductor in the DICM mode is calculated as

$$L_{p\_2}, L_{n\_2} = \left(\frac{v_{ac\_\min}{}^2}{P_{in}}\right)\left(\frac{v_{dc}}{2 v_{in\_\min} \times f_s}\right)\left(\frac{v_{dc}}{v_{in\_\min} + v_{dc}}\right) = 0.3108 \text{ mH} \tag{10.5}$$

$$C_p, C_n = \frac{v_{dc} \times D}{\gamma f_s C_p f_s R_{dc}} = 1.308\ \mu\text{F} \tag{10.6}$$

Where permitted ripple voltage of γ% (i.e. = 10%)

The DC link capacitor, C_{dc} is,

$$C_{dc} = \frac{\left(\frac{P_{in}}{v_{dc}}\right)}{2\pi f \Delta v_{dc}} = 1.37 \text{ mF} \tag{10.7}$$

To alleviate higher order harmonics, a LC filter is employed at the input providing 40 dB attenuation,

$$C_{ac} = \frac{\left(\frac{\sqrt{2}P_{in}}{v_s}\right)\tan\phi}{2\pi f \times \sqrt{2}v_s} = 2.296\ \mu\text{F} \tag{10.8}$$

The filter inductor is selected to mitigate high-order harmonics,

$$L_{ac} = \frac{1}{4(\pi f_r)^2 C_{ac}} = 1.363\ \text{mH} \tag{10.9}$$

where f_r is the resonant frequency.

10.3.4.2 Flyback converter design

$$\text{Duty cycle } D_{fb} = \frac{v_b}{\left(\frac{n_s}{n_p}\right)v_{dc} + v_b} = 0.375 \tag{10.10}$$

Transformation ratio $(n_s/n_p) = N$ is set as 9 to confer the control output voltage of 48 V.

The optimum value of magnetizing inductance L_{fb} to deliver a maximum power for battery charging in operating conditions of discontinuous operation is obtained as

$$L_{fb} \leq \frac{(v_{dc}D_{fb})^2}{2v_b \times i_b \times f_{sf}} \leq 0.1125\ \text{mH} \tag{10.11}$$

$$\text{Output capacitor, } C_b = \frac{D_{fb}v_b}{f_{sf}\left(\frac{v_b{}^2}{P}\right)\chi \times v_b} = 217.1\ \mu\text{F} \tag{10.12}$$

where χ is 3% and takes into account of the output ripple voltage.

10.3.4.3 Performance simulation of single-phase BL EV charger

The MATLAB® models of single-phase BL EV charger are developed using Simulink® and Sim Power Systems (SPS) toolboxes to simulate the performance of this EV charger in single phase.

The presented EV charger is powered by a single-phase supply of 230 V_{rms} where sinusoidal grid current i_{ac} in phase with the supply voltage is observed which shows a unity PF at AC mains. Figure 10.14(a) shows the simulated steady-state behavior of the BL EV charger at the rated load with a charging voltage (v_b) of 48 V and a charging current (i_b) of 41.67 A. The input inductors current in CCM is depicted in Figure 10.14(b), where the output capacitor maintains a constant DC of 400 V. The current waveforms of output inductor in positive and negative half-line cycles represent its conveniences for the DCM operation. The peak input current of the converters is evaluated to 14.3 A. Both converter switches (S_{w1} and S_{w2}) are

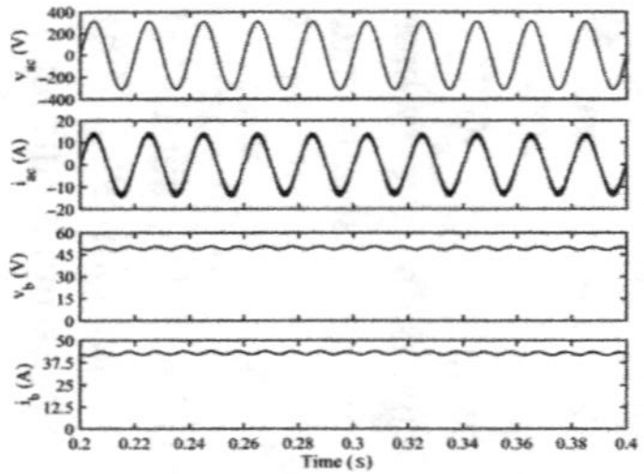

(a) *Input source voltage v_{ac}, source current i_{ac}, battery charging voltage v_b and charging current i_b*

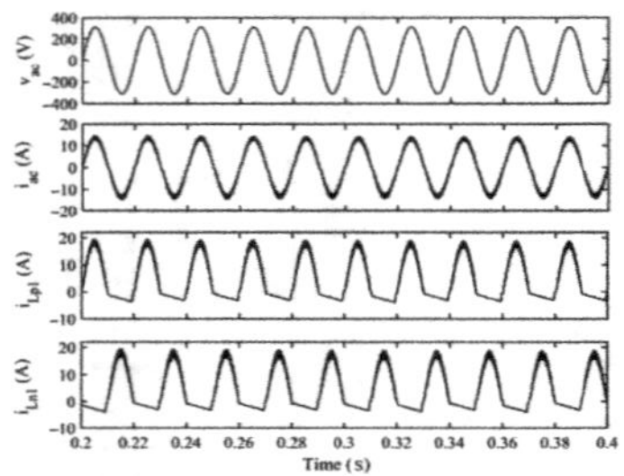

(b) *Input source voltage v_{ac}, source current i_{ac}, peak current in inductor L_{p1} and L_{n1}*

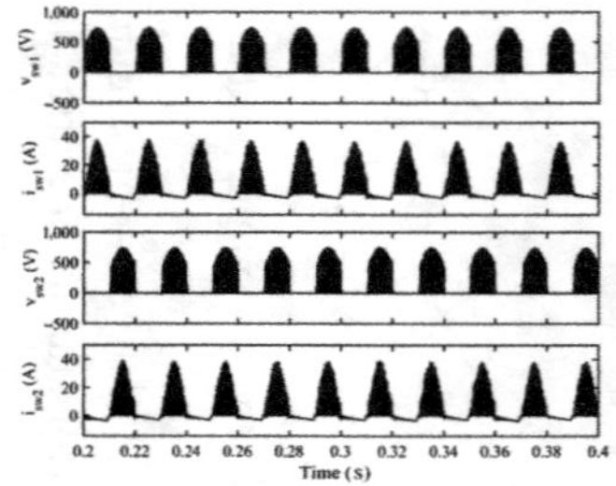

(c) *Peak voltage and current across switches S_{W1} and S_{W2} during respective half cycle of ac mains*

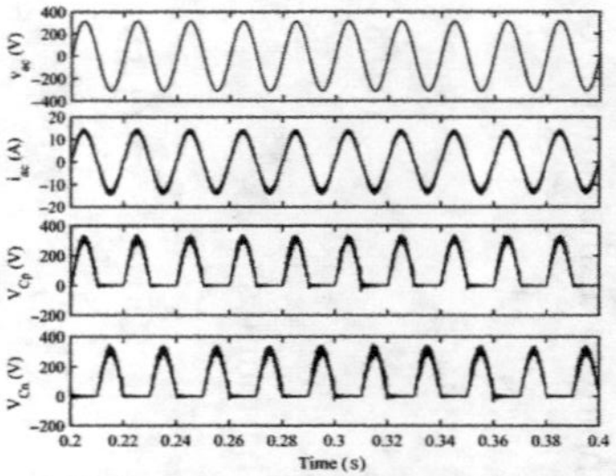

(d) *Input source voltage v_{ac}, source current i_{ac}, intermediate capacitor voltage V_{cp}, V_{cn}*

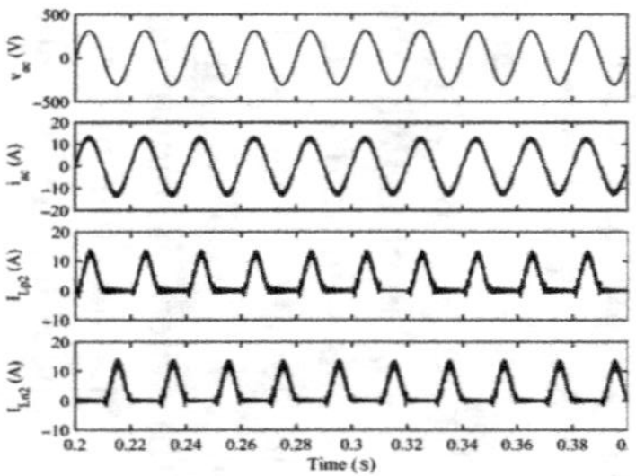

(e) *Input source voltage v_{ac}, source current i_{ac}, peak current in output inductors L_{p2} L_{n2}*

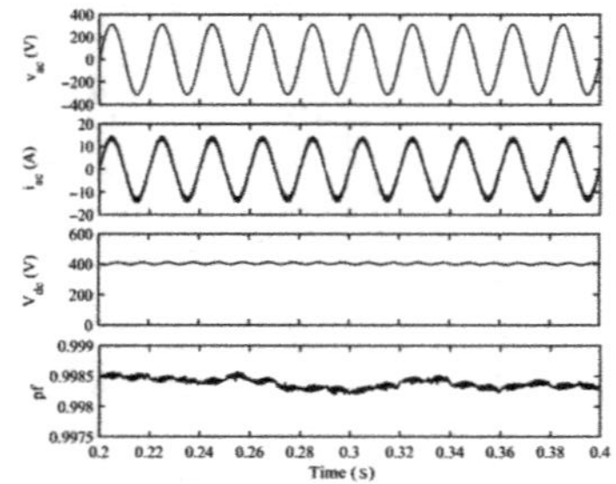

(f) *Input source voltage v_{ac}, source current i_{ac}, DC link voltage and power factor of BL SEPIC EV charger*

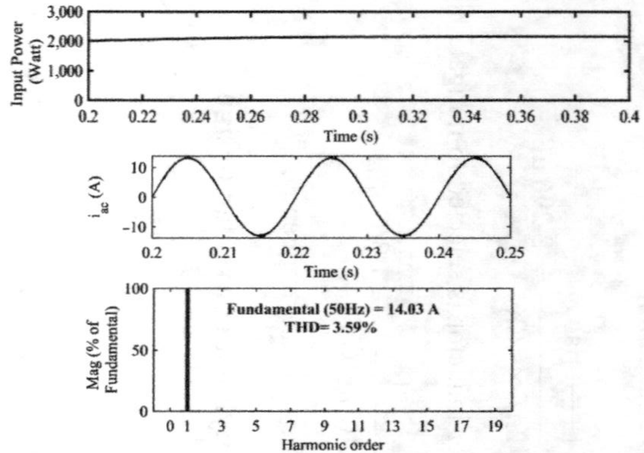

(g) *Harmonic spectrum of source current and input power of the converter*

Figure 10.14 Performance of introduced system at steady-state conditions

operated asymmetrically in each half-cycle. This performance validates the feasibility of BL configuration and its operation with reduced stress of voltage and current across all the switches.

It is observed from Figure 10.14(g) that the source voltage and current are in-phase resulting in enhanced PQ parameters with low THD of the line current which is verified through the harmonics spectrum analysis. Current THD is below 5% and the PF is 0.998, which complies with the required IEC 61000-3-2 norms. The individual dominant harmonics (3rd, 5th, and 7th) are also within IEC 61000-3-2 norms as shown in Figure 10.14(g). The efficiency of the EV charger is 91.74% because the input power drawn by the EV charger is 2.18 kW whereas the output power is 2 kW.

The improved efficiency is a result of current conduction through the lowest number of components in each cycle and, therefore, they have high efficiency due to lower losses associated with them.

10.3.5 Dual active bridge converter topology

The dual active bridge (DAB) converter topology is a specific topology which is capable of bidirectional power flow. One such topology is shown in Figure 10.15 which is used for battery charging while utilizing grid and photo-voltaic (PV) array. It consists of a single-phase AC source that feeds a nonlinear load. The DAB converter system is designed for charging low-rating batteries using solar PV and grid power.

The grid side VSC is controlled with the help of an adaptive filter algorithm and thus maintains the DC link voltage at the specified reference value. The PV source is created by connecting several modules in series and parallel, and PV voltage and current are determined for the PV array using the perturb and observe (P&O) method. The solar PV array is connected at the point of common coupling and feeds solar power utilizing maximum power point tracking (MPPT) control. MPPT control is also used to generate the reference DC link voltage for the DC link

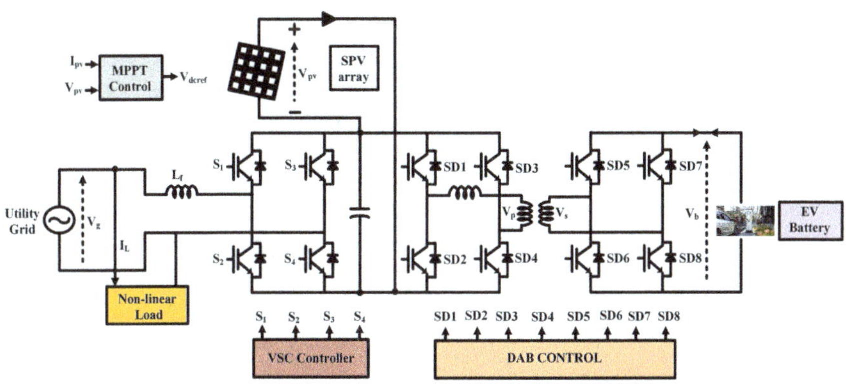

Figure 10.15 Control schematic of a DAB converter for battery charging

capacitor. After the DC link capacitor, the DAB DC–DC converter is connected, the DAB control is executed while utilizing the phase shift modulation technique. The DAB converter ensures that the voltage and the current are maintained at a specified value required for charging and discharging of the battery in the various modes of operation that may be G2V, V2G, etc.

10.3.5.1 DAB controller design

DAB controller works on the principle of phase shift. When there is an adequate phase shift between the primary and the secondary switches, then the transfer of power takes from the leading phase shift side to the lagging phase shift side. This is the same as the power transfer between two buses in a power system. For the primary and secondary voltages, V_1 and V_2, respectively, the phase shift δ is obtained as

$$\delta = \pi\left(1 - \frac{P_{\mathrm{w}}L}{V_1 V_2}\right)$$

where w is the switching frequency and L is the inductive reactance between the two sides.

To control the DAB converter, the difference between the battery current and the reference battery current is taken as

$$e(t) = I_{ba}(t) - I_{ref}(t)$$

This error signal is then passed to the PI controller whose output is given as

$$p(a) = \left(k_p + \frac{K_t}{s}\right) \times e(t)$$

A signal is produced as an output and is multiplied by the constant α for the intended design values of K_{i} and K_{p}. Where α is defined as

where Ts= 1/25,000 is the time period for the specified switching frequency of 25 kHz

$$\alpha = \left(\frac{1}{25,000}\right) \times 0.5\, p(a) \times \alpha = p_{ref}$$

This obtained signal acts as a time delay which provides the required phase shift needed between the primary and the secondary switches

10.3.5.2 Operation in charging mode (G2V)

In the charging mode of operation, the battery gets charged through the power supplied from the grid. In this mode, the primary portion of the DAB converter leads to the secondary and thus the power is supplied from the primary to the secondary. Figure 10.16 shows the control configuration of DAB charging converter.

10.3.5.3 Operation in discharging mode (V2G)

In the discharging mode of operation, the power is sent back from the battery to the grid. In this mode, the primary segment of the DAB converter leads the secondary

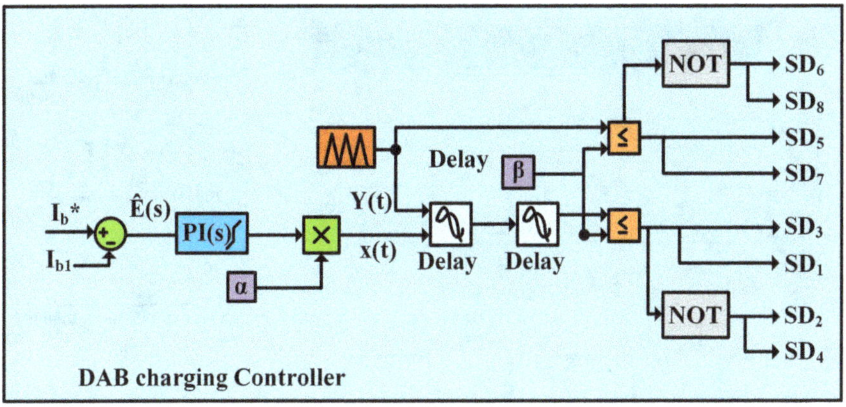

Figure 10.16 Control schematic of a DAB charging controller

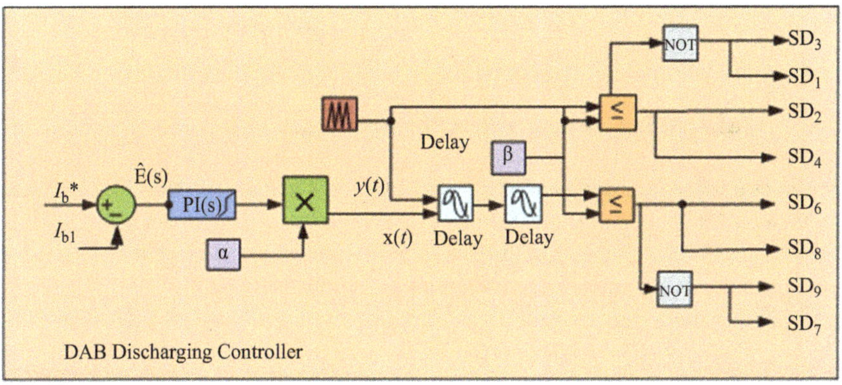

Figure 10.17 Control schematic of a DAB discharging controller

portion and the power is transferred from secondary to primary. Figure 10.17 shows the control configuration of DAB discharging converter. The operation of the DAB converter is obtained and presented in Figure 10.18 in both the modes (charging as well as discharging). DAB converter has a point of transition at 0.5 s between the charging and discharging modes. While the power is transferred from the grid to the battery in the time zone of 0–0.5 s, the primary of the DAB converter leads the secondary, resulting in power transfer from the primary to the secondary or from G2V, as observed in Figure 10.18. At the instant of 0.5 s, there is a change in operation from G2V to V2G, the primary starts to lag behind secondary after the 0.5 s and the power starts to transfer from the secondary side to the primary or V2G.

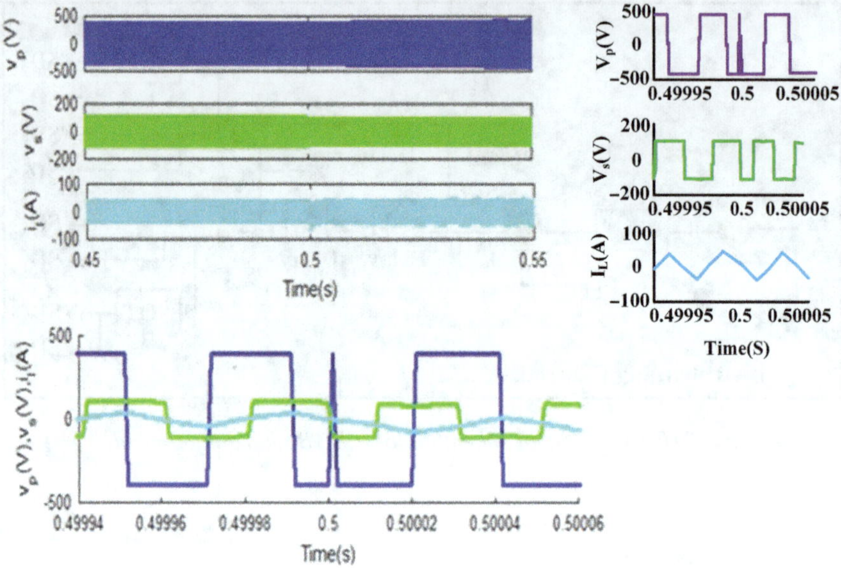

Figure 10.18 Performance simulation of a DAB controller

10.4 Multi-pulse and multi-level topologies

For power quality improvement of battery charging systems, the AC-DC converters can have multiple winding transformer-based configurations known as multi-pulse converter configurations and multiple switching pulses with variable voltage level switching known as multi-level converter configurations.

10.4.1 Multi-pulse converters

When the static rectifiers are used, the output voltage pulsates integer multiple times the input frequency and the input current is also multi-step. By popular use in the power electronics industry, a multipulse converter mean "a three-phase converter that produces more than six ripples of DC voltage per cycle" or alternately "a three-phase converter having steps in AC more than that of a 6-pulse bridge rectifier." The advantage of multipulse technique is that it acts as AC and DC side filter in a single unit.

The multipulse converters require separately wound transformers with the rectifiers mainly for:

- Stepping up or down the available utility supply voltage to desired levels.
- When the increase in the number of phases or phase shifting is needed, i.e., phase multiplication is required separately wound star/star-delta transformers are commonly used to obtain 3 to 6-phase conversion.
- For eliminating neutral conductor, star-delta transformers are used.

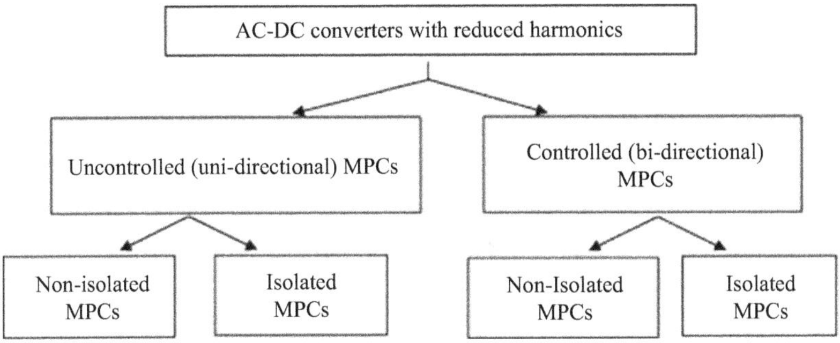

Figure 10.19 Classification of MPCs

- Saturation in the transformer core is useful in short-circuit conditions, i.e., it acts as isolation between the supply and the load side.

10.4.1.1 Classification of multi-pulse converters

The most popular AC–DC converters are diode-based or thyristor-based bridge-type converters although half-wave converters are also used in some specific applications. In the applications like DC motor drives and CSI-fed AC motor drives, the thyristor-based AC–DC converters having firing angle control are used for DC voltage/current control and bi-directional power flow. The diode-based bridges are commonly used in VSI-fed AC motor drives and power supplies. The above discussion leads to the classification of multi-pulse converters with the reduced harmonics as shown in Figure 10.19. The isolated MPCs shall be discussed here that are connected to the utility for EV charging.

10.4.1.2 Arrangement of MPCs

A number of researchers have investigated the multi-pulse AC–DC converters, many giving simulation and experimental results and also new concepts. The isolated uncontrolled rectifiers and isolated controlled rectifiers have been discussed here as isolation and harmonic reduction at input shall be possible simultaneously at the utility end.

The five important parts of these MPCs which are important for their successful operation and decide its characteristics that are: (i) transformer, (ii) interphase reactor (IPR), (iii) zero-sequence blocking transformer (ZSBT), (iv) bridge configurations, and (v) high-pass filter.

Transformers

The different three-phase transformer winding arrangements possible are shown in Figure 10.20. These isolated transformers may have more than one primary or secondary winding, arranged in Δ, Y, zigzag (ZZ), polygon, extended Δ, and hexagon shape. The full-wave (FW) MPCs generally have transformers with Y-connected secondary winding, while the primary may be Δ, Y, zigzag (ZZ), polygon, and extended Δ. At times, Tee-connected (Scott connected) windings are also employed, mainly due to the fact the even number of single-phase transformers is required.

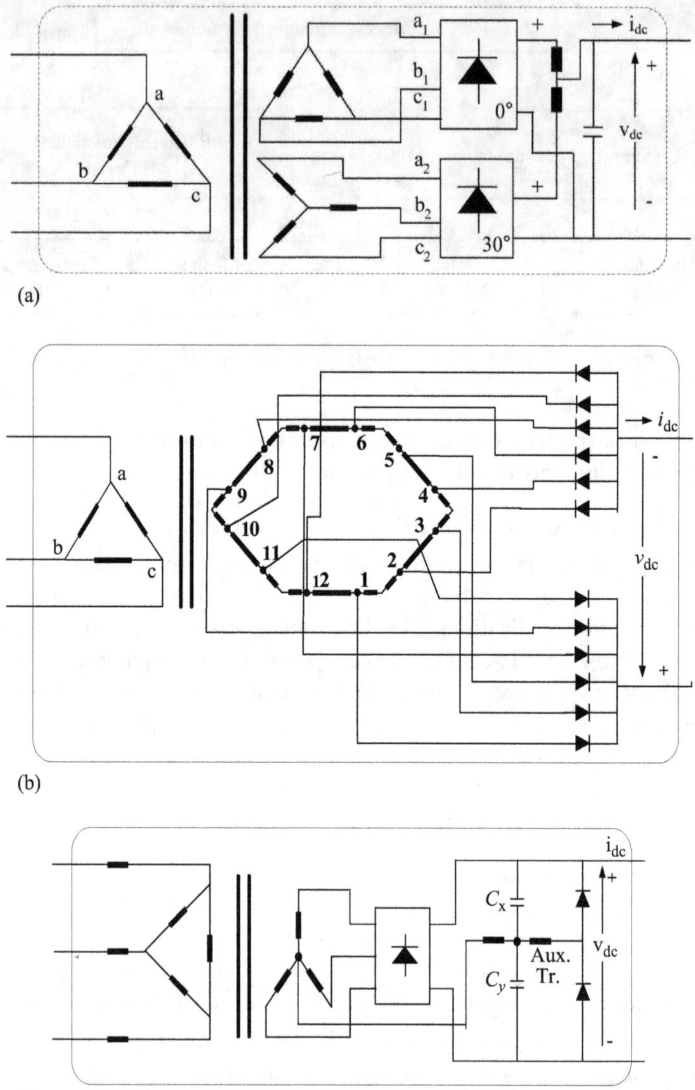

Figure 10.20 *(a) Bridge rectifier-based isolated MPCs – 12-pulse MPC. (b) 12-pulse MPC with non-isolated secondary winding. (c) Improved 6-pulse MPC. (d) 12-pulse controlled MPC. (e) 18-pulse MPC with a single primary. (f) 24-pulse MPC with two primaries. (g) 24-pulse MPC with pulse doubling. (h) 12 × 3 MPC. (i) 48-pulse (12 × 4) MPC. (j) Another 48-pulse (12 × 4) MPC.*

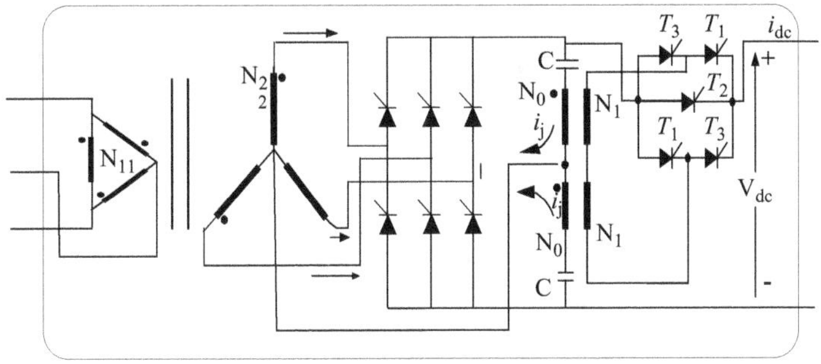

(d)

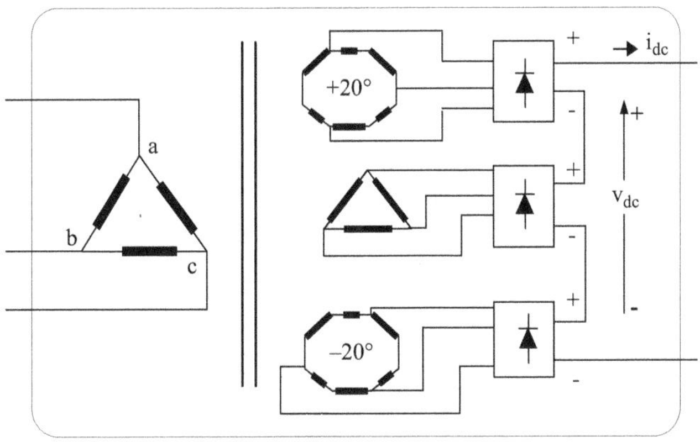

(e)

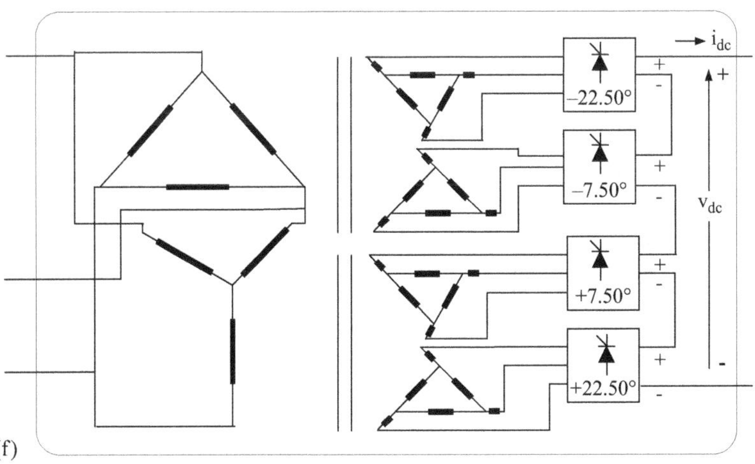

(f)

Figure 10.20 (Continued)

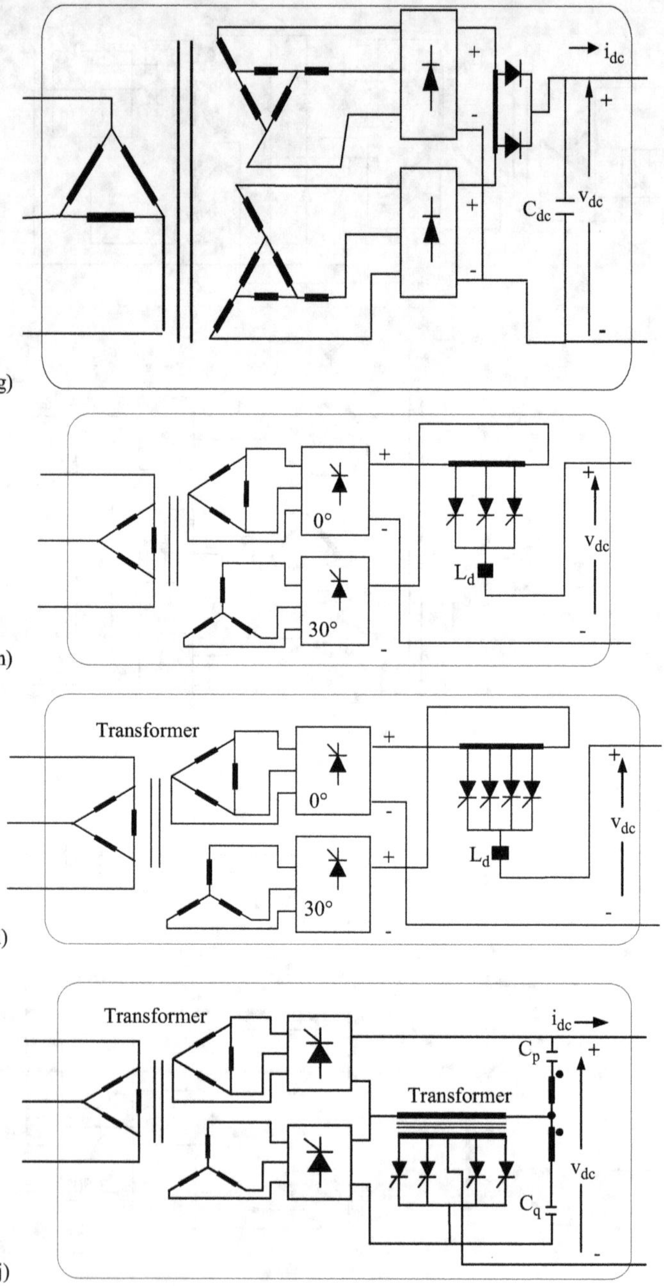

(g)

(h)

(i)

(j)

Figure 10.20 (*Continued*)

IPR

The IPRs are employed to absorb the instantaneous voltage difference between the bridge outputs. Tapped IPRs are used to help the pulse multiplication by absorbing the ripple currents and injecting it back into the input thereby reducing the harmonics. An arrangement of IPR windings for pulse doubling and tripling is shown in Figure 10.21.

The value of voltage difference at the terminals of a tapped IPR is given by

$$V_m = 0.0814\ V_0$$

And the voltages V_1, V_2, and V_3 can be written in terms of the V_m as

$$V_1/V_m = V_2/V_m = (N_o - 2N_t)/2/N_o = 0.2543.$$

$$\text{And } V_3/V_m = 2N_t/N_0 = 0.4914.$$

Therefore, the value of V_1, V_2, and V_3 is:

$$V_1 = V_2 = 0.2543,\ V_m = 0.021V_0$$

$$V_3 = 0.4914, V_m = 0.04V_0$$

$$I_{o1,\mathrm{rms}} = I_{o2,\mathrm{rms}} = 0.567I_o \text{ and so } I_{o3,\mathrm{rms}} = 0.232I_o$$

The Volt-ampere rating of the tapped IPR becomes

$$VA_{IPT} = 0.0165P_o$$

The formulas used for designing of transformer are the following:

- The voltage induced in a transformer winding with T turns,

$$Et = K\sqrt{Q}, \text{ where } K = \sqrt{(4.44f, \varphi_m.10^{-3}/(A\ T)}$$

$$T_p = V_p/E_t \text{ and } T_s = V_s/E_s$$

$$A_p = I_p/\delta \text{ and } a_s = I_s/\delta$$

- Total area of copper,

$$\text{Core area}, A_c = T_p.a_p + T_s.a_s = 2.T_P.I_p/\delta = 2AT/\delta$$

But $Q = 2.22.f.\ \varphi_m.\ A_c.\delta \times 10^{-3}$ kVA

- Maximum flux, $\varphi_m = Q/(2.22.f.\ A_c.\ \delta \times 10^{-3})$
- Core area, $A_i = \varphi_m/B_m$
- The space factor for transformer window is Kw = area for conductors in window/window area = A_c/A_w

ZSBT

For the independent operation of the two-diode bridge rectifiers fed from non-isolated secondary windings, zero-sequence currents must be blocked. This blocking transformer must offer high impedance to unwanted currents and promotes 120° conduction for each device in bridge. The arrangement of windings is

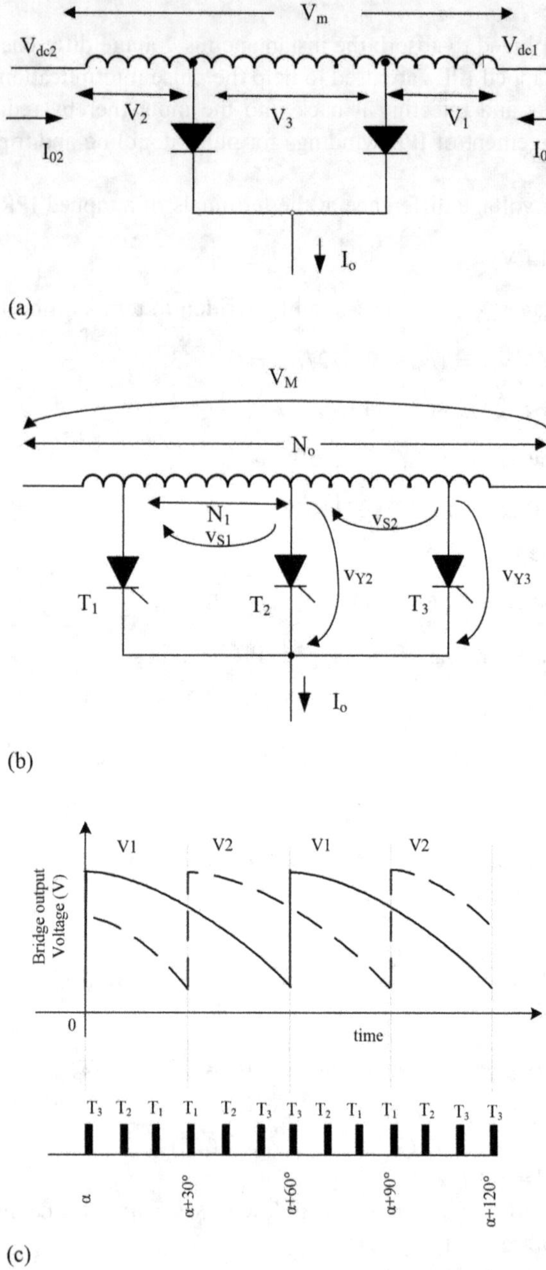

Figure 10.21 Pulse multiplication concept for Interphase Reactor – (a) An IPR tapped with two diodes for pulse multiplication 12 × 2, to get 24 Pulse AC/DC Converter (b) An IPR tapped with three thyristors for pulse multiplication 12 × 3, to get 36 Pulse AC/DC Converter. (c) The sequence of turning-ON tapped thyristors with reference to bridge output voltages

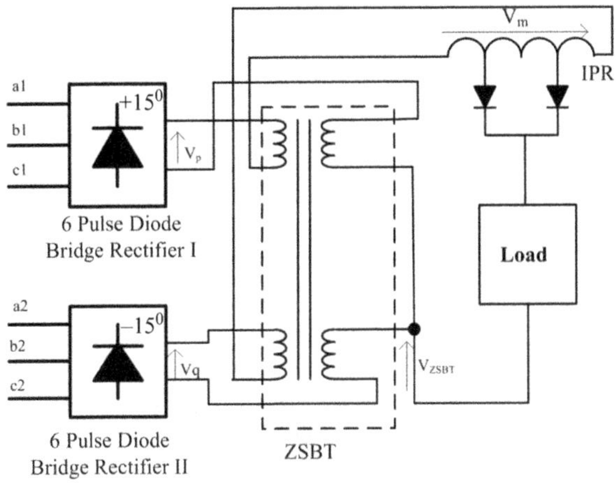

Figure 10.22 Arrangements of ZSBT and IPT windings

shown in Figure 10.22. The voltage across the ZSBT (V_{ZSBT}) is obtained as

$$V_{\mathrm{ZSBT}} = (V_p - V_q) - V_m$$
$$= (V_{a1} - V_{a2}) \text{ for } 8/12f \text{ to } 11/12f$$
$$\text{and } (V_{b1} - V_{a2}) \text{ for } 11/12f \text{ to } f$$

where f is the fundamental frequency of V_{a1}.

The ZSBT contains only triple-frequency voltage components and is expressed in the Fourier series as

$$V_{\mathrm{ZSBT}} = V_{\mathrm{LL}} * (-0.25 * \cos(3\omega t) + 0.07 * \sin(6\omega t) + 0.03 * \cos(9\omega t) + \ldots..)$$

So, we have

$$V_{\mathrm{ZSBT}} = 0.185 V_{\mathrm{LL}} \text{ and } I_{o1,\mathrm{rms}} = 0.567 I_o$$

Therefore, the VA rating of this transformer (VA_{ZSBT}) is $VA_{\mathrm{ZSBT}} = 0.075 * P_{\mathrm{out}}$ where P_{out} is the output power.

Bridge configurations

Six-pulse full-bridges may be arranged in series or in parallel with IPRs at the output. Three-phase inputs are shifted by 60 °/n where n ($n > 1$) is the number of 6-pulse bridges used. If $n = 2$, the phase shifting between inputs is 30° (or ±15°). If $n = 3$, the phase shifting between inputs is 20° (−20°, 0°,20°). If $n = 4$, the phase shifting between inputs is 15° (−22.5°, −7.5°, 7.5°, 22.5°). Figure 10.20 shows the isolated bridge-based MPCs which are most popular for uncontrolled/controlled rectification. When full-wave converters are employed, the phase shifting between the inputs to 6-pulse rectifiers is 120 °/n.

High-pass filter (HPF)

The commutation overlap is observed in the line currents and voltages when more than two devices in a rectifier bridge conduct. The duration and the magnitude of overlap vary with firing angle, load current, and source inductance. It incites six notches per cycle in the input voltage waveform of a 6-pulse thyristor converter. To mitigate this deviation, tuned passive filters and high-pass filters (HPF) at the point of common coupling (PCC) are required. In general, the first-order tuned and HPFs can be employed but these are susceptible to changes in the supply frequency besides being bulky involving more components.

The multipulse AC–DC converters inherently have higher order characteristics harmonics causing higher frequency but lesser depth commutation notches in voltage waveform. Their elimination is possible by employing a second-order HPF as the current waveform has low amplitude 23rd and 25th order (35th and 37th for 36 pulse converter) dominant harmonics for a 24-pulse converter.

The harmonic pattern without any passive filter at the input of an MPC helps determining the indices of an HPF to be employed. The HPF must offer a capacitive reactance ($X_{C,1}$) and a low-value dominant resistive impedance (Z_n) over fundamental and higher-order frequencies, respectively.

10.4.2 Multi-level converters

Owing to many advantages of multilevel converters, such as less-voltage stress across switches, low harmonics, and high efficiency, the concept is used in battery charging applications also. One such topology is presented in Figure 10.19. This configuration is created using two different configurations namely using level-shifted duty control multilevel converter and ZVS phase-shift full bridge converter. In a level-shifted duty-controlled multilevel converter, there are two active switches (S_{W1} and S_{W2}), two diodes (D_1 and D_2), one bidirectional switch (S_{W3}), and two split capacitors (C_1 and C_2) coupled through a DC link. This constant DC voltage (v_{dc}) feeds to a ZVS phase-shift full-bridge converter at the output of the level-shifted multilevel converter.

The ZVS phase-shift full-bridge converter consists of four active switches ($S_a, S_b, S_c,$ and S_d) at the primary side of the high-frequency transformer (HFT). The secondary of the transformer is connected to four diodes ($D_a, D_b, D_c,$ and D_d) and output capacitor C_f. PSFBC is controlled using a voltage controller so that the output DC voltage can be altered and controlled. The DC-link capacitor balances the charges CCCV operation.

The connections of bidirectional switch between leg y and split DC-link capacitor, as shown in Figure 10.23, provide various paths for current to achieve a five-level voltage profile at the terminals of the adopted converter, which results in low-voltage stress on the switching devices. The drawn line current by the five-level AC–DC rectifier is sinusoidal and in phase with input voltage to attain the UPF and THD of input current as per the international PQ standard IEC-61000-3-2 [13]. One important prerequisite of this control is that the DC link voltage should have a value higher than the peak voltage of AC mains, whereas lower than twice of AC mains voltage, i.e. $v_{(g(max)<v_{dc}<2v_{(g(max)}}$. The DC link voltage, v_{dc} of this

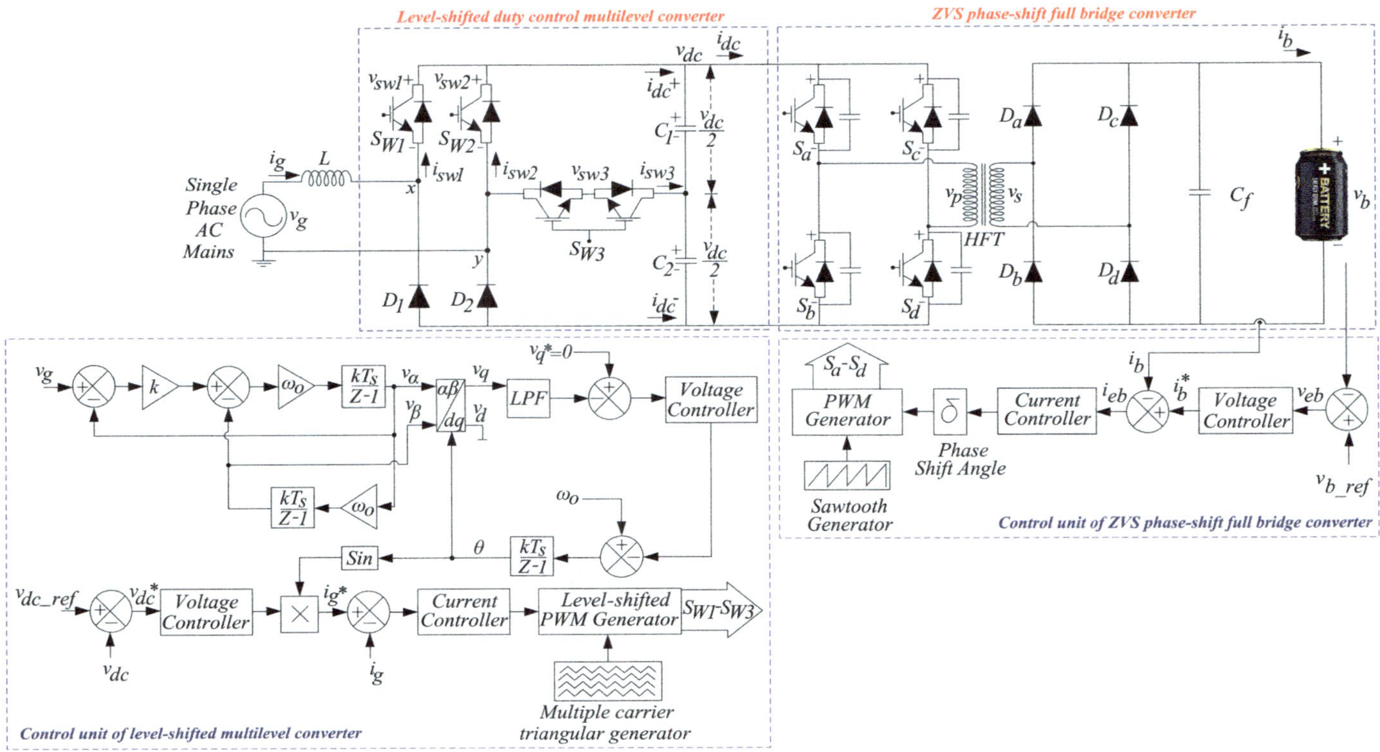

Figure 10.23 Level-shifted PWM five-level PFC rectifier-based EV battery charger

discussed topology is applied to the input of high-frequency PSFBC. The battery charging current during the switching period is maintained constant by the output filter inductor L_o and capacitor C_o. The control loops and power circuits of these converters are isolated individually.

The current directions and different operating modes are shown in Table 10.1, which are produced by semiconductor switches.

To obtain this five-level pattern at the front-end rectifier with low THD and UPF, a level-shifted PWM scheme is applied to produce appropriate switching pulses. The overall control schematic of the level-shifted duty-controlled multilevel charger, as depicted in Figure 10.24, initiates DC-link voltage sensing and voltage-error generation between the reference voltage and the sensed voltage. The obtained error is sent to the outer loop of the low bandwidth voltage controller. The output of this controller is multiplied with the unit sinusoidal signal obtained from the phase-locked loop (PLL) circuit to generate the reference sinusoidal current i_g*. To interconnect the battery charger to AC mains, the grid frequency must be accurately estimated. The grid estimation in a single-phase system cannot be determined directly, hence, phase shifting is required to construct orthogonal

Table 10.1 Operating modes of level-shifted PWM five-level rectifier

Switching states	**Sign of i_g**	**Switching signals**			**v_{xy}**
		SW_1	SW_2	SW_3	
1	> 0	ON	OFF	OFF	$+v_{dc}$
2	> 0	ON	OFF	ON	$+v_{dc}/2$
3	≥ 0 and ≤ 0	ON	ON	OFF	0
4	< 0	OFF	OFF	ON	$-v_{dc}/2$
5	< 0	OFF	ON	OFF	$-v_{dc}$

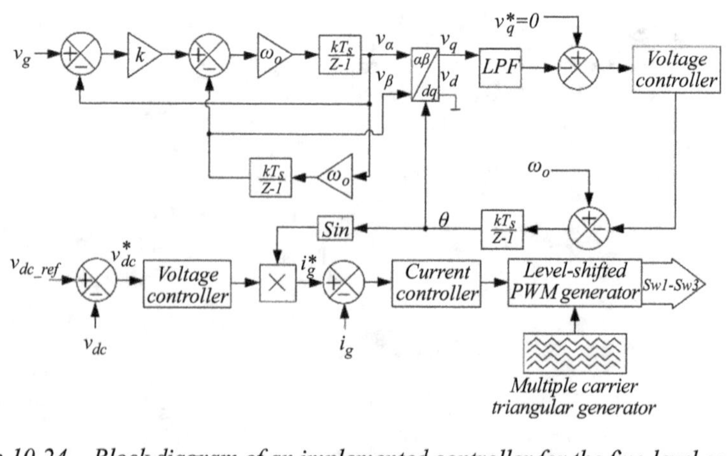

Figure 10.24 Block diagram of an implemented controller for the five-level converter

components. The frequency is calculated from the q-axis component in a single-phase PLL system using a discrete-time integrator and a PI controller. The second-order filter is used to estimate the $\alpha\beta$-components, which then leads to the computation of the dq-axis components using the line frequency as the correction factor.

The line current of the adopted five-level converter is compared with reference sinusoidal current and sent to the inner loop of the high bandwidth current controller, which is processed through a level-shifted PWM comparator to produce the switching pulses. The level-shifted PWM strategy has a low and fixed switching frequency carrier waveform of 5 kHz as shown in Figures 10.25(a) and (b), respectively, which ensures the generation of gating signals to control the power switches that reduce the switching losses, harmonics, and device stresses as compared with traditional multilevel converter.

Due to this five-level converter, the sizes of passive components are smaller, light-weighted, and cost-effective in comparison to conventional rectifiers. As shown in Figure 10.25(a), four vertically shifted triangular carrier waves with the same frequency and amplitude (C_{w1}, C_{w2}, C_{w3}, and C_{w4}) are used to modulate the sinusoidal reference waveform (U_{ref}) and generate fixed frequency switching pulses for five-level control of rectifier. Figure 10.25(b) depicts the logical operation block.

The output voltage of the level-shifted PWM converter is processed by a DC–DC converter operating in a continuous conduction mode. The gating signals with dead time for switching devices converter $S_a - S_b$ and $S_c - S_d$ are retained on for 180°. Therefore, switching pulses of $S_a - S_d$ and $S_b - S_c$ are square pulses with a 50% duty cycle. These pulses are generated by a PWM generator by comparing the voltage reference signal to a sawtooth carrier. The output of the full-bridge is connected to the HFT. This high-frequency AC is rectified by a diode bridge rectifier $D_1 - D_4$. The output inductor L_o delivers continuous current and filter capacitor C_o limits the output ripple voltage. Finally, a voltage feedback controller is used to alter and control the output DC voltage.

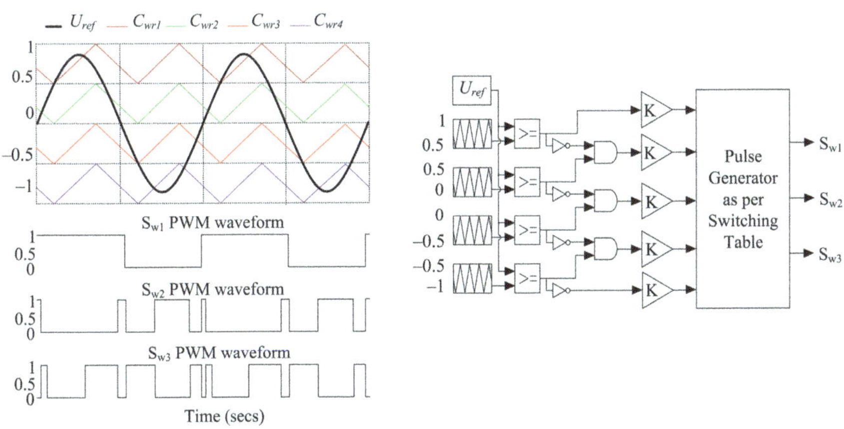

Figure 10.25 (a) Switching pattern of 4-carrier level-shifted PWM technique. (b) Fixed frequency switching pulses as per switching Table 10.1.

10.4.3 Design of single-phase multilevel EV charger

The improved PQ 3.3kW level-shifted PWM five-level EV battery charger is designed for an output of 48–72 V DC, 150 Ah battery with UPF at 220 V input AC supply. Its simulated performance is shown in Figure 10.26, which demonstrates the achievement of desired parameters.

(A) **Five-level converter design**

The input AC voltage V_{in} to the five-level converter is conferred as

$$v_g(t) = v_{g\_pk}\,\mathrm{Sin}(2\pi f_s t) = 220\sqrt{2}\,\mathrm{Sin}(314t) \tag{10.13}$$

The input inductor (L) filters out the input mains ripple current, expressed as

$$L = \frac{\frac{v_{dc}}{2}}{4 \times f_c \times \Delta ig_{p-p(\max)}} = 0.53\ \mathrm{mH} \tag{10.14}$$

The values of DC-link split capacitors, C_1 and C_2, are calculated for permissible ripple in DC-link voltage (10%)

$$C_1, C_2 = \frac{P_o}{4 \times f_c \times \{v_{dc}{}^2 - (v_{dc}{}^2 - \Delta v^2)\}} = 103.12\ \mu\mathrm{F} \tag{10.15}$$

(B) **DC–DC converter design**

The ZVS-PSPWM DC–DC full-bridge converter provides 48–72 V DC for a different range of EV battery charging. The turns ratio N, i.e., (n_s/n_p) is taken as 4.49 to obtain the range of 48–72 V output. i_b and v_b are nominal current and voltage output, respectively, at the rated power. The duty cycle D_f is calculated to have minimum switch stress in the desired voltage range

$$D_f = \frac{0.5 v_b}{(n_s/n_p) v_{dc}} = \frac{0.5 \times 72}{(1/4.49) \times 400} = 0.404 \tag{10.16}$$

The maximum average current of output inductor, L_o, is given as

$$i_{b_{\max}} = \frac{P_{o_{\max}}}{v_{b_{\min}}} = \frac{3.3 \times 10^3}{48} = 69.75A \tag{10.17}$$

The output inductor, L_o, is selected to maintain current ripple within the desired limits (i.e., 3% of average charging current)

$$L_o = \frac{v_{b_{\max}}(0.5 - D)}{\Delta i_b \times f_s} = \frac{72(0.5 - 0.404)}{2.0625 \times 25 \times 10^3} = 0.134\ \mathrm{mH} \tag{10.18}$$

The value of the output capacitor is calculated for the designated ripple in output voltage as,

$$C_f = \frac{i_{b_{\max}}}{8 f_s \times \Delta v_{co}} = \frac{68.75}{8 \times 25 \times 10^3 \times 2.4} = 143.22\ \mu F \tag{10.19}$$

where ⟦Δ*v*⟧*co* is 5% and taken into account as an output ripple voltage.

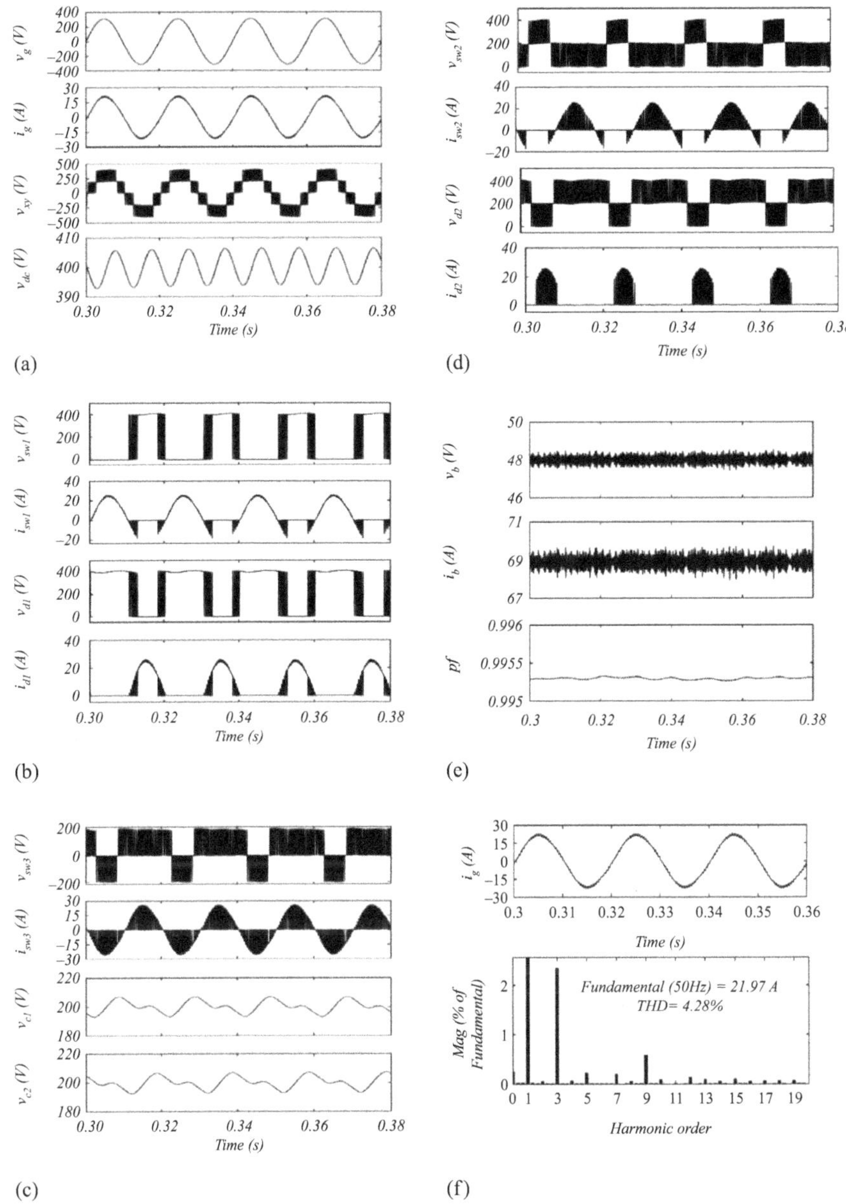

Figure 10.26 *Simulated performance of battery charging at an output voltage of 48V Dc. (a) Grid source voltage* v_{gd}*, source current* i_g*, voltage* v_{xy} *at rectifier input, and dc-link voltage. (b) Voltage and current profile of switches* S_{w1} *and* D_1*. (c) Voltage and current profile of switches* S_{w2} *and* D_2*. (d) Peak voltage, current across switches (*S_{w3}*), and voltage across capacitor (*C_1 *and* C_2*). (e) Charging voltage* v_b*, current* i_b *of battery, and measured pf of the converter. (f) Harmonic spectrum of source current.*

10.4.4 Modeling, simulation, and performance of single-phase multilevel EV charger

Another multilevel configuration shown in Figure 10.27 is a three-phase five-level AC/DC rectifier with an IBC fast EV charger. The level-shifted duty control is used in a three-phase, five-level AC/DC rectifier, which contains 6 active switches ($S_{W1}, S_{W2}, \ldots\ldots$, and S_{W6}), 12 diodes ($D_1, D_2 \ldots$ and D_{12}), and 2 split capacitors (C_1 and C_2) connected through a DC connection. The three-phase converter per phase legs is connected using three boost inductors, designated L_r, L_y, and L_b, and acts as an input filter inductor. This DC link voltage (v_{dc}) feeds to a novel parallel IBC, which is suited for the EV charger applications.

The proposed interleaved buck DC–DC converter with zero resonant current switching is suitable for low-voltage and high-current applications. The circuit can be derived from the standard buck converter where two pairs of switches are connected in parallel for sub-converter I similarly connected in parallel for sub-converter II. All the inductors L_1, L_2, L_3, and L_4 are identical ($L_1 = L_2 = L_3 = L_4 = L_b$). To eliminate harmonics and DC ripple, the carrier signals for switches (S_{W7} and S_{W8}) and (S_{W9} and S_{W10}) are 180° phase shift. The duty cycle for each switch is less than 0.5.

The buck resonant converters are used for resolving the high switching frequency losses, reducing the circuit volume, and controlling the switches with ease. Therefore, they control the output voltage via switching frequency. At the input side, a five-level voltage waveform is generated, thus reducing voltage stress on the switching power devices. The five-level voltage (v_{ab}) at converter input is decided according to AC mains current variation and line voltage. In accordance with the international PQ standard IEC-61000-3-2, the proposed EV charger draws a sinusoidal line current in phase with an input voltage to achieve UPF and THD of grid current. Parallel IBCs are integrated with DC-link voltages of the five-level AC/DC rectifier topology that keeps the battery charging current constant. Each converter's control loop is isolated from the other.

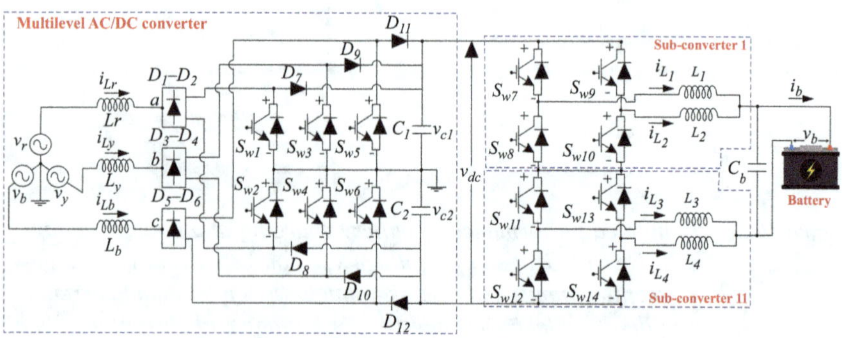

Figure 10.27 Three-phase multilevel AC/DC converter fed with two-phase parallel interleaved DC–DC converter

10.4.5 Operation of three-phase multilevel EV charger

The operation of a three-phase multi-level EV charger may be discussed in two parts: front-end AC/DC converter and interleaved DC–DC converter.

10.4.5.1 Three-phase front-end AC/DC converter

During the positive cycle of the line current, when high-frequency switches S_{w1}, S_{w2}, diode D_1, and diode D_2 conduct, i.e., zero level is generated by a short circuit between points a and b. Following a linear increase in the inductor current i_{L}r, the inductor is fed with grid power. As the demand current is continuous supplied to the battery, the energy stored in the capacitors (C_1 and C_2) discharges.

When switching device S_{w1} is on and S_{w2} is off during the same positive half cycle of the input mains, the energy is transferred from inductor L_r through diodes D_1, D_8, while D_2 is conducting to the output capacitor C_2. The voltage across an inductor L_r is $|v_r| - v_{c2}$. Split capacitors C_1 and C_2 have respective charging and discharging currents of i_{dc} and $i_{Lr} - i_{dc}$.

When switching device S_{w1} is off and S_{w2} is on, diodes D_1, D_7, and D_2 are conducting. The resulting voltage across inductor L_r is $|v_r| - v_{c1}$. Capacitor C_1 charging current is $i_{Lr} - i_{dc}$, and capacitor C_2 discharging current is i_{dc}. In this mode, the DC bus voltage v_{ab} is v_{c1}.

When high-frequency switches S_{w1} and S_{w2} are off, and diodes D_1, D_7, D_8 and D_2 are on, then capacitors C_1 and C_2 are being charged at a rate of $i_{Lr} - i_{dc}$. The voltage across inductor L_r is $|v_r| - v_{dc}$. The charging current of capacitors C_1 and C_2 is $i_{Lr} - i_{dc}$. In this state, voltage v_{ab} at the converter input terminal is $v_{c1} + v_{c2}$. Similarly, leg r has four operational modes for its negative half-cycle.

In comparison with a conventional PWM rectifier with the same switching device voltage rating, the rectifier output voltage is twice as much as the DC bus voltage. Another key benefit of the suggested design is that the DC bus-connected diodes prevent the shoot-through current even while both switches are turned on. As a result, the rectifier has a high degree of reliability, simple to operate, and produces a high output of DC bus voltage.

10.4.5.2 Interleaved buck DC/DC converter

When switches S_{w7} and S_{w10} are energized it increases the leakage inductor current i_{L1}. On the other hand, the inductor L_2 current i_{L2} decreases when S_{W8} and S_{W9} are turned off.

When switches S_{w8} and S_{w9} are turned on and i_{L2} that flows through L_2 increases, i_{L1} decreases. When switches S_{w8} and S_{w9} are activated, current i_{L2} flowing via inductor L_2 rises while i_{L1} falls.

With the activation of S_{w7} and S_{w8}, the currents $I_{\mathrm{L}1}$ and $I_{\mathrm{L}2}$ start to rise as a consequence of the increased load changes. The inductive current rises linearly as v_{d}c stores energy in L_1 and L_2.

The load variation is decreased, current i_{L1} and i_{L2} start decreasing. S_{w7} and S_{w9} are not functional. In contrast, S_{w8} and S_{w10} are enabled.

All the working modes of sub-converter II are identical as sub-converter I. The outputs of both the converter are parallelly connected, which increase the charging current and maintain the output voltage constant.

10.4.6 Hardware parameter design

The EV charger design details of inductor selection and output capacitor selection are outlined here for an output of 200–400 V wide-range DC voltage.

10.4.6.1 Three-level converter design

The input inductor (L_s), filters out the input mains ripple current, expressed as,..

The values of DC-link split capacitors, C_1 and C_2, are calculated for permissible ripple in DC-link voltage of 5%,

$$C_1, C_2 = \frac{P_o}{4 \times f_c \times \{v_{dc}{}^2 - (v_{dc}{}^2 - \Delta v^2)\}} = 826.44\ \mu F \tag{10.20}$$

10.4.6.2 Two-phase IBC

The two-phase IBC provides 200–400 V DC for different ranges of EV battery charging for an input DC voltage 1,100 V:

$$\text{Duty ratio}(D) = \frac{v_b}{v_c} = 0.181 \text{ to } 0.363 \tag{10.21}$$

$$\text{Output current } (i_b) = \frac{P_o}{v_b} = 250\ A_{\max} \tag{10.22}$$

$$\text{Inductor current } i_{L1} = i_{L2} = i_{L3} = i_{L4} = \frac{i_b}{2} = 125A \tag{10.23}$$

$$\Delta i_{L1} = \Delta i_{L2} = \Delta i_{L3} = \Delta i_{L4} = \left(\frac{i_b}{2}\right) \times 5\% = 6.25A \tag{10.24}$$

$$\text{Value of inductor } L_1 = L_2 = L_3 = L_4 = \left(\frac{v_{dc} - v_{o(\min)}}{\Delta I_L \times n \times f_s}\right) \times D = 1.045\ mH \tag{10.25}$$

The value of the output capacitor is calculated for the designated ripple in the output voltage as,

$$C_b = \frac{i_{b(\max)}}{8 \times f_s \times \Delta v_{cb}} = \frac{250}{8 \times 25 \times 10^3 \times 6} = 208.33\ \mu F \tag{10.26}$$

10.4.7 Performance simulation of three-phase multilevel EV charger

The MATLAB® models of a three-phase multilevel EV charger is developed using Simulink® and Sim Power Systems (SPS) toolboxes to simulate the performance of this EV charger.

Figure 10.28(a–d) shows the simulated performance of a multilevel AC/DC converter fed two-phase parallel interleaved DC–DC converter for fast EV charger

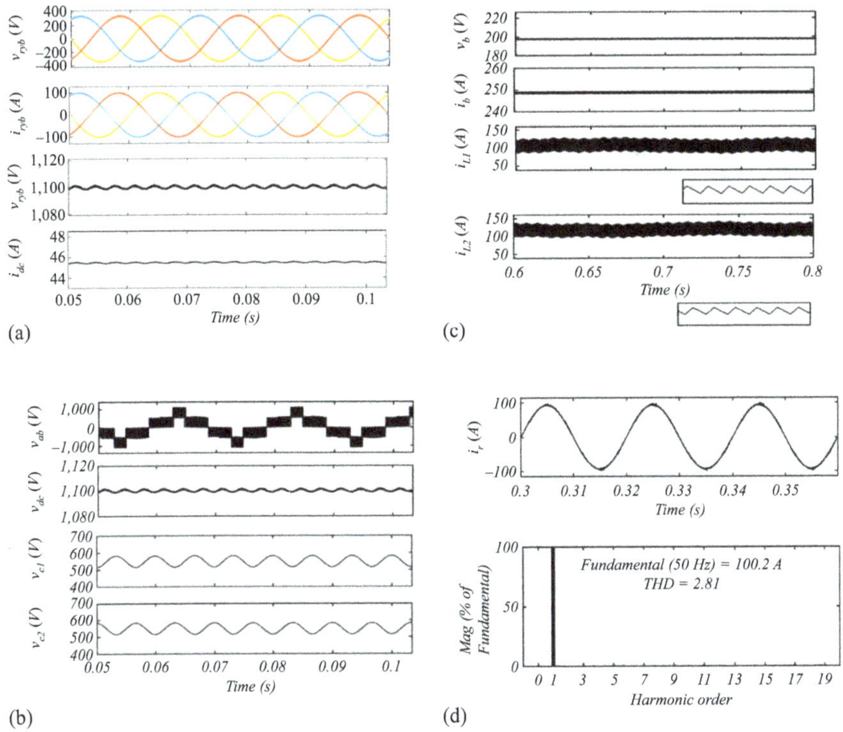

Figure 10.28 *Simulated performance of multilevel AC/DC converter-fed two-phase parallel interleaved DC–DC converter for a fast EV charger at an output voltage of 200 V dc. (a) Grid source voltage* v_{ryb}*, source current* i_{ryb}*, dc-link voltage, and dc-link current; (b) voltage* v_{xy} *at rectifier input, dc-link voltage, and voltage across capacitor (*C_1 *and* C_2*); (c) charging voltage* v_b*, current* i_b *of battery, inductor current of DC–DC converter (*L_1 *and* L_2*); (d) harmonic spectrum of* r *phase.*

operating 230 V phase to neutral at rated load condition with an output of $200V_{dc}$. As shown in these figures, the DC-ink voltage (v_{dc}) is $1,100V_{dc}$ and voltage across split capacitor (C_1 and C_2) is $550V_{dc}$. The drawn line current by the five-level AC–DC rectifier is sinusoidal and in phase with the input voltage to attain the UPF. However, peaky current from the supply system (i_{ryb}) is drawn, which does not deteriorate the PQ performance of the fast EV charger. The THD of input current is 2.57%, which satisfies the international PQ standard IEC-61000-3-2 [13].

The performance realization is deemed satisfactory and complies with the PQ criteria. The suggested rectifier with three-voltage levels also minimizes electromagnetic interference, harmonic components, voltage ratings, and device voltage and current stress when compared to standard PWM converters.

10.5 Summary

The purpose of this chapter is to present the PQ improvement aspects in a battery charger used for EV applications. Featuring a near unity input PF and low THD on the line, the input stage can have different converter-based PF correctors. The converters can be single-phase-controlled and -uncontrolled converters followed by DC–DC converters or multi-pulse converters or multi-level converters. The control strategy can be used for bidirectional power flow in these converters and a special DAB converter for V2G or G2V operation from a solar or any renewable energy-based charger. Such chargers are considered as green chargers. To improve efficiency and to increase switching frequency, a zero-voltage isolated full bridge/ interleaved DC/DC converter or zero voltage or zero current switching schemes can also be employed. It can be summarized that the PQ improvement is an essential requirement for any kind of chargers and it is implementable in many ways, as demonstrated in this chapter but not limited to these methods only. The readers may have various ideas on the basis of the presented concepts.

Symbols

L_{p2} and L_{n2}	output inductors
L_{p1} and L_{n1}	input inductors
v_{in}	input AC voltage
v_{ac_pk}	peak input voltage
	supply frequency of input mains
D	duty ratio
V_{dc}	DC-link voltage
C_{dc}	DC-link capacitance
L_{dc}	filter inductor
f_r	resonant frequency
I_l	inductor current
P_o	output power

Glossary

AC	alternating current
APF	active power filters
BEV	battery electric vehicles
BL	bridgeless
CC	constant current
CCM	continuous conduction mode
CS	charging station
CV	constant voltage

DAB	dual active bridge
DBR	diode bridge rectifier
DC	direct current
DCM	discontinuous current conduction mode
EV	electric vehicle
G2V	grid-to-vehicle
HEV	hybrid electric vehicle
HPF	high-pass filter
HFT	high-frequency transformer
IBC	interleaved buck converter
IEEE	Institute of Electrical and Electronics Engineers
MLI	multi-level inverter
MPPT	maximum power point tracking
PF	power factor
PFC	power factor correction
PHEVs	plug-in hybrid electric vehicles
P&O	perturb and observe
PQ	power quality
PSFBC	phase-shifted full-bridge converter
PV	photo-voltaic
PWM	pulse-width-modulation
SEPIC	single-ended primary inductance converter
THD	total harmonic distortion
TLB	three-level boost
UPF	unity power factor
V2G	vehicle-to-grid
VSC	voltage–source converter
ZVS	zero-voltage switching

References

[1] O. C. Onar, J. Kobayashi, and A. Khaligh, “A fully directional universal power electronic interface for EV, HEV, and PHEV applications,” *IEEE Trans. Power Electron.*, vol. 28, no. 12, pp. 5489–5498, 2013.

[2] R. Dwivedi, “Investigations on fast charging converter for electric vehicle application,” PhD Thesis, SLIET Longowal, India, May 2023.

[3] Ned Mohan, T. M. Undeland, and W. P. Robbins, *Power Electronics: Converters Applications and Design*, 3rd ed. *(An Indian Adaptation)*, Wiley, New York, NY, 2022.

[4] O. García, J. A. Cobos, R. Prieto, P. Alou, and J. Uceda, "Single phase power factor correction: a survey," *IEEE Trans. Power Electron.*, vol. 18, no. 3, pp. 749–755, 2003.

[5] C. Liu, B. Gu, J. S. Lai, *et al.*, "High-efficiency hybrid full-bridge-half-bridge converter with shared ZVS lagging leg and dual outputs in series," *IEEE Trans. Power Electron.*, vol. 28, no. 2, pp. 849–861, 2013.

[6] M. Yilmaz and P. T. Krein, "Review of battery charger topologies, charging power levels, and infrastructure for plug-in electric and hybrid vehicles," *IEEE Trans. Power Electron.*, vol. 28, no. 5, pp. 2151–2169, 2013.

[7] M. Srivastava, P. S. Tomar, and A. K. Verma, "Emphasis on switch selection and its switching loss comparison for on-board electric vehicle charger," *IET Power Electron.*, vol. 12, no. 6, pp. 1385–1392, 2019.

[8] B. Singh, S. Gairola, B. N. Singh, A. Chandra and K. Al-Haddad, "Multipulse AC-DC converters for improving power quality: a review," *IEEE Trans. Power Electron.*, vol. 23, no. 1, pp. 260–281, 2008.

[9] D. A. Paice, *Power Electronic Converter Harmonics: Multipulse Methods For Clean Power*, IEEE Press, UK, 1996.

[10] R. C. Dugan, M. F. McGranaghan, and H. W. Beaty, *Electric Power Systems Quality*, McGraw Hill, New York, NY, 1996.

[11] P. S. Jamwal, "Multilevel inverter fed induction motor drive for battery electric vehicle," *PhD Thesis, SLIET Longowal*, India, July 2023.

[12] I. Subotic, N. Bodo, and E. Levi, "Single-phase on-board integrated battery chargers for EVs based on multiphase machines," *IEEE Trans. Power Electron.*, vol. 31, no. 9, pp. 6511–6523, 2016.

[13] "IEC Standard 61000-3-2. Limits for Harmonic Current Emissions," 2018.

[14] N. Bianchi and M. Dai Pre, "Active power filter control using neural network technologies," *IEE Proc. – Electric Power Appl.*, vol. 150, no. 2, pp. 139–145, 2003.

[15] B. Singh, S. Singh, A. Chandra, and K. Al-Haddad, "Comprehensive study of single-phase AC-DC power factor corrected converters with high-frequency isolation," *IEEE Trans. Ind. Informatics*, vol. 7, no. 4, pp. 540–556, 2011.

Chapter 11

Wireless power transfer for electric vehicle

Sumit Pramanick[1] and Anandarup Das[1]

11.1 Introduction

The adoption of electric vehicle (EV) has been steadily increasing throughout the world and is expected to continue in the future. More and more countries, including developing countries, are adopting this technology at a fast rate. One of the challenges of this process is, however, building the necessary charging infrastructure for charging the EVs. This includes both adequate number of charging points in a geographical region and simultaneous fast charging options. It may not be possible to achieve both these objectives given the economic constraints in a country, availability of a strong power grid, and large area of land available. For example, there is an ever-increasing demand for the extension of a range of batteries from a single-charge demanding higher size of batteries and initial cost.

Wireless power charging can play an important role in this scenario. Instead of relying solely on development of better and compact battery technology, an alternative in the form of wireless technology is gaining attention. In this technique, batteries can be charged by electromagnetic field in air with the vehicle stationary or in moving condition [1–3]. It can be observed that many low-power devices like smartphones and tablets are already being charged wirelessly; it is expected that wireless power transfer in EVs will also play a significant role in near future.

A wireless power transfer system consists of a transmitter and a receiver. The transmitter transmits electric power and is usually connected to the electrical grid. A receiver receives the energy through an intermediate medium, usually air, and delivers to the load. In case of EVs, the transmitter is often located buried inside the road. The receiver is an electric circuit situated inside the EV and connected to the electric motor and the onboard battery. Wireless power transfer can take place while the vehicle is stationary or moving [4]. In the case of stationary vehicles parked in a spot, wireless power transfer is very similar to wired electric charging except that no physical cable needs to be connected to the vehicles. The characteristic difference in wireless charging is the ability to charge a vehicle on the go. With an adequate number of transmitters present inside the road, it may be possible

[1]Department of Electrical Engineering, IIT Delhi, India

to continuously supply energy to the vehicle putting less requirement of battery in the vehicle.

11.1.1 Wired charging and its challenges

Presently, most of the electric and hybrid EV in the world are charged through a wire. Often these vehicles (cars and buses) contain a large onboard rechargeable battery. Although battery technology has been steadily developing over the years, but so is the increasing demand for these batteries in terms of size and delivering capacity. Till a few years ago, overnight charging of the EV batteries was an expected requirement. However, all modern manufacturers stress on fast charging of EV batteries where the state of the charge (SOC) of the battery is expected to increase up to 80% within 10–15 min of charging.

The ever-increasing requirement from the batteries puts additional stress on the battery size, cost, charging time, and lifetime of the batteries. Additionally, the charging infrastructure from the power grid needs to be adequately developed for rapid charging of hundreds of EV in a region.

11.1.2 Potential gain with wireless charging

With a wireless charging infrastructure in place, the demand for the onboard battery can be drastically reduced. There is a fundamental limit to the extent of battery cost and size reduction. In addition, the total amount of Lithium that can be extracted is also finite. With wireless charging, such challenges are less severe.

With wireless charging on the go, charging time can be reduced and the anxiety to find a nearby location for charging can be substantially alleviated. In addition, safety and convenience of charging while the vehicle is in the motion can be attractive in an all-weather condition. In this way, wireless charging can be an additional boost in adopting EVs on a mass scale.

11.1.3 Challenges with WPT

There are several technical challenges that WPT is currently facing for widespread adoption. The efficiency of wireless power transfer is lesser than that of conductive power transfer. Typically, the efficiency is of the order of 80–85% in wireless power transfer where there is a scope for further improvement.

Environment often plays a crucial role in the efficiency of wireless power transfer. Operating temperature, humidity, weather (snowfall, rain, etc.) can alter the efficiency of power transfer. Note that the transmitter coils are buried inside the road, which can be substantially heated up in a hot weather scenario.

EMI regulations also play an important role while designing WPT system. Wireless power transfer involves energy transfer through magnetic and electric fields distributed in the air gap near the road, in close proximity to pedestrians. One important safety standard has been widely adopted in WPT called the International Commission on Non-Ionizing Radiation Protection (ICNIRP) guideline. In this guideline, safe values of magnetic field, electric field, and EM waves have been suggested to follow.

Wireless power transfer also involves substantial amount of initial investment. This involves building the road with transmitter coils positioned in an accurate fashion. It must be energized through a cable from the nearby substation which supplies an embedded inverter. The arrangement must be adequately protected against vibration and shock from heavy vehicles on road top. Note that, to improve efficiency it is desirable to place the transmitter coils as close as possible to the road surface. All this makes the initial cost of investment in wireless power manifold; however, over a span of 25–30 years, it has been shown that wireless power transfer can become substantially profitable.

11.1.4 Capacitive wireless charging

There are two types of wireless charging techniques: inductive type based on magnetic induction principle and capacitive type based on electrostatic principle. In inductive type, the energy exchange between the primary and the secondary takes place through magnetic field coupling them. In the capacitive type, the energy exchange takes place via the electric field that is established between them. The two techniques can be compared in terms of power level, air gap length, cost, or size of the equipment.

Over the years, however, inductive power transfer has dominated over capacitive power transfer technology. It has been found that for low-power range over short distances, capacitive power transfer is attractive. For longer distances with larger power transfer magnitude, inductive power transfer is more beneficial.

11.2 Inductive wireless power transfer

An inductive wireless power transfer (IWPT) system can be visualized as a power transfer system through an air-core transformer. Unlike a transformer, the coils in IWPT system are spatially separated as shown in Figure 11.1. A controlled high-frequency AC supply is fed to the transmitter coil (Tx) which generates magnetic flux in the air-gap. The magnetic flux lines generated by the Tx coil are linked through the second coil which is called the receiver coil (Rx). Due to the time-varying nature of the magnetic flux linked in the receiver coil, an emf is induced across the Rx coil as per Faraday's law. Thus, the Tx and Rx coils are magnetically coupled through the magnetic flux lines in the air-gap. The AC voltage induced across the Rx coil is rectified and fed to the DC output terminals where battery is connected. This magnetic coupling between the Tx and Rx coils through the air-gap enables transfer of electrical energy from the Tx to the Rx coil.

Designing an IWPT system requires quantification of the energy transfer between the magnetically coupled coils. Further, the evaluation of the performance of the IWPT system under various operating conditions would also require proper understanding of magnetically coupled electrical circuits connected to the coils. In Section 11.2.1 a methodology is explained to model the magnetically coupled coil systems in IWPT systems. The influence of the air-gap and the magnetic flux linkage between the coils on the electrical parameters of the coils has been

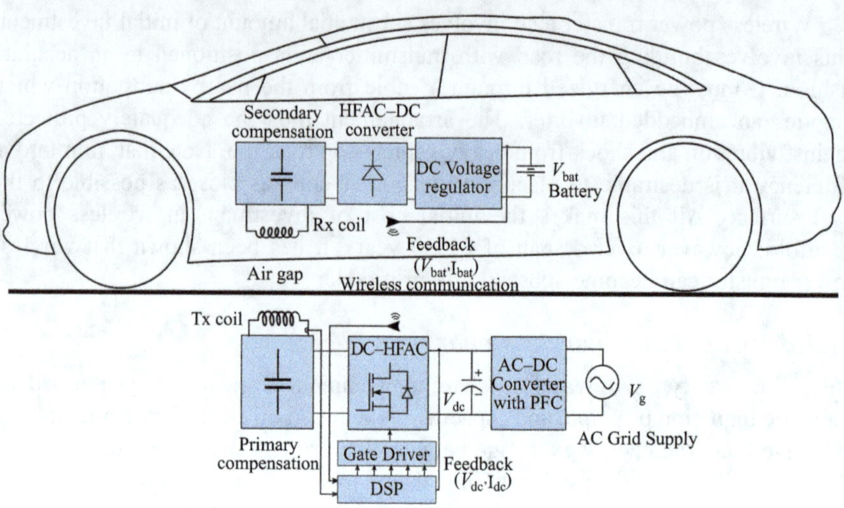

Figure 11.1 IWPT system in an EV battery charging system

explained. Thus, an equivalent electrical circuit can be derived which can be analyzed to quantify the energy transfer from Tx to Rx coils.

The magnetic flux is generated from the Tx coil when it is connected to an AC voltage source. The AC voltage source is controlled using a DC–AC power electronic converter. The current drawn by the coil from the AC source is dependent on the impedance of the coil system at any particular operating condition. To reduce the conduction losses in the coils and the power electronic converter at a particular level of power transfer, it has to be ensured that the impedance offered by the electrical system is such that minimum current is drawn from the controlled AC voltage source. In Section 11.2.2, some compensation circuits have been explained which are used in IWPT systems to ensure the above-mentioned condition.

The electrical model of the compensated coil system has been analyzed in Section 11.2.3 to derive the expressions for power transfer and efficiency of the power transfer through the coils. Further, some criteria have been derived to understand the conditions at which maximum power transfer and efficiency can be achieved through the coils. To ensure these conditions, the control of the power electronic converter to generate the AC supply at Tx coil is explained in Section 11.2.4. The control algorithm is explained by the help of MATLAB®/ Simulink® simulation in this section.

11.2.1 Modeling of coils

The coupled coil model representing the transmitter (Tx) and receiver (Rx) coils is shown in Figure 11.2. The current source connected to the Tx coil (i_p) represents the AC current being supplied to the coil. A voltage v_p is induced across the Tx coil due to flux linking through it. Similarly, the current source (i_s) connected to the Rx

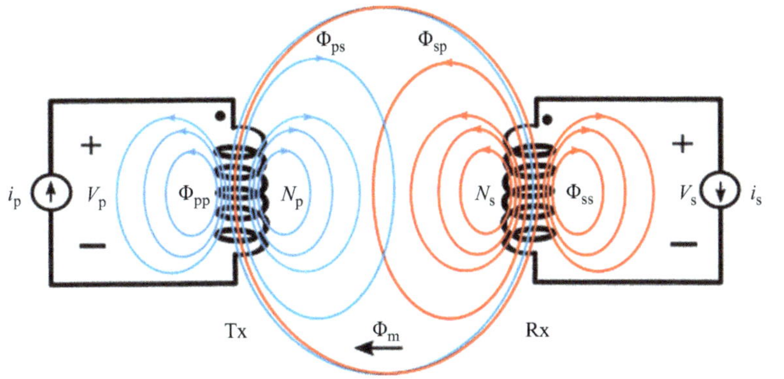

Figure 11.2 Flux linkage between magnetically coupled coils through air-gap

coil represents the load current flowing through the coil when a load is connected across the Rx coil. The voltage induced in the Rx coil is v_s In case of the EV battery charging system, the load connected across the Rx coil is the compensated AC–DC converter connected EV battery. The coupled coils are considered to be asymmetric with an unequal number of turns. The Tx and Rx coils are analyzed with N_p and N_s turns, respectively. The ratio of the number of Tx turns to Rx turns is represented by n which is defined in (11.1):

$$n = \frac{N_p}{N_s} \tag{11.1}$$

Applying Ampere's law over a closed surface around coil with N turns, with current i being flowing through it, the magnetic field intensity $(\vec{H})$ can be determined with (11.2). If the closed surface is through air, the magnetic flux density $(\vec{B})$ can be determined as per (11.2). The magnetic flux density is a measure of strength of the magnetic field around the coil. It indicates the number of flux lines per unit area perpendicular to the direction of the flow of flux lines:

$$\begin{aligned} &\oint \overrightarrow{H} \cdot \overrightarrow{dl} = Ni \\ &\overrightarrow{B} = \mu_0 \overrightarrow{H} \end{aligned} \tag{11.2}$$

It can be observed from (11.2) that the magnetic field generated through the center of a current-carrying coil depends on the product of the number of turns in the coil and the current flowing through the coil. This product (Ni) is called magnetomotive force (MMF), causing magnetic field to develop through current-carrying coil as shown in Figure 11.2. The magnetic flux (ϕ) is determined by (11.3), where $\mathbb{R}$ is the reluctance offered by the medium along the flux flow path. Similar to resistance in electrical domain which obstructs the current flow through a path, $\mathbb{R}$ offers resistance to magnetic flux in magnetic domain. It can be observed from (11.3) that $\phi \propto i$. The total flux lines linked through the coil (λ) is total flux lines linked by all the turns in the coil and is expressed as shown in (11.3):

$$\phi = \frac{Ni}{\mathbb{R}} \qquad \lambda = N\phi \tag{11.3}$$

Observing (11.2) and (11.3), it can be concluded that due to flow of currents through the coils in IWPT system flux lines are generated from Tx and Rx coils as shown in Figure 11.2. Due to flow of current i_p through Tx coil, some flux lines are linked only by Tx coil itself which is indicated as ϕ_{pp}, while some of the flux lines are linked with both Tx and Rx coils which are indicated as ϕ_{ps}. Similarly, due to flow of current i_s through the Rx coil, some flux lines get linked only with Rx coils which is indicated as ϕ_{ss}, while some flux lines are linked with both Tx and Rx coils which are indicated as ϕ_{sp}. Using (11.3), the total flux linkages in each coil are expressed in (11.4). For developing the model of the IWPT coils, it is assumed that ϕ_{ps} and ϕ_{sp} opposes each other. Expressions in (11.4) are written considering the convention that current entering the dot terminal of the coil is considered positive. From (11.3), it has been already observed that $\phi \propto i$, (11.4) has been expressed accordingly:

$$\begin{aligned} \lambda_p &= N_p\phi_{pp} + N_p\phi_{ps} - N_p\phi_{sp} \\ \lambda_s &= -N_s\phi_{ss} - N_s\phi_{sp} + N_s\phi_{ps} \end{aligned} \tag{11.4}$$

Faraday's law states that the voltage induced across a coil is equal to the rate of change of total flux linkage through the coil. Since, $\phi \propto i$ the induced voltage can be shown to be proportional to the rate of change of current through the coil, where the proportionality constant (L) shown in (11.5) is called *inductance*:

$$\begin{aligned} v = \frac{d\lambda}{dt} = N\frac{d\phi}{dt} = \frac{N^2}{\mathbb{R}}\frac{di}{dt} = L\frac{di}{dt} \\ L = \frac{N^2}{\mathbb{R}} \end{aligned} \tag{11.5}$$

From (11.4) and (11.5), the voltage induced across the Tx and Rx coils can be expressed as (11.6). Where $\mathbb{R}_{pp}$ and $\mathbb{R}_{ss}$ are equivalent reluctance to the flux paths of ϕ_{pp} and ϕ_{ss}, respectively, and $\mathbb{R}_{ps}$ and $\mathbb{R}_{sp}$ are equivalent reluctance to the flux paths of ϕ_{ps} and ϕ_{sp}, respectively. Since the magnetic path length for the flux lines linking both Tx and Rx coils are the same $\mathbb{R}_{ps} = \mathbb{R}_{sp}$:

$$\begin{aligned} v_p &= \frac{N_p^2}{\mathbb{R}_{pp}}\frac{di_p}{dt} + \frac{N_p^2}{\mathbb{R}_{ps}}\frac{di_p}{dt} - \frac{N_pN_s}{\mathbb{R}_{sp}}\frac{di_s}{dt} \\ v_s &= -\frac{N_s^2}{\mathbb{R}_{ss}}\frac{di_s}{dt} - \frac{N_s^2}{\mathbb{R}_{sp}}\frac{di_s}{dt} + \frac{N_sN_p}{\mathbb{R}_{ps}}\frac{di_p}{dt} \end{aligned} \tag{11.6}$$

Considering (11.5), the voltage equations in (11.6) can be expressed in terms of inductance and rate of change in current as shown in (11.7). $L_{lk,p}$ and $L_{lk,s}$ are leakage inductance of the Tx and Rx coil, respectively. Leakage inductance terms contribute to the voltage induced across the coils due to the flux linked exclusively by each coil (ϕ_{pp} and ϕ_{ss} in Figure 11.2) due to the current flowing through it. $L_{m,p}$ and $L_{m,s}$ are the magnetizing inductance of Tx and Rx coils, respectively. The magnetizing inductance terms contribute to the voltage induced across each coil due to the flux linked through the magnetizing path (ϕ_{ps} and ϕ_{sp} in Figure 11.2) between the two coils generated by the corresponding coil current. M is the mutual inductance, which contributes to the voltage induced in each coil due to the flux linked by it which is generated by the other coil current:

$$\begin{aligned} v_p &= L_{lk,p}\frac{di_p}{dt} + L_{m,p}\frac{di_p}{dt} - M\frac{di_s}{dt} \\ v_s &= -L_{lk,s}\frac{di_s}{dt} - L_{m,s}\frac{di_s}{dt} + M\frac{di_p}{dt} \end{aligned} \tag{11.7}$$

The instantaneous voltage equations in (11.7) capture the instantaneous behavior of the coupled coils of an IWPT system, neglecting the resistance of the coils. These equations can be used to realize the equivalent circuit model for the IWPT coil system in Figure 11.3. L_p and L_s are the self-inductance of Tx and Rx coils, respectively. Self-inductance of a coil is the sum of the leakage inductance and magnetizing inductance of a coil (e.g., $L_p = L_{lk,p}+L_{m,p}$). The model shown in Figure 11.3(a) shows the instantaneous equivalent circuit of the coupled coils; the dots above the variables indicate first derivative of a time varying quantity. If the *RMS* of the first harmonic of the voltage and current are considered, the instantaneous model shown in Figure 11.3(a) can be redrawn as shown in Figure 11.3(b).

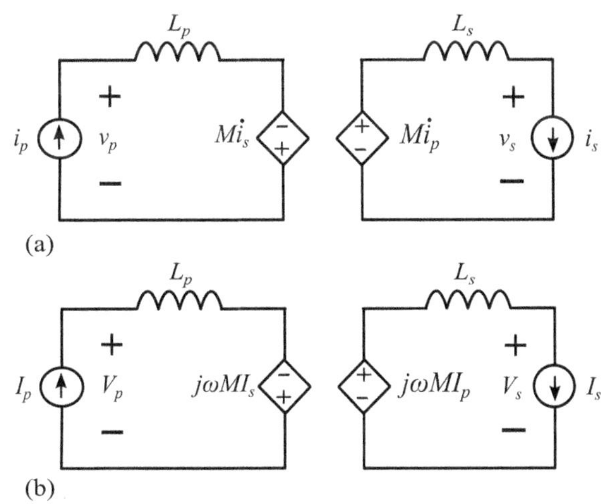

Figure 11.3 Equivalent circuit for the IWPT coil system. (a) Instantaneous model and (b) fundamental RMS equivalent model.

ω in Figure 11.3(b) is the first harmonic frequency of the voltage and current in the circuit.

The dependent voltage sources in the equivalent circuits shown in Figure 11.3 are similar to the reflected voltages from primary to secondary windings and vice versa in transformers, hence, this model is also called *transformer equivalent circuit.* Since the mutual inductance between the Tx and Rx coils is the main coupling factor for the dependent voltage sources, this model is also referred to as the *M-model.*

The inductance terms in the voltage equations in (11.7) are expressed in terms of the number of turns and reactance in (11.6). Observing the expressions in inductance terms, the inductance for the Tx and Rx coils are related to each other by (11.8):

$$L_{m,p} = n^2 L_{m,s} = nM \tag{11.8}$$

Considering the relationship between the inductance of Tx and Rx coils, and the transformation of Rx side current i_s to the Tx side, the instantaneous *M-model* in Figure 11.3(a) can be realized as an equivalent transformer model as shown in Figure 11.4.

For any coupled coil system, the magnetic coupling between the coils is quantified by a parameter called *coupling coefficient* (k). It is defined by the open-circuit voltage gain of the coil system. When the coil system is magnetized from the Tx side, the open circuit voltage gain is defined as the ratio between the Rx coil to Tx coil voltages, with no-load connected across the Rx coil. This ratio can be obtained by solving the voltage equations for the equivalent circuit shown in Figure 11.4 with $i_s=0$ as shown in (11.9):

$$\left.\frac{v_s}{v_p}\right|_{i_s=0} = \frac{L_{m,p}}{L_{lk,p} + L_{m,p}} \frac{1}{n} \tag{11.9}$$

Similarly, when the coil system is magnetized from the Rx side, the open-circuit voltage gain is defined as the ratio between the Tx side voltage to the Rx side voltage with Tx side current to be zero (11.10):

$$\left.\frac{v_p}{v_s}\right|_{i_p=0} = n \frac{\frac{L_{m,p}}{n^2}}{L_{lk,p} + \frac{L_{m,p}}{n^2}} \tag{11.10}$$

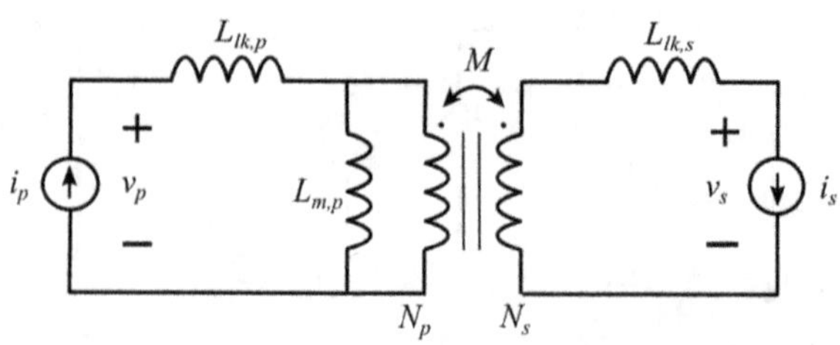

Figure 11.4 M-model equivalent circuit with Tx side transformation

From (11.9) and (11.10), the coupling coefficient can be expressed as per the following equations:

$$\kappa^2 = \left.\frac{v_s}{v_p}\right|_{i_s=0} \times \left.\frac{v_p}{v_s}\right|_{i_p=0} = \frac{L_{m,p}}{L_{lk,p}+L_{m,p}}\frac{1}{n} \times n\frac{\frac{L_{m,p}}{n^2}}{L_{lk,p}+\frac{L_{m,p}}{n^2}}$$

$$\Rightarrow \kappa^2 = \frac{\frac{L_{m,p}}{n^2}}{\left(L_{lk,p}+L_{m,p}\right)\left(L_{lk,s}+L_{m,s}\right)} \tag{11.11}$$

$$\Rightarrow \kappa = \frac{M}{\sqrt{L_pL_s}}$$

In the case of transformer, the primary and secondary coils are coupled through a core made of ferromagnetic material. Hence, the leakage inductance of coils is negligible compared to the magnetizing inductance ($L_{lk,p} << L_{m,p}$ and $L_{lk,s} << L_{m,s}$). Considering this approximation in (11.11) for transformers, coupling coefficient ≈ 1. Unlike transformers, in the IWPT system, the leakage inductance is substantial compared to the magnetizing inductance due the large air-gap between the Tx and Rx coils. Hence, the coupling coefficient is lesser than unity. Higher value of coupling coefficient indicates good coupling between the coils in IWPT system which affects the power transfer between the coils. This will be discussed in further sections of this chapter.

11.2.2 Compensation networks

In an IWPT system, the supply voltage to the transmitter coil is supplied by a DC–AC converter. Similarly, in the receiver side, an AC–DC conversion stage is used to regulate the output DC voltage or current to the DC load. The volt–ampere (VA) rating of these power conversion stages is dependent on the impedance of the network thus formed. At a specific power level, the VA rating of these power conversion stages can be maximized if the impedance connected to these converters can be minimized. Compensation networks are connected to the transmitter and receiver coils to minimize the impedances. These compensation networks are realized with resonant circuits which can be tuned to have minimal impedance at a specific resonating frequency. The transmitter-side power converter can be controlled to generate AC voltage supply (V_P) with the resonant frequency determined by the compensation network. The RMS equivalent circuit of the IWPT system is shown in Figure 11.5. The model includes the transmitter coil resistance R_p and receiver coil resistances R_s. In this section, the output impedance connected to the receiver coil Z_s, reflected receiver side impedance in transmitter side Z_{sr}, and input impedance in the transmitter side Z_p will be analyzed to study the effect of different compensation networks. The receiver coil will be connected to the load through a AC–DC converter. The load is considered to be resistive load R_L, and the equivalent load resistance as seen by the receiver coil is expressed as R_{eq} as shown in Figure 11.5. R_{eq} is dependent on the switching function of the AC–DC converter and R_L.

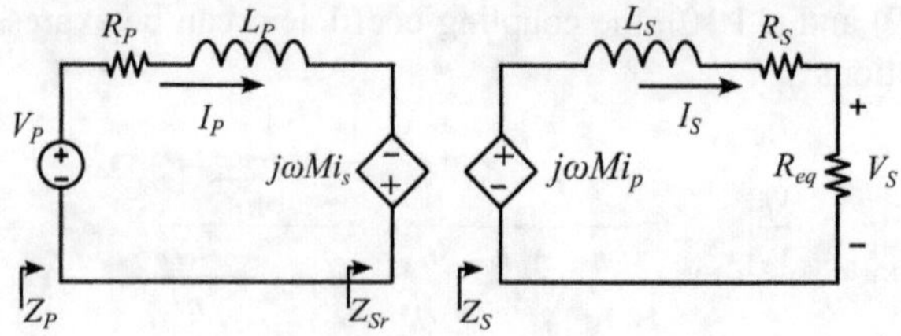

Figure 11.5 RMS equivalent circuit for the IWPT system with coil resistances. Transmitter side input impedance, output impedance of receiver coil, and reflected receiver side output impedance in transmitter side are indicated as Z_p, Z_s, *and* Z_{sr}, *respectively.*

By solving the voltage equation in the receiver side, circuit output impedance of the receiver coil is expressed in (11.12). A relation between the transmitter and receiver coil currents is derived in (11.13):

$$Z_s = (R_s + R_{eq}) + j\omega L_s \tag{11.12}$$

$$j\omega M I_p = Z_s I_s \tag{11.13}$$

Using the relation between the receiver and the transmitter coil currents I_s and I_p from (11.13) in the transmitter side voltage loop, the transmitter side input impedance is derived in (11.14). It can be observed that Z_p expression includes a term dependent on the receiver side parameters. This is the receiver side impedance reflected in the transmitter side circuit. This is expressed by the term $Z_{sr} = \frac{\omega^2 M^2}{R_s + R_{eq} + j\omega L_s}$

$$\begin{aligned} Z_p &= (R_p + j\omega L_p) + \frac{\omega^2 M^2}{Z_s} \\ Z_p &= (R_p + j\omega L_p) + Z_{sr} \end{aligned} \tag{11.14}$$

By introducing appropriate compensation to the equivalent reactances in the transmitter and receiver circuits, the *Q-factor* of the corresponding circuits can be made to zero, thus ensuring zero-stored energy in the circuits which translates to effective power transfer in the IWPT system. This reduces the VA rating of the converters to be connected in the transmitter and receiver coils. The simplest way to achieve $Q=0$ in the transmitter and receiver circuits is to connect series capacitors as shown in Figure 11.6. The transmitter and receiver side series connected capacitors (C_p and C_s, respectively) form resonating circuits by compensating for the self-inductances of transmitter and receiver coils, respectively. Thus, by considering relations $\omega_r L_p = 1/\omega_r C_p$ and $\omega_r L_s = 1/\omega_r C_s$, the transmitter and receiver side series capacitors can be sized for a specific design of the coils where ω_r is the resonant frequency of the circuits. This type of compensation network in the IWPT system is called as *SS* compensation. Observing the nature of impedances in (11.15) and the equivalent circuit in Figure 11.6 of an *SS*-compensated IWPT system, it can be concluded that the power factor of the input supply to the transmitter side will be

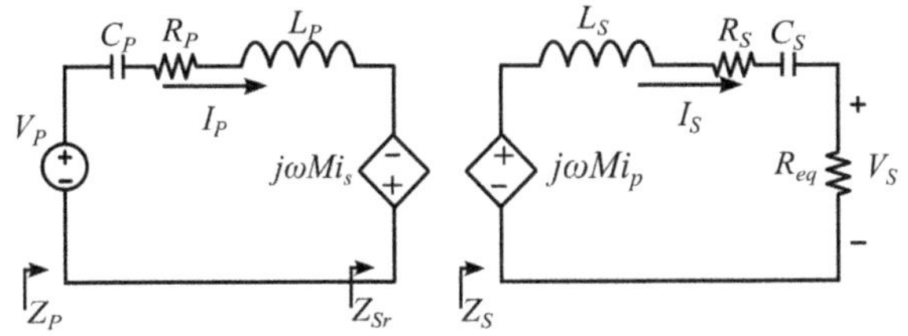

Figure 11.6 Equivalent circuit of the IWPT system with series-connected capacitor compensation in transmitter and receiver side circuits (SS)

unity, and the receiver coil current will be phase shifted by $\frac{\pi}{2}$ from the transmitter coil current. Using (11.12) and (11.13) for an *SS*-compensated IWPT system Z_p, Z_{sr}, and Z_s can be expressed as shown in (11.15). It can be observed from (11.15) that the impedances of the *SS*-compensated IWPT system are purely resistive since all the reactance terms are compensated. *SS* compensation networks are extensively used in IPWPT systems due to the simplicity in designing the resonant circuit to achieve low *Q-factor* in the circuits. In Sections 11.2.3 and 11.2.4, *SS*-compensated IWPT system will be considered for further analysis:

$$\begin{aligned} Z_p &= R_p + \frac{\omega_r^2 M^2}{R_s + R_{eq}} \\ Z_{sr} &= \frac{\omega_r^2 M^2}{R_s + R_{eq}} \\ Z_s &= R_s + R_{eq} \end{aligned} \tag{11.15}$$

The compensation networks can also be realized by series and parallel connection of passive elements to the coils in the IWPT system [5]. Some of the possible compensation networks in IWPT system are shown in Figure 11.7. The passive elements in the several compensation networks shown in Figure 11.7 can be sized by ensuring low *Q-factor* in the resonating circuits connected to the coils. The impedances of the IWPT systems thus formed can be obtained by solving the resulting compensated equivalent circuits.

11.2.3 Power transfer and efficiency

For a generic IWPT system with compensation networks explained in Section 11.2.2, the input power (P_{in}) to the transmitter coil can be determined by calculating the real power drawn from the input supply V_p. Similarly, the output power (P_{out}) transferred to the equivalent reflected load (R_{eq}) can be determined by calculating the real power dissipated through R_{eq}. These real powers can be calculated by analyzing the real components of the currents flowing through the transmitter and receiver circuits. In this section the analysis for the power transferred to the output and efficiency of the IWPT system will be conducted with *SS*

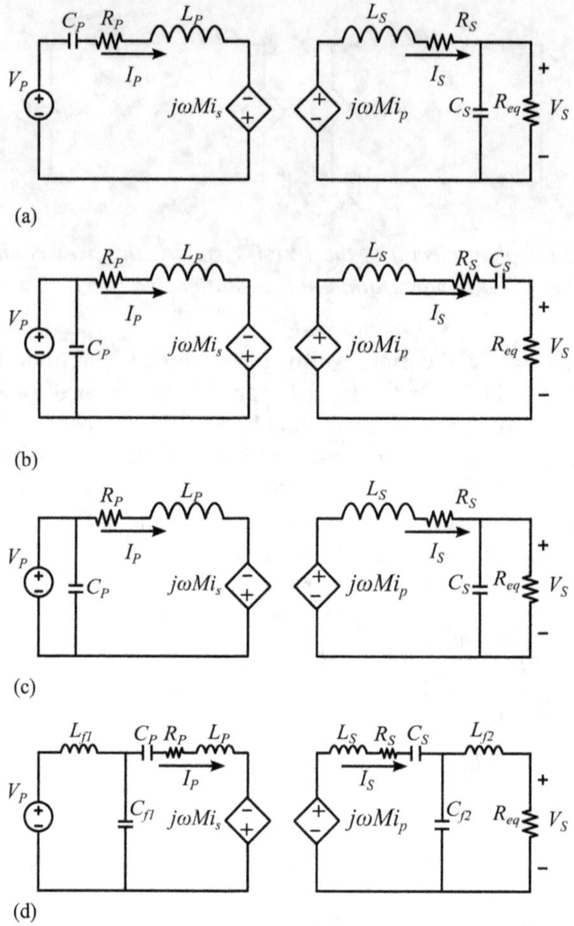

Figure 11.7 Equivalent circuit of the IWPT system with possible compensation networks. (a) Transmitter series capacitor and receiver parallel capacitor (series–parallel). (b) Transmitter parallel capacitor and receiver series capacitor (parallel–series). (c) Transmitter parallel capacitor and receiver parallel capacitor (parallel–parallel). (d) Transmitter LCC and receiver LCC (LCC–LCC).

compensation. The factors determining the efficiency of the coil-coil power transfer and the maximum efficiency possible will be analyzed in this section.

For an *SS*-compensated system, the transmitter and receiver side circuits can be solved using the impedances obtained in (11.15) to obtain P_{in} and P_{out}, respectively, as shown in (11.16). It can be noted in (11.16) that the Tx coil copper loss is deducted from the input power P_{in} to obtain the power transferred from Tx to Rx coil, and the power transmitted to the equivalent load resistance (R_{eq}) is the output power P_{out}

$$P_{in} = I_p^{\,2}\left(R_p + \frac{\omega_r^2 M^2}{R_s + R_{eq}}\right)$$
$$P_{out} = I_p^{\,2}\frac{\omega_r^2 M^2 R_{eq}}{\left(R_s + R_{eq}\right)^2} \tag{11.16}$$

For designing an IWPT system, expressing the power transfer (P_{out}) with respect to the input and output voltage ratings of the system is essential. The power transfer expressed in (11.16) considers the losses in coils, which is critical for developing an understanding of the power transfer efficiency across the coils. However, a designer may not be directly able to link the information about the mutual coupling between the coils to the voltage specifications of the IWPT system to be developed. This relation between the power transfer and the voltage ratings can be developed by neglecting the copper loss of the coils assuming the coils are of high quality factor. From (11.13), the Tx coil current (I_p) can be expressed in terms of the output voltage V_s. Thus, the apparent power transfer (S_{out}) between the input and the output of the IWPT system can be expressed by the following equations. θ_{ps} is the phase lag of V_s with respect to V_p:

$$I_p = -j\frac{V_s}{\omega M}$$
$$S_{out} = V_p I_p^*$$
$$S_{out} = \frac{V_p V_s}{\omega M} e^{j\left(90^\circ - \theta_{ps}\right)} \tag{11.17}$$
$$P_{out} = \frac{V_p V_s}{\omega M}\sin\theta_{ps}$$

The efficiency (η) of the power transfer through the coils in the IWPT system can be determined by finding out the ratio of P_{out} to P_{in}. For *SS*-compensated IWPT system, η is derived in (11.18) from the input and output power expressions in (11.16). Consider the receiver coil resistance $R_s << R_{eq}$, the approximate η is derived in (11.19):

$$\eta = \frac{1}{1 + \frac{R_s}{R_{eq}} + \frac{R_p\left(R_s + R_{eq}\right)^2}{\omega_r^2 M^2 R_{eq}}} \tag{11.18}$$

$$\eta \approx \frac{1}{1 + \frac{R_s}{R_{eq}} + \frac{R_p R_{eq}}{\omega_r^2 M^2}} \tag{11.19}$$

It can be observed from (11.18) that for a particular design of the transmitter and receiver coils, the power transfer efficiency in a SS-compensated IWPT system is dependent on the equivalent load resistance R_{eq}. To find the maximum power transfer efficiency of a particular coil system, the derivative of the efficiency with respect to R_{eq} is evaluated and equated to zero as shown in (11.20). Thus, the optimal equivalent load resistance ($R_{eq,opt}$) at which maximum efficiency can be achieved is derived in (11.20):

$$\frac{\partial \eta}{\partial R_{eq}} = \left(\frac{1}{1 + \frac{R_s}{R_{eq}} + \frac{R_p (R_s + R_{eq})^2}{\omega_r^2 M^2 R_{eq}}} \right)^2 \left(-\frac{R_s}{R_{eq}^2} + \frac{R_p}{\omega_r^2 M^2} \right) = 0$$
$$R_{eq,opt} = \frac{\omega_r^2 M^2}{R_s R_p} \tag{11.20}$$

Substituting the optimal equivalent load resistance ($R_{eq,opt}$) expression derived in (11.20) into the approximate efficiency in (11.19), the maximum power transfer efficiency ($\eta_{\max}$) is derived in (11.21):

$$\eta_{\max} = \frac{1}{1 + 2\frac{\sqrt{R_p R_s}}{\omega_r M}} \tag{11.21}$$

The maximum power transfer efficiency of the system can be expressed in terms of the coupling coefficient between coils derived in (11.11), transmitter coil quality factor ($Q_p = \frac{\omega L_p}{R_p}$), and receiver coil quality factor ($Q_s = \frac{\omega L_s}{R_s}$), by rearranging (11.21). The steps are shown in (11.22):

$$\eta_{\max} = \frac{\frac{\omega_r M}{\sqrt{R_p R_s}}}{2 + \frac{\omega_r M}{\sqrt{R_p R_s}}}$$
$$\Rightarrow \eta_{\max} = \frac{\kappa \sqrt{\frac{\omega_r L_p}{R_p} \frac{\omega_r L_s}{R_s}}}{2 + \kappa \sqrt{\frac{\omega_r L_p}{R_p} \frac{\omega_r L_s}{R_s}}} \tag{11.22}$$
$$\Rightarrow \eta_{\max} = \frac{\kappa \sqrt{Q_p Q_s}}{2 + \kappa \sqrt{Q_p Q_s}}$$
$$\Rightarrow \eta_{\max} = \frac{U}{2 + U}$$

The maximum power transfer efficiency of a *SS*-compensated IWPT system is derived in terms of a variable: $U = \sqrt{Q_p Q_s}$. This variable U is a figure of merit of the design of a transmitter and receiver coil assembly in an IWPT system. It can be observed that the figure of merit of the coil system is directly proportional to the coupling coefficient (κ) between the coils, transmitter coil quality factor Q_p, and receiver coil quality factor Q_s. From (11.22), it can be observed that higher value of the figure of merit will lead to increase in the maximum power transfer efficiency of the IWPT system. This relation of efficiency is an important design equation for the designers of coils in IWPT system. To satisfy power transfer efficiency specification of an IWPT system at a specific power level, the designer has to optimize the κ, Q_p, and Q_s. Magnetic cores are used in the design of coil assembly to increase the flux coupling between the coils. Similarly, high-quality factor coils are also designed while optimizing on the use of copper for enhancing the maximum power transfer efficiency of the IWPT system.

11.2.4 Converter control and MATLAB simulation of IWPT system

In this section, the methodology of applying a variable voltage at the resonant frequency of the compensated Tx coil will be discussed. For this discussion, an SS-compensated IWPT system will be considered. Further, the voltage and current waveforms of the IWPT system have been shown in this section by analyzing the IWPT system in MATLAB/Simulink. The AC voltage applied to the compensated transmitter coil vp is supplied from a full-bridge as shown in Figure 11.8. The frequency of vp can be forced to be the resonant frequency of the compensated coil such that the VA rating of the converter is reduced. The full-bridge is realized using MOSFETs which can be switched at high frequency to generate voltage vp with a frequency of (85 kHz). The supply voltage vp induces voltage across the coils

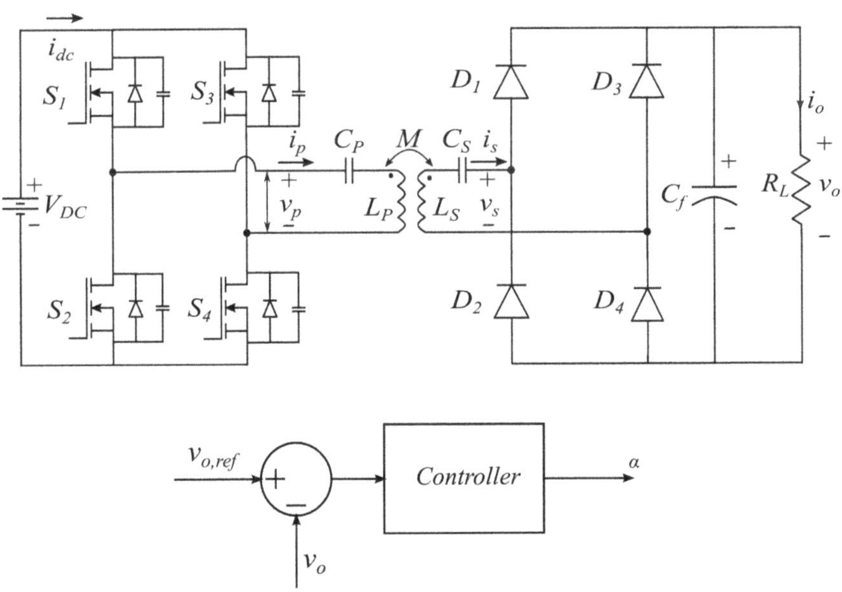

Figure 11.8 Power electronic converters connected to the SS-compensated IWPT system and the control architecture of output voltage control

(vcoil) due to flux linkages between them. A diode bridge rectifier is connected to the compensated receiver coil to rectify the induced AC voltage in the receiver coil and supply to the output load (RL). A capacitor is connected in parallel to the load such that the output voltage ripple is restricted. The formation of the simulation model of the SS-compensated IWPT system in MATLAB/Simulink is explained in the subsequent subsections. In the simulation, the IWPT system shown in Figure 11.8 will be modeled. The explanation of the model is segregated into three sections—power circuit and coil model, phase-shifted gate signal generation, and output voltage controller.

11.2.4.1 Power circuit and coil

The SS-compensated IWPT system as shown in Figure 11.8 is modeled in MATLAB/Simulink environment as shown in Figure 11.9. The model of the IWPT system consists of three stages—*Tx side full bridge, transformer equivalent model of coils, and Rx side rectifier*.

The Tx side full bridge is realized using generic MOSFET block available in Simscape → Electrical → Specialized Power Systems → Power electronics library in Simulink Library Browser. The Rx side rectifier is realized using generic diodes available in the Power electronics library. The input DC source V_p is realized using DC voltage source block available in Simscape → Electrical → Specialized Power Systems → Sources. The current and voltage measurement blocks used are *available* in Simscape → Electrical → Specialized Power Systems → Sensors and measurements. The gate signal inputs to the MOSFETs are routed through *Goto and From elements in* Simulink → Dashboard → Signal routing from the pulse generation model explained in next subsection. The symmetrical Tx and Rx coils of the IWPT system are modeled by the transformer equivalent circuit as shown in Figure 11.9. The compensation circuits comprised of C_p and C_s in the Tx and the Rx side, respectively, are connected between the coil and the converters in the respective sides. The output load is modelled as resistance. Two parallel branches of load resistances with one load resistance connected with a switch are shown in Figure 11.9 to emulate the effect of step change in load for testing the closed loop output voltage control. All these passive elements are available in Simscape → Electrical → Specialized Power Systems → Passives. The simulation considers the parameters for a symmetric coil system, which makes the coil parameters of both transmitter and receiver coils equal. The various parameters used for the simulation are shown in Table 11.1.

Since the *Specialized Power Systems* library is being used in the simulation, the *Powergui* block has been included in this model. Under *Powergui* block, the simulation type has been set to continuous which uses the simulation step type of the MATLAB model settings. Additionally, in the preferences of *Powergui* block snubber circuits, on-state voltage drop and on-state resistance of the devices have been disables. Thus, ideal power device characteristics will be used for the simulation. Under the *Modeling* tab of Simulink, the solver type and simulation step time can be selected in *Model settings*. For the model of the IWPT system, the solver type has been selected as fixed step and the step time has been selected as 10

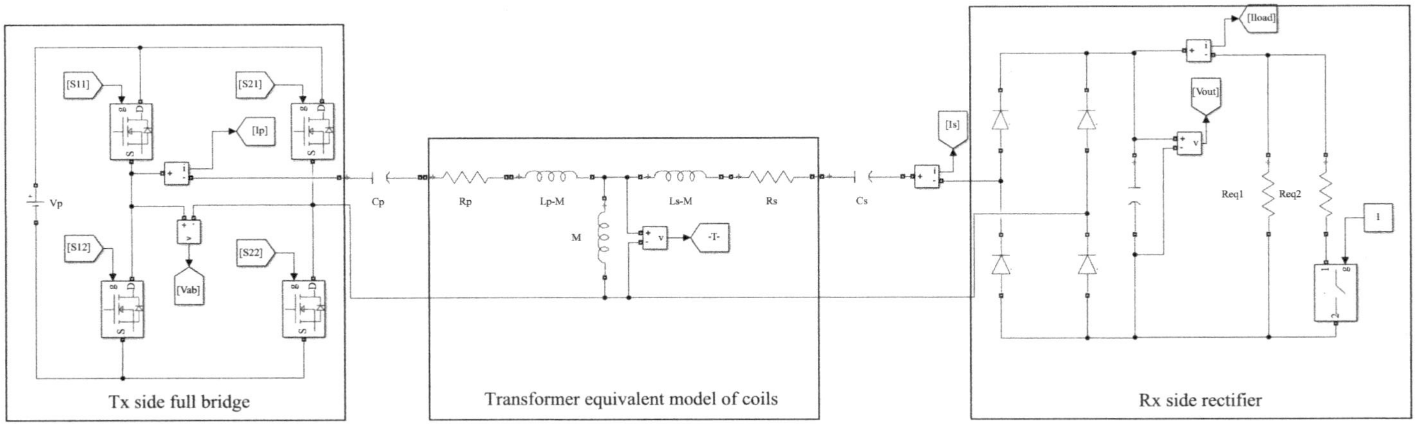

Figure 11.9 MATLAB®/Simulink® model of the power circuit and coil model of SS-compensated IWPT system

Table 11.1 Simulation parameters for the IWPT system

Power output	**P**	**3.3 kW**
DC-link voltage	V_{DC}	200 V
Output voltage reference	V_0	72 V
Resonant frequency	f_r	85 kHz
Transmitter coil resistance	R_p	0.369 Ω
Receiver coil resistance	R_s	0.369 Ω
Transmitter coil leakage inductance	$L_{lk,p}$	48.513 μH
Receiver coil leakage inductance	$L_{lk,s}$	48.513 μH
Mutual inductance	M	4.587 μH
Output filter capacitor	C_f	107.8 μF
Resonant tank capacitor	Cp, Cs	66 nF

ns, which is substantially lower than the fundamental period of the high-frequency supply to the coil.

The model for the *phase-shifted gate signal generation* block has been developed as shown in the right-hand side of Figure 11.10. The elements used for this block are taken from the *Simulink* library. The a sawtooth carrier signal is generated using the *Simulink* → *Sources* → *Repeated sequence*. The repeated sequence has been set to change from 0 to 1 linearly during the period corresponding to 85 kHz. The PWM logic is implemented using the *Simulink* → *Logic and bit operations* → *Relational operator*. A constant value of 0.5 is compared to the carrier signal to generate the gate signal for *S1*, and the complementary gate signal for *S2* is generated using a NOT logic available in *Simulink* → *Logic and bit operations* → *Logical operator*. For generating the phase-shifted gate signals, the carrier signal needs to be shifted by phase angle which is determined by the voltage controller. This is implemented by using a delay element available in Simulink → Continuous → Variable time delay. The first input to the delay element is the carrier signal generated by the repeated sequence and the second input is the absolute delay in time which corresponds to the phase shift which needs to be provided.

The variable voltage is generated by phase shifting the switching signals between two half-bridge legs in the Tx side full bridge converter. In Figure 11.8, one of the half-bridge legs is formed by MOSFETs (S_1, S_2) or S_3, S_4. The generation of the phase-shifted gate pulses in the full bridge is shown in Figure 11.11. Two carrier signals (V_{c1} and V_{c2}) are generated which are shifted by an angle α as shown in Figure 11.11. V_{c1} and V_{c2} are two sawtooth signals with equal time periods of T_s as shown in Figure 11.11. T_s is inverse of the frequency (in Hz) of the voltage supply v_p to be supplied to the compensated transmitter coil. A modulation signal V_m of half the amplitude of the sawtooth carrier signals is compared with the carrier signals to generate the gate pulses as shown by the following logic:

$$V_{S1} = \begin{Bmatrix} 1 & V_m \geq V_{c1} \\ 0 & V_m < V_{c1} \end{Bmatrix}$$

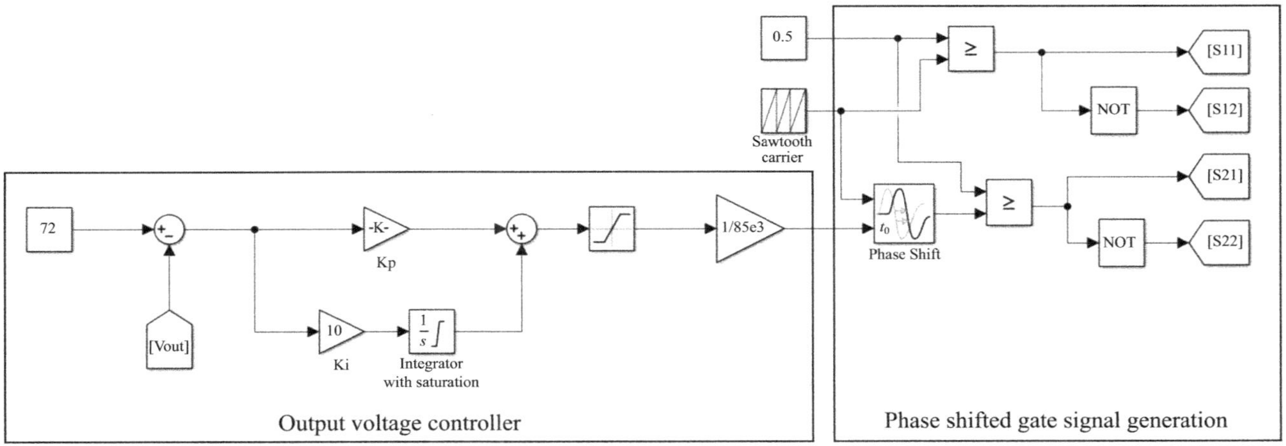

Figure 11.10 MATLAB/Simulink model of the phase-shifted signal generation and the output voltage controller blocks of IWPT system

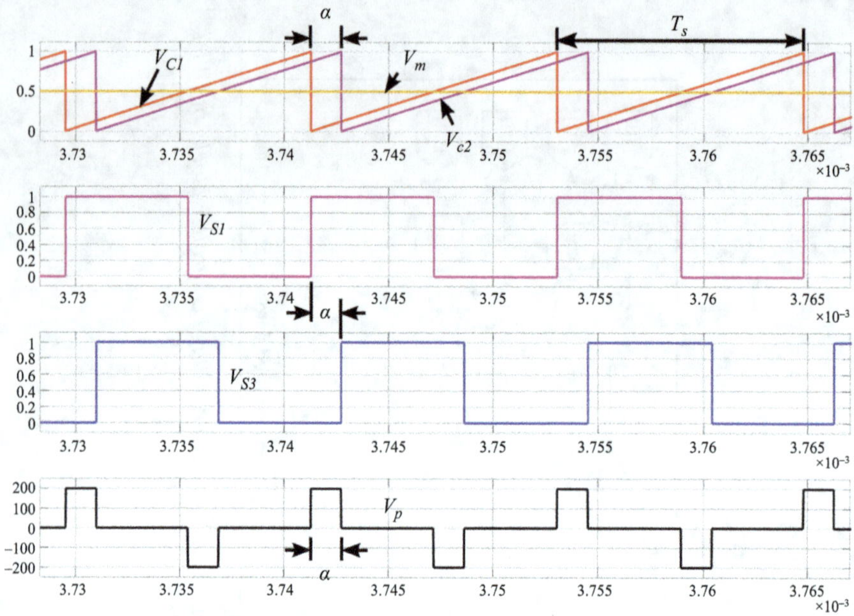

Figure 11.11 Waveforms of phase-shifted gate signal-generated obtained from simulation in $MATLAB/Simulink

$$V_{S2} = \begin{Bmatrix} 1 & V_m \geq V_{c2} \\ 0 & V_m < V_{c2} \end{Bmatrix}$$

Following the above logic, the gate signals V_{S1} and V_{S3} are generated. V_{S2} and V_{S4} are complementary to V_{S1} and V_{S3} signals, respectively. The signals V_{S1}, V_{S2}, V_{S3}, and V_{S4} are logic input signals to the gate drivers of MOSFETs S1, S2, S3, and S4, respectively, as shown in Figure 11.8. It can be observed in Figure 11.11 that the phase shift α in the carrier signals also results equal phase shift between the turn ON events of switch S1 and S3, respectively. Thus, a quasi-square wave voltage v_p is generated between the switching midpoints (switching nodes) of the half-bridges. The width of the positive and negative pulse is equal to α as shown in Figure 11.11. It can be observed from Figure 11.11 that during positive pulse, switches S1 and S4 are turned on, and, during negative pulse, S2 and S3 are turned on. Hence, the amplitude of the positive and negative pulses are V_{DC} and $-V_{DC}$, respectively, where V_{DC} is the DC link voltage of the full bridge input as shown in Figure 11.8. The fundamental component's amplitude of v_p can be expressed as shown in (11.23). Thus, it can be concluded that by changing the period T_s and full bridge phase shift α the frequency and fundamental magnitude of the transmitter side input voltage supply v_p can be controlled, respectively:

$$V_{p,fund} = \frac{4}{\pi} V_{DC} \sin \alpha \tag{11.23}$$

11.2.4.2 Output voltage controller

From (11.16), it can be seen that the power transferred to the output is dependent on the magnitude of I_p. For a series-compensated transmitter coil, the input impedance Z_p obtained in (11.15) is fixed for a particular alignment of the coils. The current I_p can then be controlled by changing the phase shift angle α, since supply voltage fundamental $V_{p,fund}$ varies with α as shown in (11.23). The output voltage V_o can thus be regulated by controlling the transmitter full bridge phase shift angle α. For the output voltage regulation, the output voltage feedback is subtracted from the reference output voltage $V_{o,ref}$ to determine the error. This error is given as an input to a linear controller to generate α, which determines the phase shift between the two carrier signals to generate switching signals for the MOSFETs in the transmitter full bridge.

The implementation of the control architecture in MATLAB/Simulink is shown on left-hand side of Figure 11.10. The elements used in the control architecture implementation are available in *Simulink* → *Commonly used blocks/ Continuous* libraries. The error in the reference input (72 V) and the output DC voltage feedback is obtained *Sum* element, where the second input has been changed to − sign. The error is then fed to a proportional gain (K_p) and integrator block after a gain of K_i as shown in Figure 11.10. The integrator output is saturated to a value of 0.5 which corresponds to the maximum phase shift possible (180°). The saturation block in the integrator emulates anti-windup action which needs to be implemented in practical implementation of an integrator. The output of the proportional gain path and the integrator path are added to obtain the controller output. The controller output is also routed through a saturation block which saturates the output to an upper bound of 0.5 and lower bound of 0. Since the controller output is equal to the fraction of the period corresponding to the phase shift angle, the controller output is multiplied with the period of voltage supply to the coil before being connected to the time delay block in *phase-shifted gate signal generation* module explained in the last subsection. For the simulation model being explained, the K_p and K_i values have been set to 0.001 and 10, respectively. The analysis of controller tuning is not explained in this chapter; however, readers can refer to [6–9] to analyze controller tuning for an IWPT system to satisfy certain stability criterion.

The steady-state simulation waveforms at full load are shown in Figure 11.12. The series resonant capacitors C_p and C_s are selected such that the self-inductances of the coils are compensated to form resonant circuit at 85 kHz. With the parameters of the coils provided, the coupling coefficient of the IWPT system can be calculated to be $\kappa = 0.0864$ following (11.11). A quasi-square wave voltage v_p is generated by phase shift control of the full-bridge converter with a fundamental frequency of 85 kHz. It can be observed that transmitter coil current i_p is purely sinusoidal and in-phase with the fundamental component of the supply voltage v_p because of resonance at 85 kHz. As a result, the voltage induced in the coil v_{coil} and the receiver coil current i_s are also sinusoidal. The controller action can also be observed in Figure 11.12, the steady-state average output voltage v_o is closely

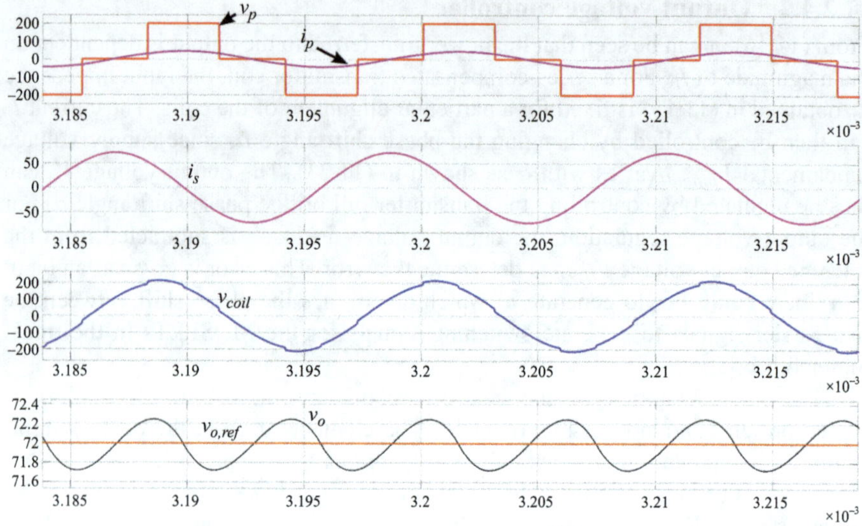

Figure 11.12 Steady-state simulation waveforms of the SS-compensated IWPT system obtained in MATLAB/Simulink

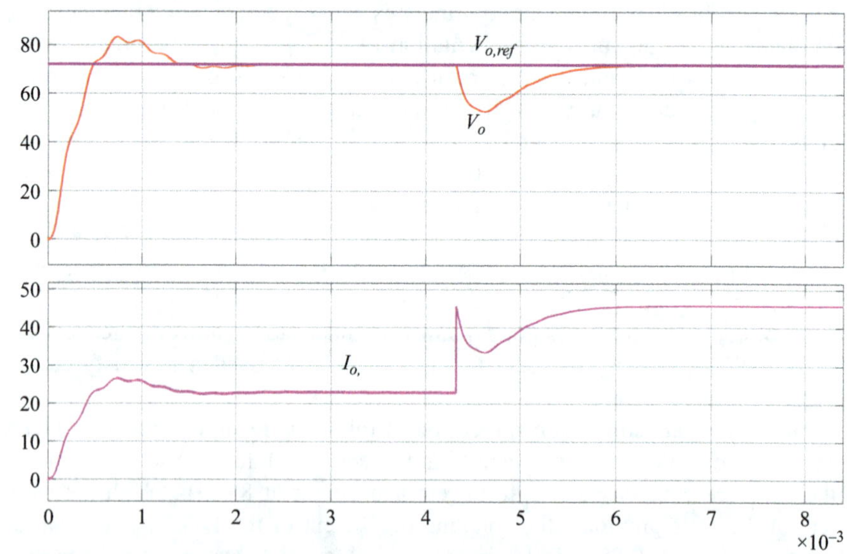

Figure 11.13 Transient simulation waveforms of the SS-compensated IWPT system obtained in MATLAB/Simulink

tracking the reference output voltage $v_{o,ref} = 72$ V. The output filter capacitor restricts the output voltage ripple within 0.6 V.

The transient response of the IWPT system is shown by the simulation waveforms in Figure 11.13. The load is increased from half-load to full load while

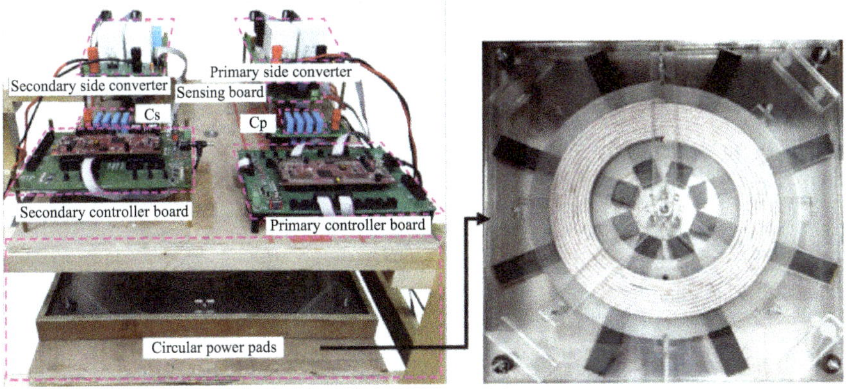

Figure 11.14 Prototype of a 3.3-kW IWPT system for charging 72 V battery, developed in IIT Delhi standards related to magnetic field wireless charger

maintaining $v_{o,ref} = 72$ V. It can be observed that the controller quickly changes the phase shift angle α such that v_o is restored to 72 V within 1.5 ms. The steady-state error is brought back to zero due to controller action.

The readers can form an idea of an actual IWPT system developed in our research laboratory from Figure 11.14. In the prototype, it can be seen a circular coil is designed for Tx and Rx side coils. The details of the design and optimization of the geometry are done in [10]. The Tx and Rx side converters, control boards, and the series compensation capacitors can be seen in Figure 11.14.

11.3 Standards of wireless power charging

A few standards have been developed in wireless power transfer to maintain safety, communication and interoperability among suppliers and end consumers. These are detailed below.

11.3.1 IEC standard

The IEC technical committee has been developing the wireless power transfer standard. One of the recommended standards for wireless power transfer is the IEC 61980:2020 standard. It replaces the first edition published in 2015. This standard applies to devices which can charge wirelessly electric road vehicles for voltage levels up to 1,000 V AC or 1,500 V DC. The standard also includes plug in hybrid vehicles.

The standard is divided into three parts. Part 1 (IEC 61980-1:2020) discusses about safety standards, efficiency and alignment requirement and specific EMC requirements. Part 2 (IEC 61980-2:2020) specifies the standards related to communication between the road vehicle and the off-board wireless charger. Part 3

Table 11.2 Classification of WPT system as given in SAE J2954 standard

	WPT1	WPT2	WPT3	WPT4
kVA input	3.7	7.7	11.1	22
Minimum efficiency at nominal alignment	85%	85%	85%	TBD
Minimum efficiency at offset position	80%	80%	80%	TBD

(IEC 61980-3:2020) specifies this standard suggests the usage of 85 kHz band (81.38–90 kHz) as the system frequency for passenger cars.

11.3.2 SAE J2954 standard

The Society of Automotive Engineers (SAE), based in the United States, is an international organization developing standards with emphasis for the transport sector. They have developed several wireless power transfer standards, collectively under SAE J2954 over the years. As of 2022, the J2954/2_202212 is the most current standard.

The aim of these standards is to enable any BEV/HEV customer to seamlessly park and charge the vehicle in a J2954 compatible parking space.

In these standards, four different classifications are made depending on the power level and the charging speed. These are termed as WPT1, WPT2, WPT3, and WPT4. The power levels are indicated in Table 11.2.

To maintain interoperability, a system complying with these standards should have its primary and secondary coils compatible with each other. For example, if the primary coil is of class WPT1, then the secondary coils should be compatible with WPT1 and WPT2; however, it is optional for the secondary coils to be of class WPT3. More detailed information is available in [11].

In addition to the power level, the standard also defines the ground clearance distances for testing and operation. The standard describes two parts: a charger station with induction pad which is called the ground assembly (GA) and a vehicle-mounted section called the vehicle assembly (VA). SAE-J2954 defines three classes of operation classified as Z-classes. In the Z-1 class, the air gap ranges from 100 to 150 mm. In the Z-2 class, the distance can be between 14 and 210 mm. Z-3 class allows a distance variation of 170–250 mm.

The communication among different subsystems in a wireless charging system is being covered under standards SAE J2847/6, SAE J2931/6, and SAE J2836/6.

11.3.3 ISO 19363 standard

The International Standard of Organization (ISO) has also developed a wireless power transfer technical standard which outlines the specifications to be followed for charging of EVs. This standard is named as "Electrically propelled road vehicles—Magnetic field wireless power transfer—Safety and interoperability requirements." This standard addresses safety requirements, efficiency, ground clearance, and test procedures for wireless power transfer. Most of the specifications are very similar to IEC and SAE standards. Bidirectional power transfer and vehicles in motion are, however, not covered in this standard.

11.4 Summary

In this chapter, a comprehensive understanding of inductive wireless power transfer (IWPT) system has been given for the readers. The IWPT system is a promising technology which can be utilized for contactless charging of EV battery. This enables the ease of usage of EV, and fully autonomous operations of EV. This technology is also being envisaged to power up drones, UAVs, electric boats, etc. In this chapter, the principle of operation of IWPT system has been discussed in detail to form the basis of further discussion which is crucial for the design of an IWPT system. Following are the different aspects which have been discussed in this chapter:

- The promise of wireless charging and its usefulness in EV charging has been discussed. Various standards that determine the design of IWPT system for EV battery charging applications have been explained.
- The physics of energy transfer through the air-gap between two systems has been explained in detail. The equivalence of IWPT system to a transformer without a magnetic core has been discussed. The flux linkages through the coils which culminates into several inductances have been described.
- The mathematical model of IWPT system has been described in this chapter. This enables the designer to form electrical equivalent circuits which is critical for the design of the compensation network and quantifying the power transfer and efficiency of the system.
- The role of compensation network to reduce the VA ratings of the converters in the IWPT system has been discussed. Various compensation methods have been shown, and analysis of the system with *SS* compensation has been described in detail.
- The analytical derivation of power transfer and efficiency of IWPT system has been given in this chapter. The dependence of efficiency on the quality factor and coupling coefficient of the coils has been established.
- Control methodology of the converter in the IWPT system has been shown. The results of steady-state and transient simulations of an IWPT system in MATLAB/Simulink have been shown in this chapter.

Symbols

i_p	Transmitter coil current
i_s	Receiver coil current
v_p	Voltage induced across transmitter coil
v_s	Voltage induced across receiver coil
N_p	Number of turns in transmitter coil
N_s	Number of turns in receiver coil
n	Transmitter to receiver coil turns ratio

ϕ_{pp}	Self-induced flux in the transmitter coil
ϕ_{ss}	Self-induced flux in the receiver coil
ϕ_{ps}	Flux linked through receiver coil due to i_p
ϕ_{sp}	Flux linked through transmitter coil due to i_s
$\mathbb{R}_{pp}$	Reluctance of path of ϕ_{pp}
$\mathbb{R}_{ss}$	Reluctance of path of ϕ_{ss}
$\mathbb{R}_{ps}$	Reluctance of path of ϕ_{ps}
$\mathbb{R}_{sp}$	Reluctance of path of ϕ_{sp}
$L_{lk,p}$	Transmitter coil leakage inductance
$L_{lk,s}$	Receiver coil leakage inductance
$L_{m,p}$	Transmitter coil magnetizing inductance
$L_{m,s}$	Receiver coil magnetizing inductance
L_p	Transmitter coil self inductance
L_s	Receiver coil self inductance
M	Mutual inductance between the coils
ω	Electrical angular frequency of coil voltage supply
κ	Coupling coefficient
R_p	Transmitter coil resistance
R_s	Receiver coil resistance
R_L	Load resistance
R_{eq}	Equivalent load resistance
Z_s	Output impedance connected to receiver coil
Z_{sr}	Receiver side impedance reflected to transmitter side
Z_p	Input impedance of transmitter coil
C_p	Series compensation capacitor in transmitter side
C_s	Series compensation capacitor in receiver side
ω_r	Resonant electrical angular frequency
P_{in}	Input power to the IWPT system
P_{out}	Output power to the IWPT system
S_{out}	Apparent power transferred by the IWPT system
θ_{ps}	Phase shift between the transmitter and receiver coil voltages
η	Efficiency of the IWPT system
η_{max}	Maximum efficiency of the IWPT system
$R_{eq,opt}$	Optimal equivalent load resistance to achieve maximum efficiency
Q_p	Transmitter coil quality factor
Q_s	Receiver coil quality factor
U	Figure of merit for maximum efficiency

Glossary

WPT Wireless Power Transfer from charging circuit to the battery through electronic circuits
SOC State of the Charge in the battery
EMI Electromagnetic Interference to the nearby circuits
IWPT Inductive wireless power transfer for charging a battery
UAV Unmanned Aerial Vehicle

References

[1] J. Gozalvez, "WiTricity – The Wireless Power Transfer [Mobile Radio]," *IEEE Vehicular Technology Magazine*, vol. 2, no. 2, pp. 38–44, 2007.
[2] J. M. Miller, P. T. Jones, J.-M. Li, and O. C. Onar, "ORNL Experience and Challenges Facing Dynamic Wireless Power Charging of EV's," *IEEE Circuits and Systems Magazine*, vol. 15, no. 2, pp. 40–53, 2015.
[3] G. A. Covic and J. T. Boys, "Modern Trends in Inductive Power Transfer for Transportation Applications," *IEEE Journal of Emerging and Selected Topics in Power Electronics*, vol. 1, no. 1, pp. 28–41, 2013.
[4] I. Hwang, Y. J. Jang, Y. D. Ko, and M. S. Lee, "System Optimization for Dynamic Wireless Charging Electric Vehicles Operating in a Multiple-Route Environment," *IEEE Transactions on Intelligent Transportation Systems*, vol. 19, no. 6, pp. 1709–1726, 2018.
[5] W. Zhang, S. Wong, C. Tse, and Q. Chen, "Analysis and Comparison of Secondary Series- and Parallel-Compensated Inductive Power Transfer Systems Operating for Optimal Efficiency and Load-Independent Voltage-Transfer Ratio", *IEEE Transactions on Power Electronics*, vol. 29, no. 6, pp. 2979–2990, 2014.
[6] Y. R. Kumar, D. Nayak, M. Kumar, and S. Pramanick, "The Effect of Cross-Coupling on the Dynamic Performance of a Dual Phase-Shift Controlled BiWPT System," *IEEE Transactions on Transportation Electrification*, vol. 9, no. 3, pp. 1–1, 2023.
[7] H. Li, J. Li, K. Wang, W. Chen, and X. Yang, "A Maximum Efficiency Point Tracking Control Scheme for Wireless Power Transfer Systems Using Magnetic Resonant Coupling," *IEEE Transactions on Power Electronics*, vol. 30, no. 7, pp. 3998–4008, 2015.
[8] A. Berger, M. Agostinelli, S. Vesti, J. A. Oliver, J. A. Cobos, and M. Huemer, "A Wireless Charging System Applying Phase-Shift and Amplitude Control to Maximize Efficiency and Extractable Power," *IEEE Transactions on Power Electronics*, vol. 30, no. 11, pp. 6338–6348, 2015.
[9] W. Zhang, S.-C. Wong, C. K. Tse, and Q. Chen, "Design for Efficiency Optimization and Voltage Controllability of Series–Series Compensated

Inductive Power Transfer Systems," *IEEE Transactions on Power Electronics*, vol. 29, no. 1, pp. 191–200, 2014.

[10] Y. R. K. S. Pramanick, D. Nayak, and M. Kumar, "Optimization of Circular Coil Design for Wireless Power Transfer System in Electric Vehicle Battery Charging Applications," *Transactions on Indian National Academy of Engineering*, vol. 6, pp. 765–774, 2021.

[11] Wireless Power Transfer for Heavy-Duty Electric Vehicles. https://www.sae.org/standards/content/j2954/2_202212/

Chapter 12

Grid integration of electric vehicles

Bindeshwar Singh[1], *Pankaj Kumar Dubey*[1], *Varun Kumar*[1] *and Mukesh Kumar*[2]

In urban areas, emissions of greenhouse gases from oil-based fuel cars with internal combustion engines (ICEs) are still a challenge. Electric vehicles (EVs) will ultimately be the most cost-effective method of transportation due to their significant benefits of using versatile fuels, convenience, secure charging, excellent performance, and cost benefits. The location and the size of aggregated EVs, on the other hand, could compromise power system reliability. As a consequence, to guarantee power network dependability, efficient charging point planning is vital. On the other hand, potential charging station placements should include city traffic conditions as well as electricity distribution facilities. This study provides a thorough examination of several optimization strategies for the appropriate location and sizing of EVs in smart power grid networks (SPGNs) with load models.

For EV planning, the optimization strategies are classified into different types: traditional, optimization, artificial intelligence, hybrid, and current optimization strategies. Various network characteristics are taken into account for EV planning with various objective function perspectives, such as minimizing the system's actual and imaginary power deficits. This comprehensive survey chapter also includes an overview of current frameworks and strategies for organizing, evaluating, and dividing up current and upcoming research advancing in the field of electric cars. The editors are confident that this chapter will be of great assistance to researchers, industry professionals, academics, and scientists looking for relevant references on the best positioning and scalability of EVs in intelligent power grid infrastructure.

12.1 Introduction

EVs have become a potential answer in the fight to reduce carbon emissions and open the door for a future of sustainable mobility. The power grid load produced by

[1]Electrical Engineering Department, Kamala Nehru Institute of Technology, Sultanpur, India
[2]Electrical Engineering Department, Maulana Azad National Institute of Technology (MANIT) Bhopal, India

EV charging scenarios potentially exceeds acceptable limits when multiple EVs fill their batteries at once. EVs, their components, classification, charging schemes and infrastructure, control systems, and grid integration will all be discussed here.

One or more electric motors provide the power for an EV. Due to the type of car, the mobility is provided by wheels, spinners, or, in the case of track-follower cars, straight motors. This could be powered by a bank of cells or a collector system that draws power from the vehicle's exterior (this is recharged from time to time by PV arrays or transformed to electricity by fuel cell (FC) technology or generators). It has an electric motor for propulsion and a battery pack. When you press the pedal, the battery is used to power the controller, the controller supplies the correct quantity of electrical energy to the motor, electric engines turn electricity into mechanical power, and the cars move as the wheels turn.

In Figure 12.1, the basic element of EVs is discussed:

(a) *Battery:* It is used to drive electric motors. The capacity is measured in a unit called Ah. A battery's design entails a series of complex calculations that assess charge distribution, charging ability, power density, lifespan, security, form factor, price, and adaptability for several types of batteries.
(b) *Power converter:* The electric power in a battery should be transformed to adjustable DC or changeable AC depending on the kinds of electric engines used to move the tires.
(c) *Electric motor:* DC series and induction motors were formerly used. Now the concentration is on specialized electrical machines.
(d) *Clutch:* The clutch disconnects the engine from the wheels, allowing it to change from low-speed to high-speed ratios.
(e) *Transmission:* The gearbox, often known as the transmission, is a device that permits power to be transferred from the engine to the wheels.
(f) *Drivetrain:* The electric motor, clutch, and gearbox make up the drivetrain.

Chatzikomis *et al.* [1], for driverless EVs, route monitoring and torque vectoring actuators are contrasted. Wang *et al.* [2] presented an application of the extension Taguchi method to stand-alone power system capacity planning.

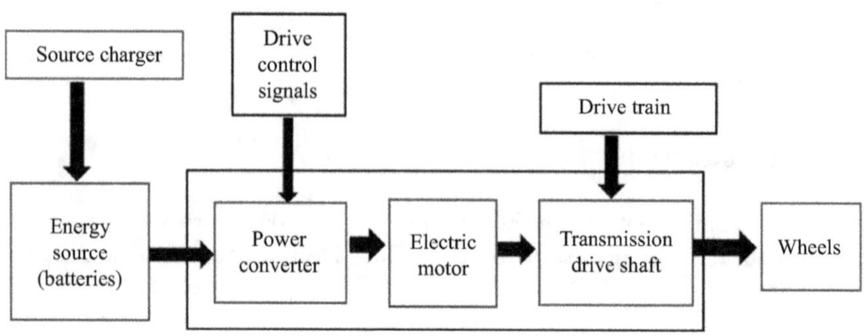

Figure 12.1 Block diagram of EVs

Hur *et al.*'s [3] proposed the representation of the product and desirability functions of electric car models for flow dispersion improvement in multi-channel conditioning circuits of power inverters. Duan *et al.* [4] proposed an optimal EV dispatching using smart contracts and the IoT. Saski [5] suggested the current and future IoT big data analytic systems survey. Chiang *et al.* [6] addressed next-generation networks and smart data pricing. Cai *et al.* [7] have devised a stream function-based 3D obstacle detection system for portable Automatic unmanned vehicles (AUVs), and the Internet of underwater things has been suggested.

The fundamentals of multi-access edge computation in 5G and later technologies were discussed by Pham *et al.* [8], along with technological innovation and conditions. Pohan *et al.* [9] investigated the fusion of the swiftly expanding random tree and ant colony platforms as a route routing method. Park *et al.* [10] utilizing a dispersed ADMM technique, a non-myopic path routing for multi-target monitoring was proposed. Li *et al.* [11] suggested route-planned assessment technologies for self-driving underwater vehicles. Drake *et al.* [12] addressed the path planning by a mobile robot with a moving goal. Ding *et al.* [13] have formulated an EV recharging alert and course-planning mechanism based on sparking. Yu *et al.* [14] suggested that for a driverless electric bus, a path planning and navigation control system should be designed. Cabrera *et al.* [15] designed and implemented a cheap wireless charging facility for electric bicycles.

Wang *et al.* [16] presented China's smart grid technology and development. Haus *et al.* [17] proposed centroid vectoring for trajectory tracking of floating-based robotics for applications ranging from marine to aviation. Wang and Wang [18] proposed that to reduce grid power peaks and fill valleys, vehicle-to-grid systems are used. Ogaili *et al.* [19] discussed the constraints and suggestions for planning, clustering, and prediction systems for EV charging. Yagmur *et al.* [20] presented the DSO's view on blockchain-based energy applications. Hu *et al.* [21] suggested that to evaluate battery-swap stations for EVs, as well as cooperative optimization of decentralized planning, blockchain consensus protocols are used. Huang *et al.* [22] proposed that the LNSC is a blockchain-based safety method for managing EVs and charger piles. Lasla *et al.* [23] address the smart city's blockchain-based trading system for EVs recharging.

12.1.1 Categorization of EVs

EVs are divided into five categories:

(a) *Battery EVs (BEVs)*
High-potential batteries, one or more electric drivetrains (either AC or DC), and a controller to regulate the power electronics are the essential components of a BEV as shown in Figure 12.2. Electric motors, unlike ICEs, can deliver steady and greater torque beyond a range of velocities beginning at 0 mph; thereby, most BEVs do not require a minimization gearbox or the extremely sophisticated engine control systems that have been incorporated into ICEs and their transmissions meet emission regulations while still allowing them to operate.

(b) *Hybrid EVs (HEVs)*
A hybrid electric car has an electric motor as well as traditional ICEs as shown in Figure 12.3. In comparison to a traditional car, this one is thought to have higher performance and fuel economy.

(c) *Plug-in hybrid EVs (PHEVs)*
In PHEVs, the battery operates a motor, while the other energy, such as petrol, drives an engine. The car usually operates on electric power till the batteries are nearly depleted, at which moment, it changes to an ICE. Plug-in hybrids combine a battery and an electric motor with gasoline, diesel, or an alternative fuel engine as shown in Figure 12.4. When you connect to the mains, you might get a wider range of electric-only power. The batteries have a larger storage capacity than a current hybrid vehicle.

(d) *Range extension type of EVs (Rx-EVs)*
Range-extended EVs are EVs with a gas or diesel generator to replenish the battery when the charge runs low as shown in Figure 12.5. Range extension, on the other hand, can be performed using either series or parallel hybrid layouts. In a series-hybrid arrangement, the petrol engine powers an electric generator rather than powering the tires immediately. The generator charges batteries to supply power for the electric motors.

(e) *FC EVs (FCEVs)*
An FCEV is an electric car that uses an FC to operate its electric engine onboard, sometimes in confluence with a cell or ultracapacitor. FCs in automobiles produce energy by compressing oxygen from the air. It has a propulsion system that is similar to that of EVs, in which H energy is stored and converted to electric power using an FC (shown in Figure 12.6). The Table 12.1 presents a comparative analysis of the above categories of EVs.

12.1.2 Charging station

The equipment utilized to replenish plug-in electric cars is called a recharging terminal. It is also referred to as an EV charger or an EV providing basic hardware. The bulk of EV power packs have an integrated AC-to-DC converter that enables them to be plugged into a regular household AC outlet even though EV battery

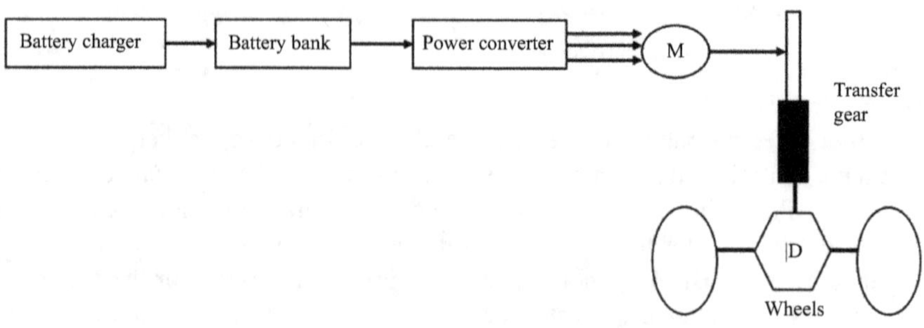

Figure 12.2 Block diagram of BEVs

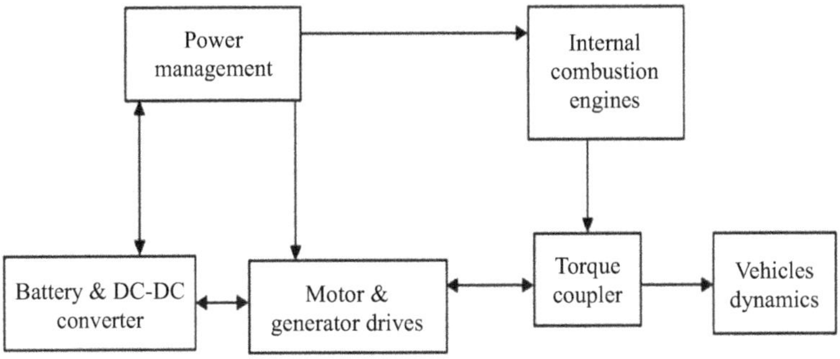

Figure 12.3 Block diagram of hybrid EVs

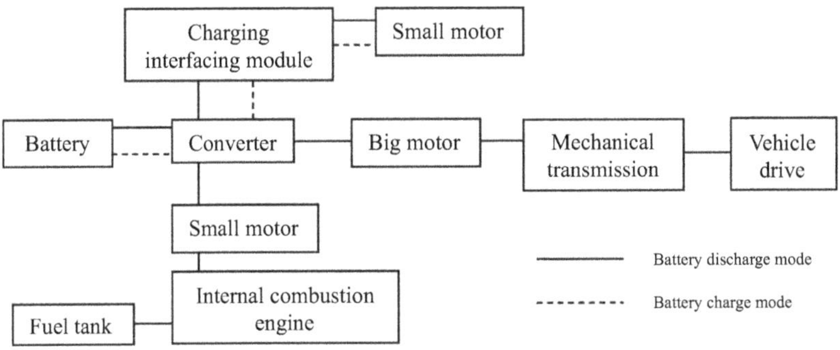

Figure 12.4 Block diagram of PHEVs

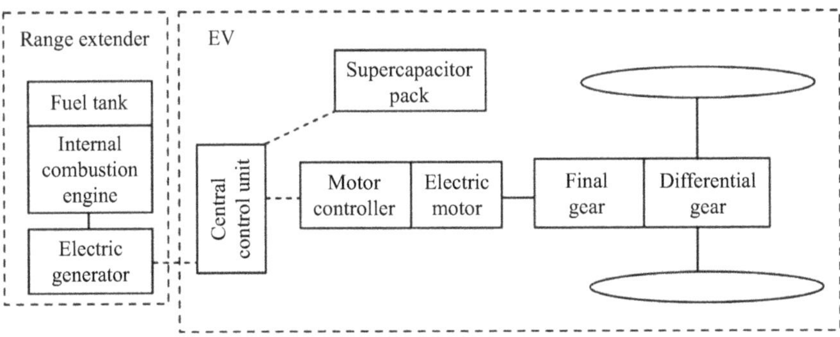

Figure 12.5 Block diagram of range extension type EVs

packs can only be recharged with a DC supply. AC power will also be provided by AC charging points, which are inexpensive and low-public charging stations. To enable larger power charging, which requires significantly larger AC-to-DC converters, the inverter arrangement is put within the power outlet rather than the car.

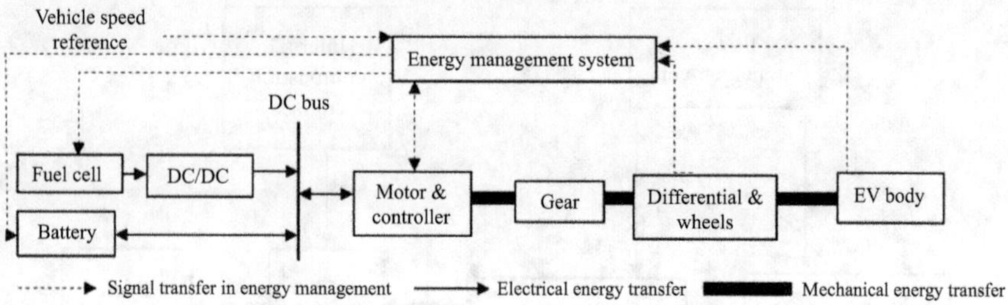

Figure 12.6 Block diagram of FC EVs

Table 12.1 Comparison between different EVs

S. no.	Parameter	HEV	PREV	BEV	FCEV
1	Capacity of battery (kWh)	5–15	8.0–18.0	40.0–100.0	1.0–2.0
2	Traveling with a seriously low battery	Possible	Possibly	Never feasible	Feasible
3	Peak reach with full tanks.	On fossil fuel, 300.0–500.0 km	Mobility of a battery-powered vehicle is 50.0–180.0 km, and on fossil fuels, 250–500 km	Peak battery mobility is 100.0–300 km	On hydrogen fuel, 350.0–700.0 km

The infrastructure now directly supplies DC power to the automobile, bypassing the automobile's built-in converter. "DC charging stations (CS)" is the name given to them (Figure 12.7).

12.1.2.1 Charging technologies for EVs

Level 1 EV charging

The voltage for level 1 EV charging is 120 V. Every electric car comes with the necessary hardware, which consists of a cord with an attached control box. Simply put the device into a three-pin (grounded) electrical outlet. Depending on the capacity of the vehicle's batteries, this charging method takes 16–20 h to fully charge them. The benefit of this charging method is that it does not necessitate the addition of any additional hardware. To use the charging cable, simply park near a three-pin wall outlet and plug it in. The drawback of this type is that charging the battery takes so long.

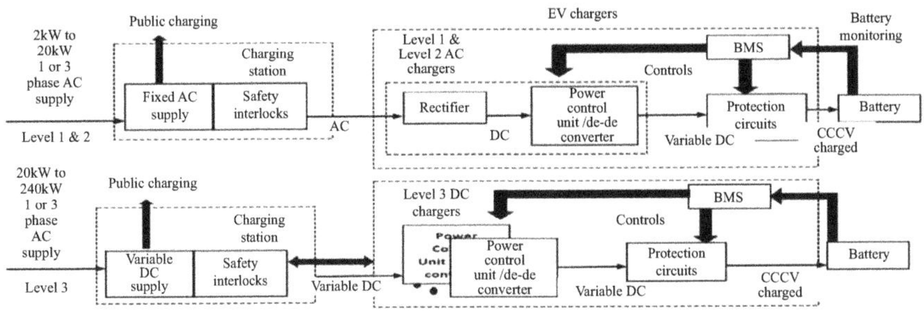

Figure 12.7 Block diagram of electrical parts used in levels 1, 2, and 3 charging

Level 2 EV charging

240 V is the standard for level 2 charging for EVs. This form of charging necessitates the use of extra hardware. When a customer buys an electric car, certain manufacturers will install an AC wall-box charger at their home and, in some cases, at their workplace, either for free or at a reduced price, to enable level 2 charging. An EV can be fully charged in as little as 6 h or a little more with this method, depending on the battery capacity. In comparison to level 1 charging, level 2 charging is much faster. Not only that but it's also supposed to be more energy efficient. However, due to the utilization, this charging approach is pricey.

Level 3 EV charging

Level 3 charging is available at public charging stations for EVs. It transforms AC into DC for direct storage in EV batteries and is known as DC rapid charging. Normally, it is rated at 480 V. An electric car can be charged to 80% capacity in less than an hour using a DC fast charger. Within half an hour, Tesla superchargers may reach the same charging capacity. The necessary hardware is relatively costly and can usually be found at public charging stations. It is necessary to pay a fee to the service provider to use them.

12.1.3 Grid integration of EVs

An electric grid is a synchronized system of electricity producers and consumers linked by transmission and distribution systems and managed by one or even more command centers. In this integration process, EVs are linked with the grid network as a load and as a piece of equipment to improve various system parameters. Two forms of grid integration occur with EVs.

G2V: grid-to-vehicle = employing energy from the grid to charge an automobile.

V2G: vehicle-to-grid = returning much of the power that has been preserved to the grid.

12.1.3.1 V2G integration

V2G refers to "vehicle-to-grid," and it is a technique that lets power from an EV battery be sent back to the power grid. A key fact to remember when using V2G in practice is that EV drivers have quite enough power in their battery packs whenever they need it. When they go to work early, the automotive batteries must be properly charged to drive passengers to and from the office. This need must be reached by V2G or any charging technique. Whenever an EV operator wishes to remove the vehicle, they ought to be able to communicate how charged up the batteries are. V2G has several benefits. Electric cars are a less expensive way of travel for customers because power costs less than gasoline. It generates increased income for the car's proprietors.

It lowers emissions by limiting the usage of renewable resources. Residences, parking lots, employers' workplaces, and community charging points can all be part of the V2G ecosystem. V2G has several drawbacks, including the fact it is not an inexpensive form of energy when compared to huge energy plants. The widespread use of these cars may cause technical issues. When small-scale generators are used with huge power generators, compatibility issues may occur.

12.1.3.2 G2V integration

It is a method wherein the electricity is transferred from the grid to EVs, with the latter acting as a demand. A network operator must assume that most automobiles are parked and possibly connected at any given time of the day to arrange electricity delivery. The ordinary private car is on the street for 4–5% of each day, which indicates that the automobiles remain parked for the rest of the time. Even during rush hour, about 90% of the total private cars are stationed.

Customers might gain from the usage of EVs because electrical is more inexpensive than gasoline for equal spacing driven, according to the advantages of G2V. Customers who want to resell electricity back to the grid could benefit from EVs in a V2G setup. The power grid could design and develop a great deal of the power required to fuel the country's automobiles at just the minimal price of fuel, other than during peak demand hours.

Due to the smaller level of production by each car, integration with current systems depending on huge power generation may be hard.

12.1.3.3 G2V and V2G rechargeable configuration for solar-powered EV charging station

Three bidirectional DC-to-DC charging converters, a bidirectional DC-to-AC converter, a distribution transformer, and a PV system MPPT boost converter make up the charging station [24] (as shown in Figure 12.8). An energy storage unit with a bidirectional DC-to-DC converter and a distribution transformer connects the microgrid to the AC utility grid.

Its block diagram model is shown in Figure 12.9. The roles of each element in the solar-powered CS are listed below.

Solar PV framework

It consists of solar units that are wired together either in parallel or in series. With the help of DC-to-DC converters, the PV module converts solar energy into DC power, which is then linked to the DC bus. In this project, four 250 W panels are connected in a 1 × 4 arrangement to create a 1,000-W total output. A variable-step

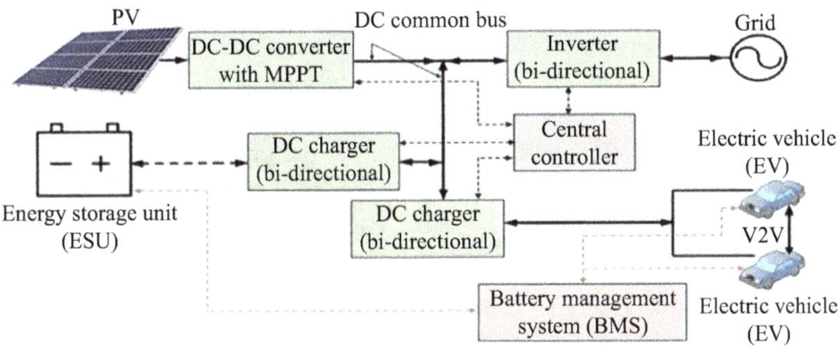

Figure 12.8 Grid interconnection with EVs

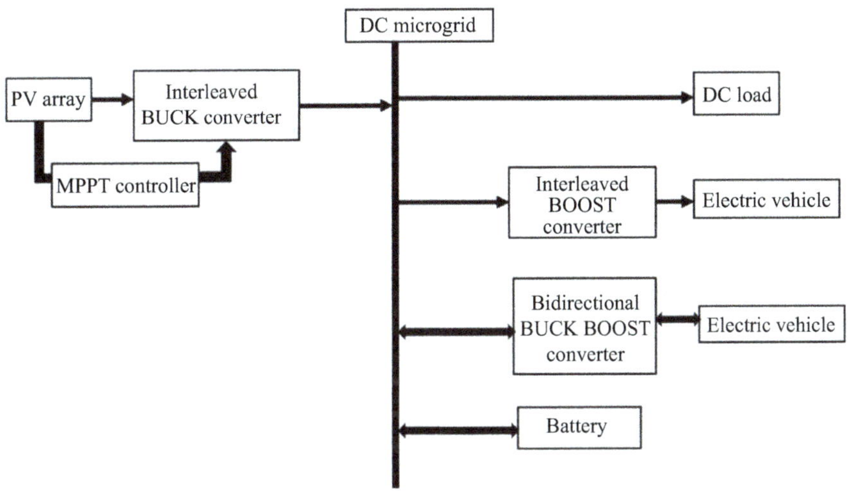

Figure 12.9 Block diagram of a grid interactive microgrid system

P&O MPPT controller is utilized to maximize the output of the PV system. The primary power source for the CS is a solar PV (SPV) system.

Electric drivetrains

One EV has a G2V system that is unidirectional, while the other has both G2V and V2G arrangements that is bidirectional and can be used for electric vehicle-to-vehicle (EVV) power transfer also. The main objective of this study is to use an SPV system to charge an EV and, if the SPV system is unable to provide enough power, to store energy in the EV battery. EVV functions as a standby reserve in a DC microgrid [25].

Bidirectional DC-DC Converter

The two-phase interleaved buck converter that powers the SPV technology has a 122.8-V input and a 48-V DC microgrid attached to its output. Interlaced buck converters reduce power to the required amount. Based on the circumstances, the bidirectional buck–boost converter will either charge or discharge the EV battery (in boost mode), while the two-phase interleaved boost converter will raise the grid voltage to 72 V.

DC load

The SPV scheme, the reserve batteries, or a mixture of the two will provide the DC load.

12.1.3.4 Converters and controllers for grid integration

A grid-connected EV charger facility, as illustrated in Figure 12.10, was made up of three primary components: a variety of EV chargers, a power converter for the grid side, and a regulator for systems development. The EV-side converter is a PWM DC/DC voltage source converter that is found in each EV charging controller. A 3 three-level rectifier/inverter converts the three-phase voltage level to a DC voltage (see Figure 12.10).

EVs are provided by the DC via EV chargers and are placed in parallel. The EV-side controllers, grid-side interface regulators, and centralized charging regulator levels make up the charging point control scheme. To conduct the charging /discharging operation of battery systems, every EV is charged/discharged through a DC/DC converter at the EV-side regulator. The 3-level AC/DC converter is regulated at the grid-side controller by maintaining a steady DC-bus potential and adjusting reactive power provided to/from the grid via the 3-phase three-level (TPTL) rectifier.

Three-level PWM bidirectional rectifier/inverter

Reverse AC–DC converters are the fundamental link across networks and EVs. In the EV charging process, this converter acts as a rectifier, transforming the AC power from the utilities to a DC bus. It works as an inverter in the V2G mode,

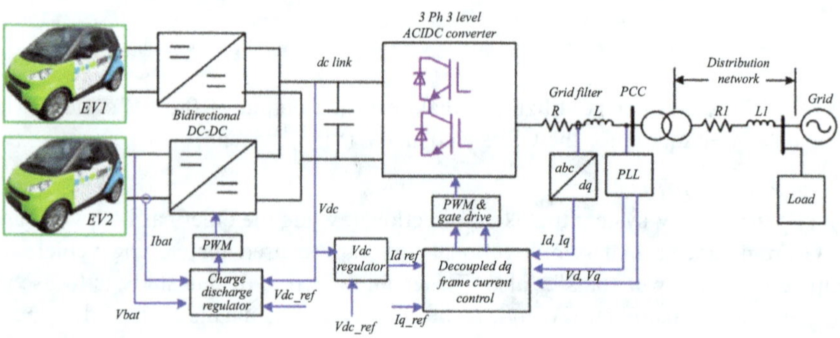

Figure 12.10 Grid-connected charging station

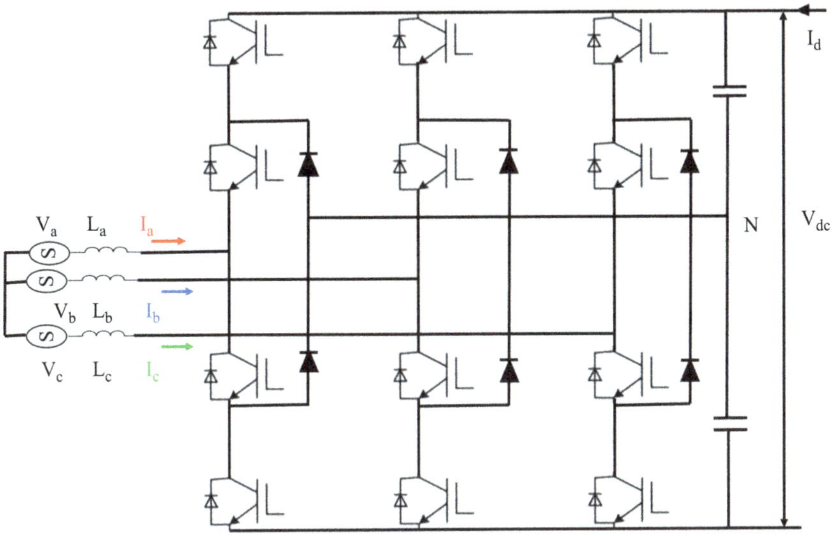

Figure 12.11 TPTL AC–DC converter

changing DC power to AC voltage and resupplying power to the grid. A multiple three levels (TPTL) voltage origin inverter (VSI) premised on a diode-clamped technique is employed to further reduce harmonic components of the inverter-mode/mains voltages, as shown in Figure 12.11.

A grid-side filter of a tiny quantity can decrease harmonic currents caused by the operating frequency. Aside from the positive and negative DC-bus voltages, the neutral point voltage is also essential for voltage creation in TPTL. As an outcome, all power semiconductor converters are exposed just to halves of the line voltage, resulting in a two-fold decrement in energy switching losses. As an outcome, the capacity of the system is increased, the heat sink is reduced, and the converter energy density is increased even more.

EV side converter

The charging point investigated here is capable of charging/discharging EV energy storage systems for both G2V and V2G modes, implying the use of bi-directional converters. Some charging points only charge an EV battery in the G2V mode, which may require a unidirectional converter. Even though the V2G method is projected to be widely used in electric cars and HEVs in the future smart grid, it is unlikely that people can accept energy recycling at a wayside charging point. In a 3-level converter, the load voltage on active switches and diodes is 1/2 that of a half-bridge converter.

The current load of energy in active and passive devices is also comparable to that of three-level and half-bridge converters. The operating frequency of the TPTL converter can be substantially lower than in the half-bridge power converter in terms of maintaining the same inductance ripple-current ratio (e.g., 30%). As a result, the inductance dimension in the TPTL converter is substantially smaller than in the half-bridge converter, for example, 1/3. The suggested bidirectional PWM DC/DC converter can be used to charge and discharge the EV battery, with a high voltage level of V_d.

The effectiveness of that power converter will deteriorate if V_d is minimal. Since the duty ratio is low, the current ripple in the inductor will grow, and the converter effectiveness will be worse because the output power is reduced but the power losses do not fall accordingly. An adjustable frequency control system is used to solve this problem. In both buck and boost different modules, it is clear that the TPTL converter has better performance effectiveness. However, when all of the evaluations of these two converters are taken into account, the three-level DC/DC converter outperforms the two-level DC/DC converter on practically every condition.

As a result, it will be used to charge and discharge EVs. The TPTL power converter has a distinct disadvantage in that it needs four switches, whereas the half-bridge power converter just wants two. In the half-bridge arrangement, however, the voltage impact on switching devices is double that of the TPTL converters.

Control approach for charging infrastructure

In EVs, the battery is a critical component. Through recharging, batteries can collect electricity into kinds of chemical energy, which can then be released through inner chemical processes during the discharge phase. Many parameters, such as concentration of reactants, temperatures, and reaction ranges, can determine the charging and draining operation of a battery system. As a result, a charging or discharging power management plan should always be followed and created to enhance the lifespan of the EV battery and keep it running at a higher level. Three main techniques are commonly utilized for recharging controls of an EV's battery for change management. Fixed current, constant voltage (CV), and taper-current recharging are three of these recharging strategies.

Constant current (CC) recharging essentially means that the charger delivers current at a roughly constant rate, independent of battery temperatures or charge conditions. This charge methodology aids in the reduction of cell and series-connected battery instabilities. Determining the battery charging current rate, on the other hand, can be difficult. A charged current that is too short cannot fulfill the recharge time rate need, and a particularly high charging current can quickly cause massive harm to the battery's electrolyte throughout the gasification charging period. CV recharging guarantees that the battery receives nearly the same input voltage all through the charging cycle, irrespective of its charge state.

Due to the obvious large voltage differences between the batteries and the EV chargers, CV chargers generally provide a large starting current to the EV battery. The charging current reduces according to the EV battery's terminal voltage rise in a taper-current charging technique. In most cases, the CC and CV recharging techniques are implemented during the recharging phase. CC recharging is used till its EV battery voltage exceeds the charge reference voltage, after which CV charging is used, enabling the charge flow to taper until it achieves a very tiny amount. During the charging period, it is suggested to maintain the charging current, which reflects the ideal current where the battery must be recharged first (to around 770% state-of-charge) underneath the CC recharging system before converting to the CV charging strategy.

Control of EV side converter

This ESC charger may be fitted into each EV to allow for individual charging methods based on the EV's needs. The CC and CV control techniques were used in this part, and they were then integrated to meet the criteria of the EV battery charging/discharging operation. The ESC processor has two control cycles: a rapid current inner loop (PI_c) and a slow voltage outer loop (OSV) (PI_v). Only the CV management method requires this nested-loop charging control mechanism, presuming that the current going into the EV battery is positive.

The voltage of the battery V_{bat} is matched to the value of the battery voltage reference V_{bat}, and the outcome is sent to the voltage loop regulator PI_v. The battery is charged if $I_{batt} > 0$; otherwise, it is depleted if $I_{batt} = 0$. The switching cursor in Figure 12.12 goes via the beginning or the end depending on the control input. The SOC current standard $I_{ref(soc)}$ and the CC modes reference current $I_{ref(sat)}$ in the center define this input. As a result, depending on the SOC and indications, the system operates in one of the three different ways (I_{ref}):

Case 1: The switches are in the lowest position in the condition of $I_{ref}<0$, and the circuits are regulated to work in the CC discharging process. The batteries send the power to the power grid or other EVs. By controlling and modifying the present baseline, this can be accomplished. The voltage signal is produced by the current controller, which is retained between $-V_{tri}$ and $+V_{tri}$ through a saturation column, where to prevent the DC–DC converter from operating in a non-linear modulating state, V_{tri} corresponds to the magnitude of the triangle carrier PWM waveform.

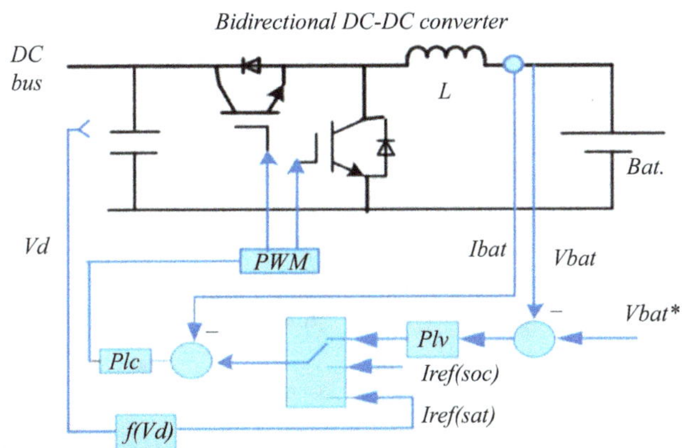

Figure 12.12 A CC and CV control approach is used to charge and discharge an EV battery

Case 2: The switching is in the lowest position, and the EV-side converter runs in a CC charging cycle if the current standard $I_{ref} > 0$ and the EV battery SOC is less than 70%.
Case 3: The switches are shifted to the leading position and the EV-side converter changes to CV charging mode if the current standard $I_{ref} > 0$ and the EV battery SOC is greater than 70%. As a result, the current-loop controller receives a reference current from the voltage regulator loop. In this example, the current-loop controller regulates battery charging by operating the DC/DC power converter at the specified current standards.

Fuzzy logic controller
Because of the non-linear nature of EV chargers and distribution systems, a standard control method must be replaced with the use of creative passion fuzzy logic controllers (FLCs). The controller is critical in distribution systems because it ensures that EV chargers operate efficiently, reliably, and with greater power quality. A fuzzification section, a base rule facility, and a defuzzification component are the three major elements of an FLC controller. FLC is used to manage the charging/discharging rates. In this case, the charge controller's input variables include the SOC for each EVs and the voltage at the nodes that connect the charging stations to them.

The EV battery will charge if the SOC is small and the node voltage is large, but it will discharge if the SOC is large and the node voltage is small. However, there are situations in which both the nodal voltages and the SOC are strong or weak. In these circumstances, a particular charge/discharge pace is used. As a result, the FLC should be changed to account for important situations to keep voltage variations at the distribution energy node below acceptable levels. The truth values of the fuzzy logic parameters might range from 0 to 1, which is known as their degree of membership.

Through a simple IF–THEN rule-based method, fuzzy logic can deal with the ambiguities in the EV charging system, removing the need for a statistical model of the regulated environment. This is especially beneficial in complex mechanisms that are challenging to quantitatively measure [26]. The amount of inputs and outputs, on the other hand, significantly changes the nature of a fuzzy logic-based system (Figure 12.13).

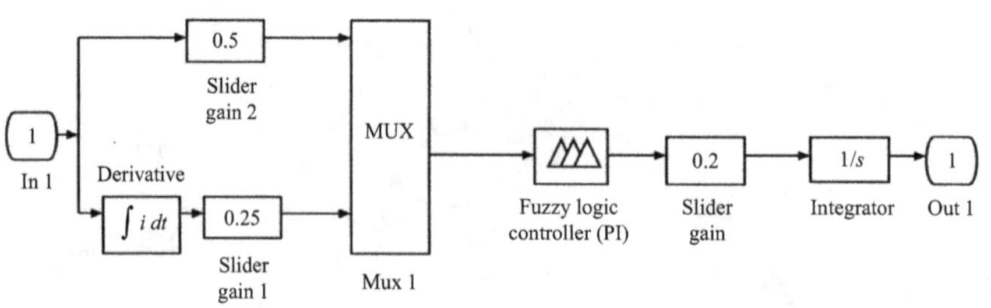

Figure 12.13 Block diagram of a fuzzy logic based PI controller

12.2 Case study of analysis of EVs planning in 16 bus systems for ZIP-LDMs

The basics of the planning structure for EVs were covered in the section above, but this section describes the mathematical modeling of the planning and results for EVs. Here, an investigation of how EVs interact with the distribution system to improve performance characteristics is conducted using simulation.

12.2.1 Mathematical modeling of EVs planning

The functioning of EVs is crucial in the power grid procedure. Sections 12.2.1.1–12.2.1.2 cover the design of planning problems, objective functions, and restrictions for these EVs [27].

12.2.1.1 Problem formulation

The following are the mathematical problem formulations for EV planning offered in this segment:

The total MVA intake without EVs (OP_{WOEVs}) for the main substation is provided by (12.1):

$$OP_{WOEVs} = \sqrt{AP_G{}^2 + RP_G{}^2} \tag{12.1}$$

The total MVA intake of the main substation with real-power EVs (OP_{WEVs}) is given by (12.2):

$$OP_{WEVs} = \sqrt{(AP_G \pm AP_{EVs})^2 + (RP_G \pm RP_{EVs})^2} \tag{12.2}$$

Four quadrants are used to drive various sorts of EVs. Table 12.2 shows the actual and reactive power values of several EVs.

Goal functions

Two-goal functions are presented in (12.3) and (12.4) and can be used to formulate optimum EV scheduling.

This target function lowers the system's overall actual energy loss from the standpoint of active power loss minimization. Eq. (12.3) represents the AP_L in the

Table 12.2 Actual and reactive power values of various EVs

EVs	Actual power	Reactive power
BEVs	AP_{BEVs}	-
PHEVs	AP_{PHEVs}	RP_{PHEVs}
Ex-PHEVs	$AP_{Ex-PHEVs}$	$RP_{Ex-PHEVs}$
FCEVs	AP_{FCEVs}	-

system:

$$AP_L = \frac{AP^2 + RP^2}{|V_{i\_bus}|^2} r_{ij\_bus} \quad \text{for } i,j \in N \tag{12.3}$$

The AP_L is determined by the total network active power loss and resistances in lines (r_{ij_bus}).

This objective serves to decrease the device's total reactive energy loss from the aspect of reactive power loss reduction (RP_L). Eq. (12.4) represents the RP_L in the scheme:

$$RP_L = \frac{AP^2 + RP^2}{|V_{i\_bus}|^2} x_{ij\_bus} \quad \text{for } i,j \in N \tag{12.4}$$

The RP_L is determined by the total system bus potential (V_{i_bus}), and resistances in lines (x_{ij_bus}). The total loss is primarily determined by the voltage profile. The total loss is primarily determined by the voltage profile. The total loss is primarily determined by the voltage profile.

Restrictions on equality and restrictions on inequality
The preceding is the EV scheduling equality constraints:

- The combined power generation of conventional and EV must be sufficient to provide both the whole expected load requirements and any power disruptions.

The following are the inequality requirements for EV planning:

- The bus voltage at the bus I is constrained by its maximum and lowest limits for all buses.
- The bus voltage angle at bus m is constrained for all buses by its upper and lower restrictions.
- The power of a conventional generator must be constrained by its lowest and greatest limits.
- The lower and maximum limits of each EV set a restriction on how much power it can produce.
- The system line deficit must be less with EVs than the overall line losses.

The formulas for the ZIP-LMs' real and reactive power [22] are given in (12.5) and (12.6):

$$AP = AP_0\left[Z_p\left(\frac{V_i}{V_0} + I_p\frac{V_i}{V_0} + AP_p\right)\right] \tag{12.5}$$

$$RP = RP_0\left[Z_q\left(\frac{V_i}{V_0} + I_q\frac{V_i}{V_0} + AP_q\right)\right] \tag{12.6}$$

The actual and reactive powers at working potential are *AP* and *RP*, respectively (Vi); the actual and reactive powers at nominal voltage are AP_0 and RP_0, respectively (V_0); Z_p, I_p, and P_p are the ZIP-LMs coefficients for actual power; and Z_q, I_q, and AP_q are the ZIP-LMs coefficients for reactive power. Tables 12.3 and 12.4 discuss the ZIP-LDMs [25] and their coefficients for actual and reactive powers, accordingly.

Table 12.3 ZIP-LMs and their behaviors

ZIP-LDMs	Equipment	ZIP-LMs are contrasted
LDM1	Air Conditioner	Low inductive load
LDM2	Lift	Highly capacitive load
LDM3	LED Bulb	Dynamic resistive load
LDM4	Micro Owen	Highly inductive load
LDM5	Fridge	Middle Inductive load

Table 12.4 Coefficient values for ZIP-LDMs that are actual and reactive (100 V cut-off voltage) [25]

ZIP-LDMs	Equipment	*AP*			*RP*		
		Z_p	I_p	AP_p	Z_q	I_q	AP_q
LDM1	Air conditioner	1.159	−1.8489	1.660	15.669	−27.150	12.469
LDM2	Lift	0.409	−0.7489	1.320	3.7700	−5.740	2.9799
LDM3	LED bulb	0.589	1.1489	−0.7089	1.7900	−0.8009	0.0199
LDM4	Micro Owen	1.379	−1.950	1.5689	50.0599	−93.500	44.50
LDM5	Fridge	1.159	−1.820	1.6600	7.06899	−10.940	4.8699

With EVs scheduling, the following system performance indices are described:

- *Actual power loss index (ILAP):* The index of actual power loss is defined by (12.7):

$$ILAP = \frac{|AP_{WEV}|}{|AP_{WOEV}|} \times 100 \tag{12.7}$$

Where AP_{WEV} represents the actual power loss with EVs and P_{WOEV} is the actual power loss without EVs. The lower value of this index gives a greater advantage with respect to actual power loss minimization due to EV placement and size.

- *Reactive power loss index (ILRP):* In (12.8), the coefficient for a reactive power outage is calculated:–

$$ILRP = \frac{|RP_{WEV}|}{|RP_{WOEV}|} \times 100 \tag{12.8}$$

Where RP_{WEV} is the loss of reactive power with an EV, and RP_{WOEV} is the loss of reactive power without an EV. Because of the positioning and scale of EVs, this score has a small value, indicating better performance in terms of reducing reactive power loss.

- *Voltage deviation index (VDI)*: One of the advantages of the proper size and location of different EVs kinds is an enhancement in the voltage level. This index is penalizing the S_{DG}–L_{DG} pair that gives the nominal higher voltage deviations (V_1=1.03 p.u.). As a result, whenever the indices are near 0, the network performance will improve. It is the maximum voltage drop for each node as well as the root node. The *VDI* could be described in (12.9):

$$VDI = \max\left(\frac{|\overline{V}_1| - |\overline{V}_{m\_bus\_EVs}|}{|\overline{V}_1|}\right) \times 100 \; for\ m\_bus = 2\ to\ N_B \quad (12.9)$$

Where $\overline{V}_1$ is the voltage of a slack bus and $\overline{V}_{m\_bus\_EVs}$ is the voltage with EVs at the *i*th bus. The voltage constraints ($V_{\min} \leq V_{m\_bus} \leq V_{\max}$) for a certain bus are commonly referred to as technical limitations; therefore, the IVD value is typically modest and within acceptable bounds.

- *Short circuit current (ISC*): Electricity flow can be reduced in some elements of the network, allowing for additional capacity by bringing power closer to the demand. The *ISC* index is given by (12.10):

$$ISC = \max\left(\frac{|\overline{S}_{mn\_bus\_EV}|}{|\overline{CS}_{mn\_bus}|}\right) \times 100 \; for\ mn\_bus\ set = 1\ to\ N_L \quad (12.10)$$

Where $\overline{CS}_{mn\_bus}$ is the capability of the MVA line without DG and $\overline{S}_{mn\_bus\_EV}$ is the capability of the MVA line with EVs. Due to the obvious load end power source, MVA rates may decrease in certain parts of the circuit, resulting in more capacity, while in others, they may rise to levels exceeding distribution line constraints.

The index gives essential information about the distribution lines' maximum capacity for the numbering of power flows that pass through the network. It contains information on a network line that requires to be upgraded. Greater than unity index values (estimated MVA flow value higher than capacity MVA) indicate a capacity sum breach inline low, while smaller values indicate the available space.

The benefit of using different types of EVs in a network for discharged line capacity is evaluated by comparing the IC of the system with and without EV. The reduction of flow at the flow boundary is an essential parameter since it indicates how much time order should be adjusted, and, hence, how much period should be increased cost. On the subject of boosting the current transfer rate in the transmission system, the IC value may not be successful. In general, these limits ($S_{(m,n\_bus)} \leq S_{(m,n\_bus)\max}$) at the specified line are regarded as limitations, so the IC level ought to be positive as well. The lower the index value is, hence, the more the space is accessible. Overloaded lines have index values greater than 100%.

- *Actual power penetration of EVs:* The ratio of EV actual power injected into the system to the sum of EV real power and without EV is indicated in (12.11):

$$\%PI_{EVAP} = \frac{PI_{WEV}}{PI_{WOEV} + PI_{WEV}} \times 100 \quad (12.11)$$

- *Reactive power penetration of EVs:* In (12.12), the number of EV reactive power put into the system is divided by the quantity of reactive power with or without EV:

$$\%PI_{EVRP} = \frac{PI_{WEV}}{PI_{WOEV} + PI_{WEV}} \times 100 \tag{12.12}$$

12.2.1.2 GA implementation

The multiple steps for GA techniques for optimum position and dimension for EVs utilizing ZIP-LDMs, starting from the platform's minimal total real power, are as described in the following:

1. Describe as from bus system, load data from ZIP-LDMs, and statistics from EVs.
2. Execute load flows for the base scenario, i.e., the starting fitness choice, and evaluate parameters like (percent ILAP, percent ILRP, percent VDI, percent ISC, percent PI_{WEVAP}, and percent PI_{WEVRP}). Make a note of the properties of the basic instance.
3. Statistics from the bus network, ZIP-LDMs, and EVs are binary-coded.
4. Establish a financial rate for the basic demographic and fitness component: obtain a randomized populace with n chromosomes (an appropriate solution to problems): in a specified range of S_{EV}–L_{EV} couples of EVs, dynamically acquire S_{EV}–L_{EV} pairings of EVs using ZIP-LDMs.
5. Calculate fitness mechanism values [$f(x)$] for each S_{EV}–L_{EV} in the populace of different types of EVs (chromosome) x: for each S_{EV}–L_{EV} pair, run load flow and evaluate indices like percent ILAP, percent ILRP, percent VDI, percent ISC, percent PI_{WEVAP}, and percent PI_{WEVRP} in consistent load cases. The variables and their relevant S_{EV}–L_{EV} pairings are included in the program.
6. Repeat the steps below to create another population until it is complete.
   - Choose a pair of chromosomes from a team based on how fit they are (the better fitness; the greater probability of becoming chosen).
   - A new progeny is created by the parent due to crossover likelihood (children). The offspring will be a replica of the parent if there is no overlap.
   - Using this mutation probability procedure, new offspring mutates (children) on each locus (in chromosome location).
   - In the new population, fresh offspring are used.
   - Examine whether this satisfies limitations such as the thermal max capacity, the power flow preservation constraints, and the voltage deviation limit. Otherwise, proceed to Step 6.

7. Use the recently created populace to run the program. With every S_{EV}–L_{EV} combination for EVs, execute for load flow and assess a new fitness alternative (chromosome). Further, evaluate the respective indices. such as % *ILAP*, % *ILRP*, % *VDI*, % *ISC*, % PI_{WEVAP}, and % PI_{WEVRP}. A comparison of the current fitness solution with the characteristics of the base case is performed.

8. If all of the conditions for stopping are met, pause and decide which possibility is best for the current population.
9. The parents and children come from a more recent generation, making up the most recent population size. Does exist a minimized MOPIF (multi-objective performance indices function)? If not, set I = i+1 for a generation. Go on to step 6 now.

A flowchart is shown in Figure 12.14 for an approach such as GA-based multi-objective optimum of EV scheduling in distribution networks with ZIP-LDMs from the least total real and reactive power losses for the network.

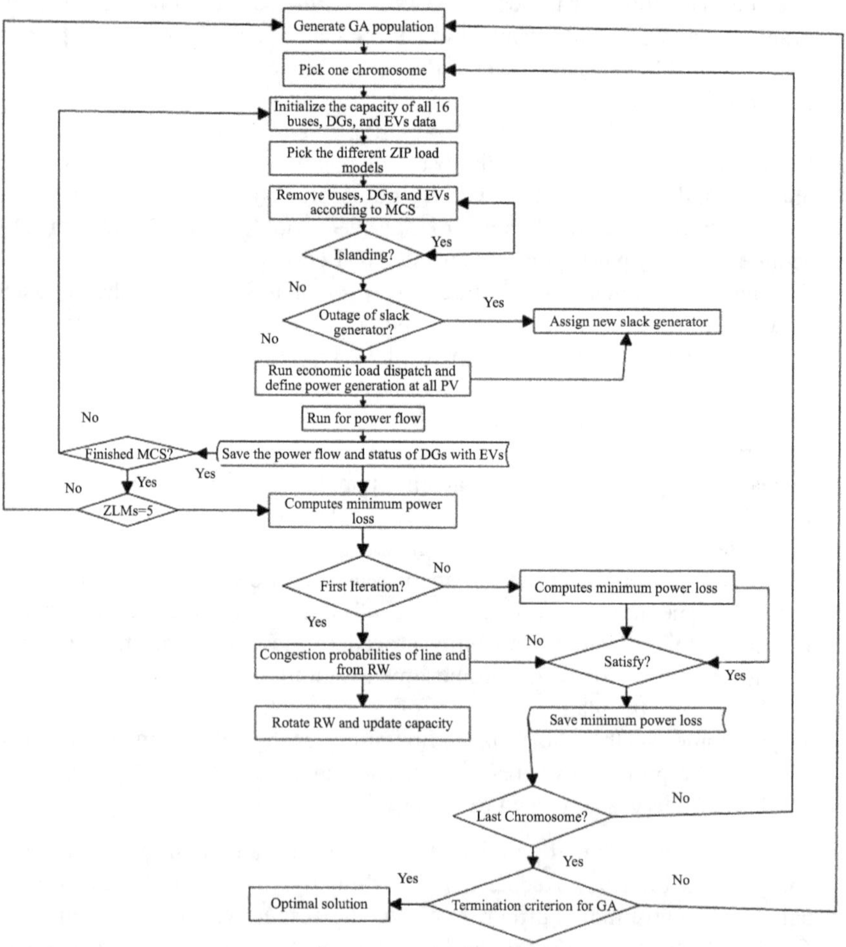

Figure 12.14 Illustrates the steps for multi-objective GA-based EV planning in distribution systems using ZIP-LDMs

12.3 Result and discussion

12.3.1 Equations relying on several objectives functions

To achieve an optimum organization, the advancement of a digital model necessitates the use of the corresponding computational modeling, the evaluation of the display destinations, and the repetitive modification of the conceptual model elements. Streamlining challenges with many targets are common in the design sector. Even though it would be ideal to streamline the desired capacities simultaneously, this is not always possible, thus the enhancement procedure must seek the best possible trade-off configuration. The main objectives of a multi-target challenge configuration are as follows:

(a) to continue gaining computational ground in the target area as it approaches the Pareto front; to maintain non-attention ruled on the target space while connected accord's concentration is on protecting the favored space.
(b) the Pareto front and Pareto ideal frameworks should be focused on different (option space);
(c) Pareto prefers to focus on estimating to offer the leadership, the designer, a suitably large but fixed number.

The multi-index for measuring the performance of distribution networks considers all repeatedly described indicators by tactical weight when preparing EV location and size with LDMs. All impact output indices have been standardized, thus it should be possible (values ranging from 0 to 1). In this sort of issue, (12.13) provides the GA-based multi-objective output index function (MOF) (12.14):

$$MOF = \beta_1(ILAP) + \beta_2(ILRP) + \beta_3(VDI) + \beta_4(ISC) + \beta_5(PI_{WEVAP}) + \beta_6(PI_{WEVRP}) \tag{12.13}$$

where

$$\sum_{r=1}^{6} \beta_r = 1 \wedge \beta_r \in (0\ 1) \tag{12.14}$$

The parameters are compared by their importance in the output indexes dispersion scheme. Distinctiveness is valued highly, and the imports of a particular output index get the top priority among the rest. The objective action indicated above will be employed with GA in the latest project. The weights used in this investigation are close to the values found in standard operators. Such values, however, may change depending on an engineer's priorities.

The ILAP and ILRP got appropriate 1st and 2nd weights of 0.40 and 0.30, correspondingly, in this experiment. Because of its impact on power efficiency, the VDI action receives the 3rd biggest weight of 0.10. Because it provides essential data regarding the current flow in the distribution system around the circuitry, including the entire thermal performance of the conductors, the ISC was assigned the 4th substantial weight of (0.10). The 5th substantial weight was given to the actual power EV penetration index (PI_{WEVAP}) (0.05). The PI_{WEVRP} (reactive power DG penetration index) was given the 6th important weighting (0.05).

To satisfy the electricity needs for the dispersion network, the multi-objective function given in (12.13) is minimized and exposed to multiple operating parameters. There are a few restrictions that are discussed:

(i) *Power flow conservation limits:* The totality of all entering and exiting power, comprising distribution system line losses, ought to be zero, as well as the power generated by the EV units is provided by (12.15):

$$P_{ss}(m, V) = \sum_{m=2}^{N_B} P_D(m, V) + \sum_{N_B=1}^{N_L} P_{loss}(V) - P_{EVm} \tag{12.15}$$

Where there are several buses in the scheme, N_L is the number of lines, $P_D(m, V)$ is the need for power (MW), $P_{loss}(V)$ is the system's loss of power, P_{EVm} is the amount of electric power given to the system.

(ii) *Short circuit current capacity:* Any power flow through a distributed feeder must meet the line's thermal performance, as calculated by (12.16):

$$S_{(m,n)} \leq S_{(m,n)\max}\,. \tag{12.16}$$

(iii) *Voltage deviation limits*: The voltage lowering capacity is determined by the distributing company's voltage regulatory limits, which are defined by (12.17):

$$|V_1 - V_n| \leq \Delta V_{\max}\,. \tag{12.17}$$

However, when the voltage and MVA restrictions for a particular S_{EV}–L_{EV} pairing in buses in the system are satisfied, approve the couple for the future generation's populace; it eliminates the S_{EV}–L_{EV} duo, which would struggle to satisfy voltage and MVA constraints in the upcoming generation.

For a minimum MOF, obtain a S_{EV}–L_{EV} combination. On all feasible generations, the optimal S_{EV}–L_{EV} couple verifies with operating restrictions, ensuring that the dimension and positioning correspond to the minimal MOF. Figure 12.15 and Table 12.5 illustrate the 16-bus test network single-line schematic and data in order.

Now various comparison charts are present in Figures 12.16–12.23; such as a comparison of the credentials of AP_L (p.u.) without as well as with EVs for ZIP-LDMs, a contrast of the characteristics of RP_L (p.u.) without but also with EVs for ZIP-LDMs [28], a resemblance of the credentials of ILAP (percent) without but instead with EVs for ZIP-LDMs, a contrast of the characteristics of ILRP (percent)

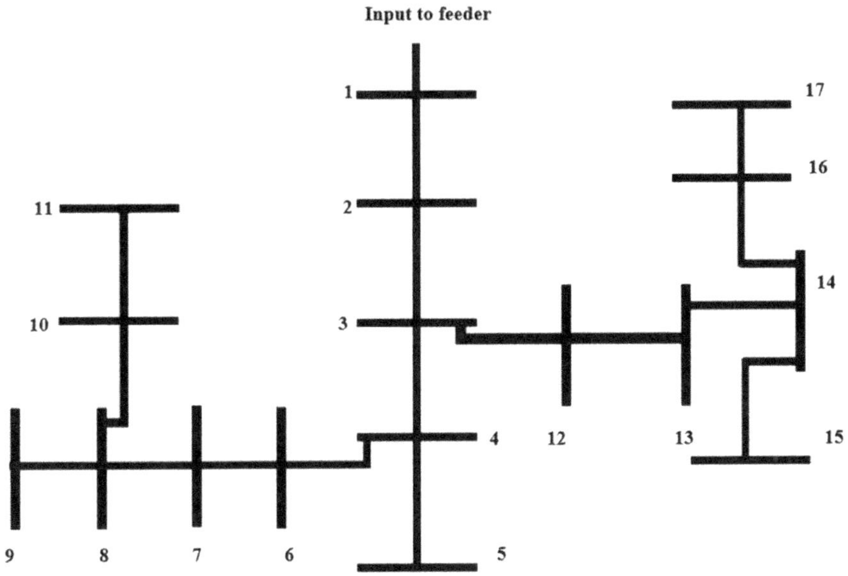

Figure 12.15 A single-line representation of a 16-bus test program (is a sub-system of the 16-bus system)

Table 12.5 Information from the 16-bus test program (which is a sub-system of the 16-bus network) [18,19]

From	To	Line impedance in p.u.		Load on to node (p.u.)			
		$R_{p.u.}$	$X_{p.u.}$	L	S_L	AP_L	RPL
1	2	0.0005729	0.0002939	1	4.59	0.101	0.0599
2	3	0.003069	0.0015639	6	4.09	0.0901	0.0402
3	4	0.0022800	0.0011609	11	2.899	0.1199	0.0799
4	5	0.00237299	0.001210	12	2.899	0.059	0.0302
5	6	0.005099	0.0044019	13	2.899	0.059	0.0201
6	7	0.0011659	0.0038529	22	1.49	0.201	0.101
7	8	0.004429	0.0014639	23	1.049	0.201	0.101
8	9	0.0064129	0.0046079	25	1.049	0.059	0.0201
9	10	0.0065009	0.0046079	27	1.049	0.059	0.0201
10	11	0.0012239	0.0004049	28	1.049	0.0451	0.0299
11	12	0.0023309	0.0007709	29	1.049	0.059	0.0349
12	13	0.0091409	0.0071919	31	0.499	0.059	0.0349
13	14	0.0033719	0.0044389	32	0.4499	0.119	0.0799
14	15	0.00370	0.0032749	33	0.301	0.059	0.0101
15	16	0.0046469	0.0033939	34	0.249	0.059	0.0201
16	17	0.0080259	0.0107159	35	0.249	0.059	0.0201

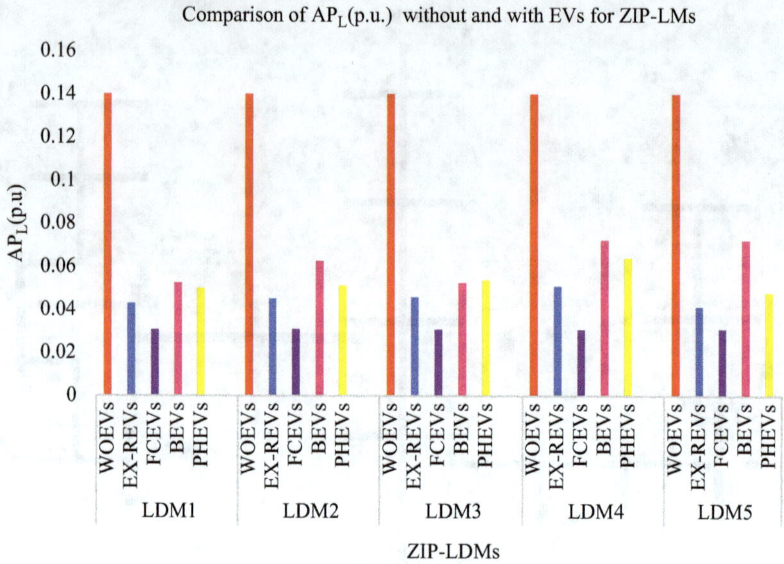

Figure 12.16 For ZIP-LDMs, a comparative of the pattern of AP_L (p.u.) without and with EVs

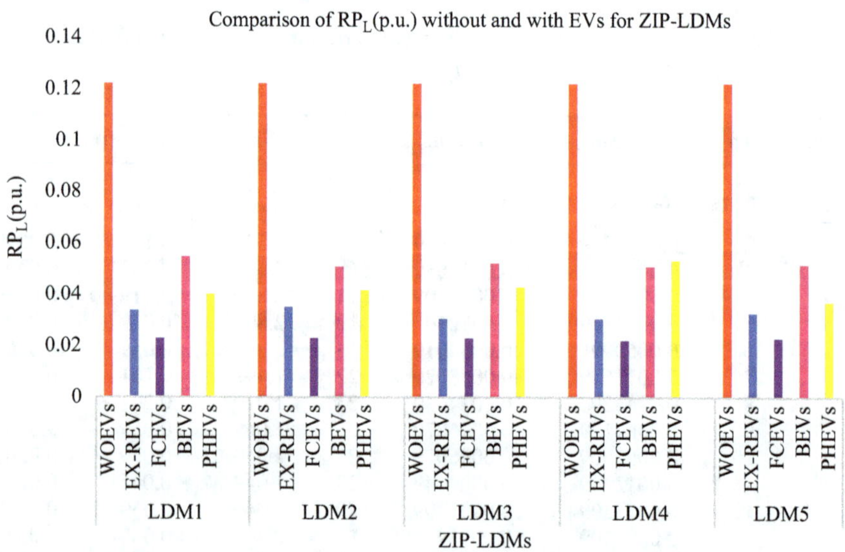

Figure 12.17 For ZIP-LDMs, a contrast of the contour of RP_L (p.u.) without and with EVs

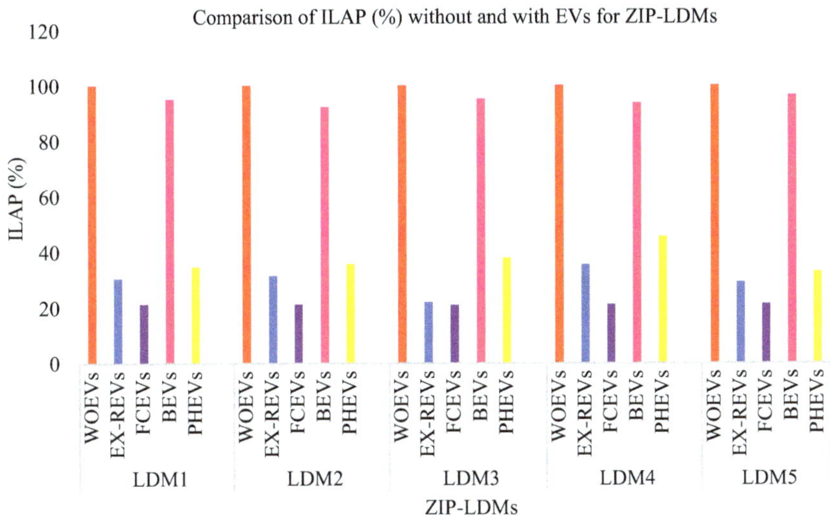

Figure 12.18 Regarding ZIP-LDMs, a comparison of the ILAP (p.u.) pattern without as well as with EVs

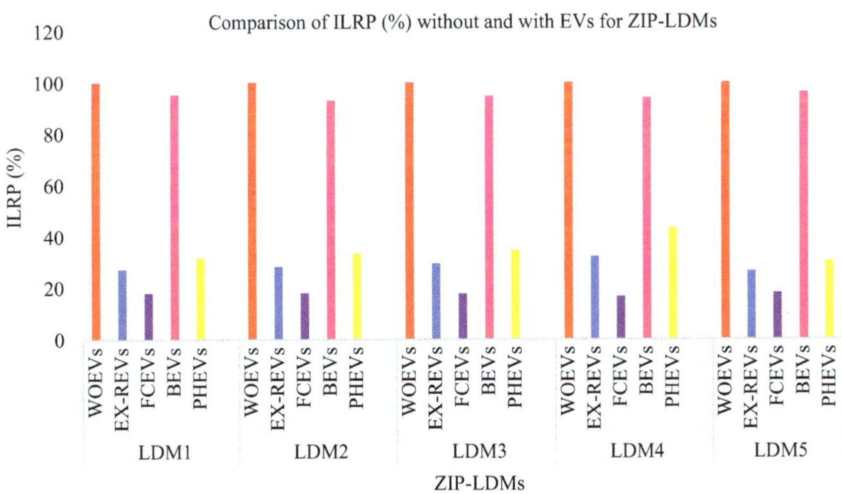

Figure 12.19 Analysis of the ILRP profiles (percent) for ZIP-LDMs without but instead with EVs

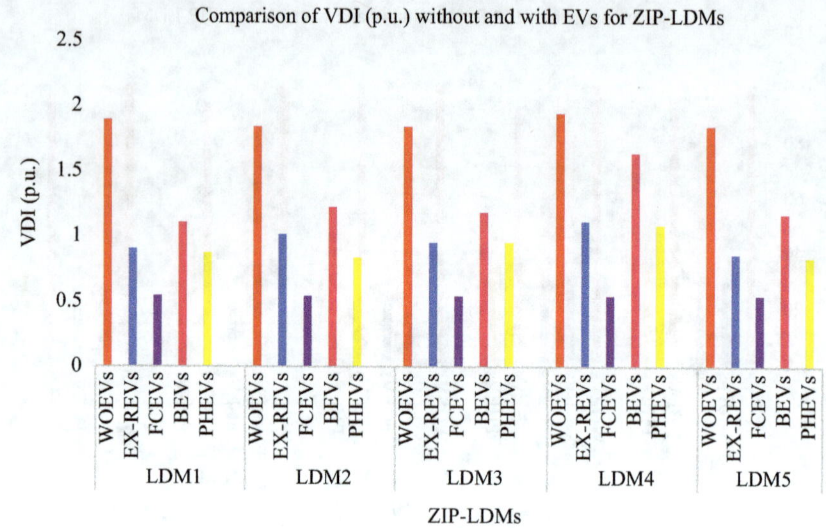

Figure 12.20 At ZIP-LDMs, the pattern of VDI (percentage) without but instead with EVs is compared

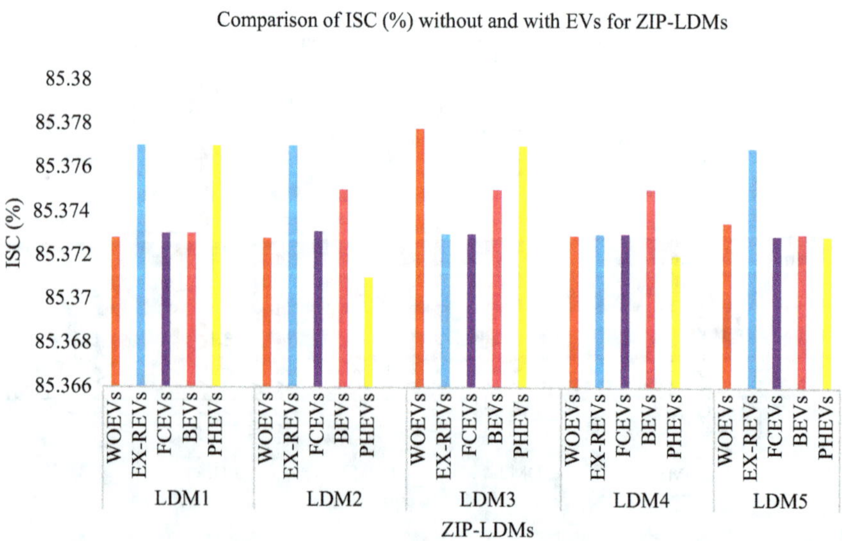

Figure 12.21 Evaluation of the ISC (percent) pattern for ZIP-LDMs without as well as with EVs

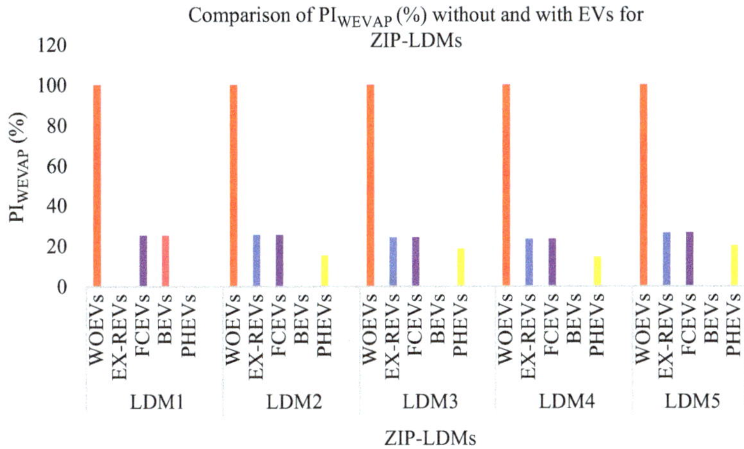

Figure 12.22 Correlation of percent PI_{WEVAP} *profiles without but also with EVs using ZIP-LDMs*

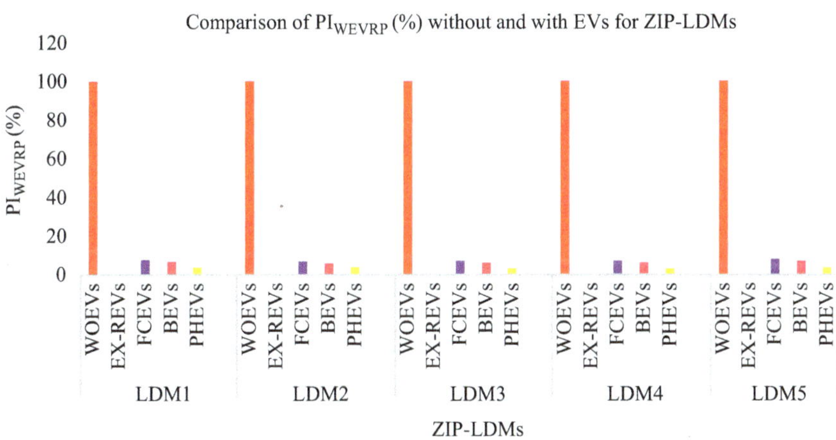

Figure 12.23 Comparing of percent PI_{WEVRP} *profiles without or with EVs in ZIP-LDMs*

without as well as alongside EVs for ZIP-LDMs, a correlation of the profile of ILAP (percent), comparative of the VDI (percent) characterization without as well as with EVs for ZIP-LDMs, contrast of the ISC (percent) characteristics without as well as with EVs for ZIP-LDMs, compared of the characteristics of percent PI_{WEVAP} without but instead with EVs for ZIP-LDMs, and comparing of the credentials of percent PI_{WOEVRP} without as well as with EVs for ZIP-LDMs. EV1 corresponds to Ex-EVs, EV2 corresponds to FCEVs, EV3 corresponds to BEVs, and EV4 corresponds to PHEVs (Table 12.6).

Table 12.6 GA and systems efficiency metrics for various forms of EV scheduling using ZIP-LDMs

ZIP-LDMs	WOEVs/ WEVs	EVs	OP_{EV} (p.u.)	EVs PF	EVs Loc.	AP_L (p.u.)	RP_L (p.u.)	% *ILAP*	% *ILRP*	% *VDI*	% *ISC*	% PI_{WEVAP}	% PI_{WEVRP}
LDM1	WOEV	–	–	–	–	0.1400	0.1219	100	100	1.8900	85.3728	–	–
	WEV	Ex-EVs	0.5055	1.00	8	0.0429	0.0337	30.55	27.37	0.9021	85.3770	25.27	–
		FCEVs	0.7140	0.86 ld	7	0.0307	0.0228	21.32	18.05	0.5432	85.3730	25.28	7.73
		BEVs	0.4530	0.00	8	0.0524	0.0546	95.09	95.27	1.1018	85.3730	–	6.74
		PHEVs	0.4426	0.86 lg	7	0.0500	0.0400	34.68	31.71	0.8743	85.3770	20.00	3.62
LDM2	WOEV	–	–	–	–	0.1400	0.1219	100	100	1.8400	85.3728	–	–
	WEV	Ex-EVs	0.5052	1.00	7	0.0451	0.0350	31.52	28.44	1.0141	85.3770	25.60	–
		FCEVs	0.7253	0.86 ld	8	0.0309	0.0229	21.22	18.00	0.5413	85.3731	25.58	7.00
		BEVs	0.4146	0.00	7	0.0625	0.0507	92.33	93.10	1.2200	85.3750	–	6.00
		PHEVs	0.4028	0.86 lg	8	0.0511	0.0415	35.71	33.33	0.8374	85.3770	15.33	3.94
LDM3	WOEV	–	–	–	–	0.1400	0.1219	100	100	1.8400	85.3778	–	–
	WEV	Ex-EVs	0.4762	1.00	8	0.0458	0.0304	22.00	29.44	0.9500	85.3730	24.21	–
		FCEVs	0.7149	0.86 ld	8	0.0308	0.0229	21.02	17.85	0.5422	85.3730	24.22	7.12
		BEVs	0.4195	0.00	7	0.0524	0.0521	95.2	94.81	1.1794	85.3750	–	6.11
		PHEVs	0.4105	0.86 lg	7	0.0537	0.0427	38.01	34.73	0.9520	85.3770	18.74	3.22
LDM4	WOEV	–	–	–	–	0.1400	0.1219	100	100	1.9400	85.3729	–	–
	WEV	Ex-EVs	0.4568	1.00	8	0.0509	0.0304	35.53	32.33	1.1131	85.3730	23.45	–
		FCEVs	0.7228	0.86 ld	8	0.0305	0.0221	21.25	16.72	0.5413	85.3730	23.44	7.19
		BEVs	0.3199	0.00	5	0.0721	0.0508	93.65	94.09	1.6345	85.3750	–	6.21
		PHEVs	0.3121	0.86 lg	8	0.0639	0.0532	45.53	43.46	1.0833	85.3720	14.55	2.99
LDM5	WOEV	–	–	–	–	0.1400	0.1220	100	100	1.8400	85.3735	–	–
	WEV	Ex-EVs	0.5028	1.00	8	0.0414	0.0326	29.18	26.49	0.8601	85.3769	26.55	–
		FCEVs	0.7165	0.86 ld	7	0.0308	0.0229	21.35	18.00	0.5432	85.3729	26.66	8.00
		BEVs	0.4589	0.00	7	0.0721	0.0516	96.44	96.22	1.1685	85. 3730	–	7.00
		PHEVs	0.4562	0.86 lg	8	0.0479	0.0369	32.94	30.42	0.8374	85.3730	20.22	3.55

12.4 Future aspects of grid integration EVs

Modernizing the load power chart, grid connection governance, overload minimization on distribution systems, voltage regulatory oversight on distribution grids, a decrease of "over a generation" by renewables, backstage amenities for consumers, and the positive impact of ultra-adaptors for heavy vehicles are among some of the major advantages suggested by recharging administration [29].

By properly and quickly interfacing PEVs with the grid, electricity providers may "reduce peak energy requirements, regulate system frequency and voltage, and control the unpredictable output power of distributed energy resources." The metering and communicative infrastructure of PEVs, which is necessary to supply power to the grid, are well, standardized, and constructed with robust cybersecurity. As the world's population of automobiles grows, careful diagnosis and management will increase the value of each alternative.

- Encouraging grid innovation: A modernized grid promotes long-term system reliability, efficiency, robustness, and steady levelized pricing, all of which increase the value of electricity as a source of energy and increase trust in a dependable electricity supply.
- Information systems for power usage: The current tendency is to create built-in wireless national PEV recharging systems that cut across utility lines. Large amounts of data about electric mobility in a specific place are maintained by these systems. While this information is transferred to electric providers, they can contribute to enhancing grid effectiveness, load prediction accuracy, the value of vehicle electrification, and customer perception.
- Smart cities: Smart cities are created metropolitan regions that excel in economics, transport, environment, population, lifestyle, and governance to produce sustainable economic growth and a good standard of living. Municipalities all over the world are working to become smart cities. The effectiveness of a smart city is dependent on electricity, insight into traffic flows, database administration, and data science. Electric utilities can be important partners in PEV grid integration, with the added load and data visibility potentially compensating some of the grid upgrading expenses.
- Unregulated PEV recharging is expected to result in a small rise in power usage. PEV charging, if properly managed, can give the electrical power network the versatility it needs for active demand response and address a range of difficulties, such as the integration of renewable energy sources.
- PEVs also are part of a vast database because of onboard connectivity systems like General Motors' OnStar. Customers' data will also be collected and managed by electric providers to calculate voltage requirements, distribution system assessment, and consumer billing. Allowing these two systems to interchange vital information will help PEVs, the grid, and customers. The management of this "big data" presents possible market opportunities for energy providers and others. Electric utilities are required by law to monitor customer and framework operating data in an organized, private, and secure

manner, which makes them potentially well enough to retain PEV charging data.

- The PEV may be a crucial element of any power management when linked to the HAN network because it is one of the greatest energy consumers in the home.
- Grid modernization is strongly linked to the grid's potential to become clean, which has both medical and economic advantages. Clean energy generation minimizes contaminants and greenhouse gases, resulting in considerable health and environmental advantages. According to an Argonne assessment published in June 2016, electrification cars would cut the grams of carbon dioxide comparable per mile by 26% considering modern technology and as high as 91% assuming renewable energy as the power source. To realize this potency, local, state, and federal energy policymakers must work together.
- Vehicle-to-home backup energy requirement: In the event of a power outage due to weather, the PEV can offer backup generators to the home, particularly to power smaller digitized loads in contemporary advanced homes.

12.4.1 EVs scenario in India

NMEM and NEMMP are two plans developed by the Government of India. The Nationwide Plan on E-Mobility was established by the Government of India in 2011. As part of a mission, the Ministry of Industries has designed a plan.

12.4.1.1 FAME-India

Hybrid and EVs will be embraced and developed at a faster rate in India. From 2015 to 2017, Stage 1 of the NMEM will be deployed over two years. The four main areas of attention are technological research, market generation, piloting, and charging points. The Ministry of India's National Electric Transportation Mission Plan 2020, which was introduced in 2013, seeks to have 6.0–7.2 million hybrid and EVs in India by 2020.

To boost eco-friendly automobiles, the administration has indeed been offering discounts on electric and hybrid vehicles consisting of up to 29,000 rupees for motorcycles and 1,38,000 rupees for automobiles under the FAME India scheme. In the 2017–2018 budgets, 175 crore rupees were allotted to the FAME various government schemes. On February 16, 2017, BHEL contracted a deal for the construction of a bus launch vehicle with Ashok Leyland Ltd and Tata Motors Ltd. Within this strategy, the first phase of FAME began in April 2015 and was expanded until March 31, 2019.

By April 1, 2019, FAME II entered into force for 3 years. FAME is a requirement incentive scheme (60% of overall spending), focusing on technology innovations, infrastructure building, and need stimulating via incentives and pilot initiatives. This regulation applies to all-electric and hybrid vehicles, namely moderate hybrids, powerful hybrids, plug-in hybrids, and battery electrics. FAME started with a capital of INR 8,950 million and gave a one-third discount on the difference in cost between an electrical automobile and a corresponding gasoline automobile in metropolitan areas. Incentives ranged from 1,800 to 2,000 INR.

Table 12.7 Electric bikes in India

S. no.	Brand and model	Range	Maximum speed	Charging time	Cost (Rupee)
1	LOHIA Oma star series	60 km	25 km/h	6–8 h	26,000–35,000
2	Hero-electric Optima series	70 km	25 km/h	6–8 h	38,000–55,000
3	YO bikes	55–60 km	55 km/h	6 h	36,000–55,000
4	BSA motor	70 km	25 km/h	6 h	25,000–36,000

The cost for FAME II is INR 100,000 million, including 86% going to consumption subsidies and 10% going to chargeable infrastructure spending. This stage of the project will include incentives for 7,000 electric buses, 55,000 four-wheeled passenger cars (including powerful hybrids), 500,000 tri-wheeled vehicles, and 1,000,000 two-wheeled bikes. This policy also includes around 2,700 recharging points around the country, with a minimum of 1 charging point in every 3 km × 3 km grid in the nation's biggest cities, other towns with populations over a million, smart cities, and urban areas in mountainous regions. In addition, recharging facilities will be constructed every 20 km on highways (Table 12.7).

12.5 Summary

The Summary in terms of conclusions and the future scope of the presented design are dealt with in Sections 12.5.1 and 12.5.2.

12.5.1 Conclusions

This study presented the effect of properly located kinds of EVs with ZIP-LDMs using GA from the standpoint of the system's least overall active power loss. The focus was on reducing this during the given time limit. The feasibility of this approach is determined using the 16-bus testing system.

The total outcomes of this research have shown that comparing these four EVs and five ZIP-LDMs separately is beneficial to society. The following are the overall results of the comparisons for EVs and ZIP-LDMs separately:

(a) EVs: Generally, the system performance indicators improved when PHEVs supplied active and reactive power to the grid at leading power factors of 0.80–0.99. For example, diesel and IC engines. While BEVs supply only active power to the grid at zero power factor, the stem's overall performance measures improved. The system performance indices order of FCEV > Ex-EVs > PHEVs > BEVs with ZIP-LMs.

(b) ZIP-LMs: Within the system, EVs with ZIP-LDMs (such as LDM1, LDM2, LDM3, LDM4, and LDM5) perform differently. Whenever it relates to EV scheduling, various ZIP-LDMs act differently. ZIP-LDMs such as LDM1 (low inductive load), LDM2 (highly capacitive load), LDM3 (dynamic resistive load), LDM4 (very inductive load), and LDM5 (medium inductive load) are considered in DG scheduling. The list consists of network performance parameters in a set pattern: LDM4>LDM2>LDM5>LDM1>LDM3.

(c) Varied kinds of EVs with ZIP-LDMs display various behaviors for variables like percent ILAP, percent ILRP, percent VDI, percent ISC, percent PI_{WDGAP}, and percent PI_{WDGRP}. Finally, it is discovered that in distribution systems, FCEVs perform better than BEVs, with ZIP-LDMs performing the poorest. The following are the performance indices in order: FCEVs are superior to Ex-EVs, PHEVs, and BEVs.

(d) New aspects of EV and grid interface with ZIP-LDMs are highlighted:
   - In the case of EVs with ZIP-LDMs, FCEVs with ZIP-LDMs provide excellent system performance measures, whereas BEVs with ZIP-LDMs provide lower system performance parameters.
   - According to the ZIP-LDMs scenario, FCEVs with LDM4 provide superior network performance parameters, but BEVs with LDM3 provide lower system performance indicators.

(e) For extremely limited DG allocating issues using ZIP-LDMs, GA-OPF is recommended.

(f) The GA approach is more efficient than the GA-OPF approach in terms of computation time speed.

12.5.2 Future scope

In the future, a suggestion for the extent of research design on this topic will be included.

(a) Static load models (SLDMs), also known as realistic load models (RLDMs) for superior performance measures, are frequently used for properly organized surveillance of EVs and FACTS controllers and their optimal position.

(b) For correctly integrated administration of EVs for their optimal location, computationally methodologies with SLDMs and RLDMs within the network for performance boost indices are frequently used.

(c) For practitioners, experimenting with integrating renewable resources and creating future electric grids is generally recommended for the shape of future projects, as well as numerous performance measures for financial assistance, boosting technical difficulties, decreasing environmental pollution levels, enhancing safety perceptions, and lowering economic cost.

(d) Specialists should strive with both renewable resource incorporation and, as a consequence, the creation of potential power grids for upcoming work, as well as a variety of performance indicators for correctly structured multiple EV control and optimum system position.

(e) Because of the employment of specific kinds of EVs for system load demands, the various types of EVs that are absorbed/delivered to the system supporting actual and reactive power are characterized. As a result, specialists must concentrate on combining renewable energy sources in the same way that they are creating future power systems.

Symbols

AP_L	actual power loss per unit
RP_L	reactive power loss per unit
ILRP	reactive power loss index
ILAP	actual power loss index
VDI	voltage deviation index
ISC	short-circuit current

Glossary

Name	Description
ICEs	internal combustion engines
EVs	electric vehicles
SPGNs	smart power grid networks
IoT	Internet of Things
5G	fifth generation
ADMM techniques	the alternating direction method of multipliers is an algorithm that solves convex optimization problems by breaking them into smaller pieces, each of which is then easier to handle
BEVs	battery electric vehicles
HEVs	hybrid electric vehicles
PHEVs	plug-in hybrid electric vehicles
Rx-EVs	range extension type of electric vehicles
FCEVs	fuel cell electric vehicles
PEMFC	proton exchange membrane fuel cell
V2G	vehicle to grid = returning much of the power that has been preserved to the grid
G2V	grid to vehicle = employing energy from the grid to charge a vehicle
PV	photovoltaics is the conversion of light into electricity using semiconducting materials that exhibit the photovoltaic effect
MPPT converter	maximum power point tracking is a technique used with variable power sources to maximize energy extraction as conditions vary
VSI	a voltage source inverter is a device that converts a unidirectional voltage waveform into a bidirectional voltage waveform, in other words, it is a converter that converts its voltage from DC form to AC form
PWM	pulse width modulation
TPTL converter	three-phase three-level DC/DC converter

FLC	fuzzy logic controllers
NMEM	Nationwide Plan on E-Mobility
FAME	faster adoption and manufacturing of hybrid and electric vehicles is a scheme launched by the Government of India to give a boost to the development of electric vehicles
GA	a *genetic algorithm* is a search heuristic that is inspired by Charles Darwin's theory of natural evolution
PWEVQ	reactive power DG penetration index
PWEVP	true power EV penetration index
ZIP-LMs	constant impedance, constant current, and constant power load models
P_L	real-power loss in per unit
Q_L	reactive power loss in per unit
ILQ	reactive power loss index
ILP	real power loss index
IVD	voltage deviation index
IC	short-circuit current

References

[1] C. Chatzikomis, A. Sorniotti, P. Gruber, M. Zanchetta, D. Willans, and B. Balcombe, 'Comparison of Path Tracking and Torque-Vectoring Controllers for Autonomous Electric Vehicles'. *IEEE Transactions on Intelligent Vehicles*. 2018, vol. 3(4), pp. 559–570.

[2] M.-H. Wang, M.-L. Huang, Z. Zi-Yi, and C.-J. Huang, 'Application of the Extension Taguchi Method to Optimal Capability Planning of a Stand-alone Power System'. *MDPI Access*. 2016, vol. 9, pp. 996–1073.

[3] D.-J. Hur, S.-H. Jeong, S.-II Song, and J.-H. Non, 'Optimization Based on Product and Desirability Functions for Flow Distribution in Multi-Channel Cooling Systems of Power Inverters in Electric Vehicles'. *MDPI Access*. 2019, vol. 9, pp. 2076–2086.

[4] B. Duan, K. Xin, and Y. Zhong, 'Optimal Dispatching of Electric Vehicles Based on Smart Contract and Internet of Things'. *IEEE Access*. 2020, vol. 8, pp. 9630–9639.

[5] Y. Sasaki, 'A Survey on IoT Big Data Analytic Systems: Current and Future'. *IEEE Internet of Things Journal*. 2022, vol. 9(2), pp. 1024–1036.

[6] M. Chiang, R. El-Azouzi, L. Gao, J. Huang, C. Joe-Wong, and S. Sen, 'Guest Editorial: Smart Data Pricing for Next-Generation Networks'. *IEEE Journal on Selected Areas in Communications*. 2020, vol. 38(4), pp. 641–644.

[7] W. Cai, Q. Xie, M. Zhang, S. Lv, and J. Yang, 'Stream-Function Based 3D Obstacle Avoidance Mechanism for Mobile AUVs on the

Internet of Underwater Things'. *IEEE Access*. 2021, vol. 9, pp. 142997–143012.

[8] Q. Pham, 'A Survey of Multi-Access Edge Computing in 5G and Beyond Fundamentals, Technology Integration, and State-of-the-Art'. *IEEE Access*. 2020, vol. 8, pp. 116974–117017.

[9] M. A. R. Pohan, B. R. Trilaksono, S. P. Santosa, and A. S. Rohman, 'Path Planning Algorithm Using the Hybridization of the Rapidly-Exploring Random Tree and Ant Colony Systems'. *IEEE Access*. 2021, vol. 9, pp. 153599–153615.

[10] S. Park, Y. Min, J. Ha, D. Cho, and H. Choi, 'A Distributed ADMM Approach to Non-Myopic Path Planning for Multi-Target Tracking'. *IEEE Access*. 2019, vol. 7, pp. 163589–163603.

[11] D. Li, P. Wang, and L. Du, 'Path Planning Technologies for Autonomous Underwater Vehicles – A Review'. *IEEE Access*. 2019, vol. 7, pp. 9745–9768.

[12] D. Drake, S. Koziol, and E. Chabot, 'Mobile Robot Path Planning with a Moving Goal'. *IEEE Access*. 2018, vol. 6, pp. 12800–12814.

[13] D. Ding, J. Li, P. Tu, H. Wang, T. Cao, and F. Zhang, 'Electric Vehicle Charging Warning and Path Planning Method Based on Spark'. *IEEE Access*. 2020, vol. 8, pp. 8543–8553.

[14] L. Yu, D. Kong, X. Shao, and X. Yan, 'A Path Planning and Navigation Control System Design for Driverless Electric Bus'. *IEEE Access*. 2018, vol. 6, pp. 53960–53975.

[15] A. Triviño-Cabrera, J. M. González-González, and J. A. Aguado, 'Design and Implementation of a Cost-Effective Wireless Charger for an Electric Bicycle'. *IEEE Access*. 2021, vol. 9, pp. 85277–85288.

[16] Y. Wang, 'Guest Editorial Smart Grid Technologies and Development in China'. *IEEE Transactions on Smart Grid*. 2016, vol. 7(1), pp. 379–380.

[17] T. Haus, M. Orsag, A. P. Nunez, S. Bogdan, and D. M. Lofaro, 'Centroid Vectoring for Attitude Control of Floating Base Robots: From Maritime to Aerial Applications'. *IEEE Access*. 2019, vol. 7, pp. 16021–16031.

[18] Z. Wang and S. Wang, 'Grid Power Peak Shaving and Valley Filling Using Vehicle-to-Grid Systems'. *IEEE Transactions on Power Delivery*. 2013, vol. 28(3), pp. 1822–1829.

[19] A. S. Al-Ogaili, 'Review on Scheduling, Clustering, and Forecasting Strategies for Controlling Electric Vehicle Charging: Challenges and Recommendations'. *IEEE Access*. 2019, vol. 7, pp. 128353–128371.

[20] A. Yagmur, B. A. Dedeturk, A. Soran, J. Jung, and A. Onen, 'Blockchain-Based Energy Applications: The DSO Perspective'. *IEEE Access*. 2021, vol. 9, pp. 145605–145625.

[21] W. Hu, W. Yao, Y. Hu, and H. Li, 'Collaborative Optimization of Distributed Scheduling Based on Blockchain Consensus Mechanism Considering Battery-Swap Stations of Electric Vehicles'. *IEEE Access*. 2019, vol. 7, pp. 137959–137967.

[22] X. Huang, C. Xu, P. Wang, and H. Liu, 'LNSC: A Security Model for Electric Vehicle and Charging Pile Management Based on Blockchain Ecosystem'. *IEEE Access*. 2018, vol. 6, pp. 13565–13574.

[23] N. Lasla, M. Al-Ammari, M. Abdallah, and M. Younis, 'Blockchain-Based Trading Platform for Electric Vehicle Charging in Smart Cities'. *IEEE Open Journal of Intelligent Transportation Systems*. 2020, vol. 1, pp. 80–92.

[24] A. R. Bhatti, Z. Salam, M. J. B. A. Aziz, and K. P. Yee, 'A Critical Review of Electric Vehicles Charging Using Solar Photovoltaic'. *International Journal of Energy Research*, 2015, vol. 40 (4), pp. 439–461.

[25] J.-T. Liao H.-W. Huang, H.-T. Yang, and D. Li, 'Decentralized V2G/G2V Scheduling of EV Charging Stations by Considering the Conversion Efficiency of Bidirectional Chargers'. *Energies*. 2020, vol. 14(4), pp. 1–16.

[26] K. Sayed and H. A. Gabbar, 'Electric Vehicle to Power Grid Integration Using Three-Phase Three-Level AC/DC Converter and PI-Fuzzy Controller'. *Energies*. 2016. vol. 9(7), pp. 1–16.

[27] P. K. Dubey, B. Singh, and V. Kumar, 'A novel approach for comparative analysis of distributed generations and electric vehicles in distribution systems'. *Electrical Engineering*. 2023, pp. 1–20.

[28] B. Singh and P. K. Dubey, 'Distributed Power Generation Planning for Distribution Networks Using Electric Vehicles: Systematic Attention to Challenges and Opportunities'. *Journal of Energy Storage*. 2022; vol. 48(1), pp. 1–44.

[29] B. Singh, P. K. Dubey, and S. N. Singh, 'Recent Optimization Techniques for Coordinated Control of Electric Vehicles in Super Smart Power Grids Network: A State of the Art'. *IEEE 9th Uttar Pradesh Section International Conference on Electrical, Electronics, and Computer Engineering (UPCON)* Prayagraj, India, 2022, pp. 1–7.

Chapter 13

Regulations and standards of electric vehicles

Rahul Arora[1], Paramjeet Singh Jamwal[2] and Ujjwal K. Kalla[3]

13.1 Introduction

For the commercialization of any product, the ease of operation and maintenance and availability of its constituent components are essential, which can be ensured by common standards across the country and the globe as well. Therefore, common standards are required for the sustainable growth of the electric vehicle (EV) sector. The EVs must support infrastructures, associated peripherals, and user-friendly software globally [1]. EVs must comply with international standards and codes. A summarized review is available in the literature [2] regarding the progress of EV charging infrastructure, grid integration of EVs, impact of EV charging on the grid with smart grid interaction performing two functions, i.e., vehicle-to-grid (V2G) and grid-to-vehicle (G2V). An investigation on fast charging converters for xEVs is reported in [3].

Based on the available literature, the EV standards, charging standards, grid integration standards, and safety standards are summarized in this chapter. The EV charging infrastructure, such as power, control, and communication infrastructure, along with the impacts of EV integration on grid are discussed in view of the developmental challenges of EV sector.

The decarbonization of the environment has led to a rapid growth of EVs. This resulted in new ways of transportation in the form of different types of EVs such as hybrid EVs (HEVs), all-EVs (AEVs), battery EVs (BEVs), fuel cell EVs (FCEVs), and plug-in HEV (PHEV). These vehicles need refueling, i.e., charging after a regular interval and the only charging source for these vehicles is the grid. Therefore, the electric power industry shall also be subjected to enormous demand in view of increasing number of EVs. To overcome any unwanted problem due to EV technology, it is essential to have standards and operation codes, uniformly across the globe.

[1]Department of Electrical Engineering, MANIT Bhopal, India
[2]Department of Electrical Engineering, NIT Hamirpur, India
[3]Department of Electrical Engineering, NIT Delhi, India

There are various standards related to EVs and the associated infrastructure [4]. The Government of India also introduced a type of approval procedure for EV and HEV for pilot/demonstration projects intended for government schemes [5]. The regulatory bodies across the globe and their established standards are summarized in the following sections. The standardization of EVs is taken care of by International Organization for Standardization (ISO). Other standards are on component-level specifications. These can be classified into different categories such as EV charging component standards, EV grid integration (EVGI) standards, and safety standards. Among the charging component standardization of EV, the codes are country specific as well as global.

The standards related to EV charging/discharging with the involvement of the grid are included in grid integration standards. The EVs act like a distributed energy resource (DER) during charging/discharging from the grid, therefore, the standards of DERs for grid interconnection also apply to EVs. The grid interconnection standards are mainly prepared by the Underwriters' Laboratories (UL) and the Institute of Electrical and Electronics Engineers (IEEE).

13.2 EV batteries standards

There are several standards issued by ISO, the Society of Automotive Engineers (SAE), and the United Nations (UN) for the batteries used in the EV (Table 13.1). The ISO 12405 standard sets the specifications for the lithium-ion-based battery packs and systems used in EVs. Various categories of ISO 12405 standards are listed in the Table. 13.1. The SAE J2464 and J2929 standards ensure the safety of EV batteries during charging. The UN 38.3 standard ensures the transportation suitability of EV batteries.

Table 13.1 Standards for EV batteries

S. no.	Standard	Focus area	References
1	ISO 12405-1:2011	High-power applications	[6]
2	ISO 12405-2:2012	High-energy applications	[7]
3	ISO 12405-3:2014	Safety performance requirements	[8]
4	ISO 12405-4:2018	Performance testing	[9]
5	SAE J2464	Rechargeable energy storage system (RESS) safety and abuse testing	[10]
6	SAE J2929	Battery system safety standard using lithium-based rechargeable cells	[11]
7	UN 38.3	Transportation testing for lithium-ion batteries	[12]

13.3 Grid interface standards

The IEEE1547 and UL1741 are two broad standards that are used for grid interfacing. The summary of all these standards and codes is given in the following section. The IEEE1547 standard is used for the interconnection of distributed resources (DR) and electric power systems (EPS). This standard is applicable to all distributed energy resources technologies. This standard has a collective capacity of 10 MVA or less at the point of common coupling. This standard covers the requirements relevant to the performance, operation, testing, safety considerations, and maintenance for the interconnection of DERs. This standard emphasizes on the installation of distributed energy resources on primary and secondary network distribution systems. There are several categories of IEEE1547 and UL standards, which are listed in Tables 13.2 and 13.3.

Table 13.2 Categories of IEEE1547 standard

S. no.	Category	Details	References
1	IEEE 1547.1-2020	Conformance test procedures for Equipment interconnecting.	[13]
2	IEEE 1547.2-2008	Application guide.	[14]
3	IEEE 1547.3-2007	Guide for monitoring, information exchange, and control.	[15]
4	IEEE 1547.4-2011	Guide for design, operation, and integration of island systems.	[16]
5	IEEE 1547.6-2011	Recommended practice for interconnecting secondary networks.	[17]
6	IEEE 1547.7-2013	Guide for conducting distribution impact studies.	[18]
7	IEEE 1547.9-2022	Guide for interconnection of distributed energy storage systems.	[19]

Table 13.3 Categories of UL standard

S. no.	Category	Details	References
1	UL 1741	Safety of inverters, converters, controllers, and interconnection system equipment.	[20]
2	UL 1741SA	Renewable energy inverters.	[21]

13.4 Charging standards

Several charging standards are available worldwide for the charging infrastructure of EVs. In the United States, SAE- and IEEE-based charging standards are used. In Europe, IEC-based charging standards are used. In Japan, they have their own charging standards named CHAdeMO. In China, Guobiao (GB/T) charging standard is used. This GB/T charging standard is issued by the Standardization Administration of China and Chinese National Committee of ISO and IEC for AC and DC charging. The GB/T AC charging standards are similar to IEC standards. In this section, IEC- and SAE-based charging standards are discussed. Tables 13.4 and 13.5 show a summary of the voltage and current levels of IEC and SAE standards. The terms used in IEC61851 and SAE J1772 are almost the same. The term that differs in both these standards is the level of power. This level of power is called as level and mode in SAE and IEC, respectively.

13.4.1 IEC standards

The IEC standards are developed by the British standardization organization, which develops standards for electrical, electronic, and other related technologies.

13.4.2 SAE standards

There are seven charging standards issued by SAE which are summarized in Table 13.5.

Table 13.4 Summary of IEC standards

S. no.	Standard	Details	References
1	IEC 61851-1:2017	✓ Covers overall standard operation for EV conductive charging systems. ✓ Applies to onboard and off-board equipment for charging EVs/PHEVs. ✓ Supply voltages up to 1,000 V AC and 1,500 V DC.	[22]
2	IEC 61980-1:2020	✓ A standard for WPT system. ✓ Applicable for a supply voltage up to 1,000 V AC and 1,500 V DC. ✓ Also applies to WPT system supplied by the on-site storage systems.	[23]
3	IEC 62196-1:2022	✓ A standard for plugs, socket outlets, vehicle connectors, and vehicle inlets that are used for conductive charging of EVs.	[24]

Table 13.5 Summary of SAE standards

S. no.	Standard	Details	References
1	SAE J2293	✓ Establishes the requirement of OBC and OfBC equipment. ✓ Has two sections: J2293-1 and J2293-2. ✓ J2293-1 discusses the power requirements and system architecture for conductive AC, conductive DC, and inductive charging. ✓ J2293-2 discusses the communication requirement and network architecture for EV charging.	[25]
2	SAEJ1772	✓ Discusses all the equipment ratings for EV charging including CB current rating, charging voltage rating, and so on. ✓ Defined for both AC and DC where each of them has three levels.	[26]
3	SAEJ1773	✓ Specifies the minimum requirements of inductively coupled charging scheme for EVs. ✓ Establishes explicitly the requirement for manually connected inductive charging systems ✓ Elaborates the requirements of software interface.	[27]
4	SAEJ2847	✓ Establishes requirements and specifications for communication between PEV and DC OfBC.	[28]
5	SAEJ2836	✓ Establishes the instructions for the documents required for the variety of potential functions for PEV communications, energy transfer options, interoperability, and security.	[29]
6	SAEJ2931	✓ Establishes the requirements for digital communication between EVs, EVSE, utility, energy service interface, advanced metering infrastructure, and home area network. To set up a communication network in a smart grid environment for EV.	[30]
7	SAEJ2954	✓ World's first WPT specification and recommended practice (RP) for EVs. ✓ Specifies wireless charging up to level 2 (7.7 kW) but recently published RP version declared up to level 3 (11 kW). ✓ The updated version also provides a standardized testbed for performance measurement and validation of new products from EV manufacturers and infrastructure companies. ✓ Includes driving assistance for seamless EV parking, payment establishment, and autonomous charging.	[31]

13.5 Safety standards for charging infrastructure

The safety standards for EV charging infrastructure and grid integration are set by the National Electric Code (NEC) and the National Fire Protection Association (NFPA). However, other above-mentioned organizations have also defined the safety standards (Table 13.6).

Table 13.6 Summary of safety standards for charging infrastructure

S. no.	Category	Details	References
1.	ISO 6469-1:2019	✓ Specifies safety requirements for rechargeable energy storage systems (RESS) of EV for the protection of persons.	[32]
2.	ISO 17409:2020	✓ Specifies electric safety requirements for conductive connections of EV to an external electric power supply using a plug or vehicle inlet.	[33]
3.	IEC 61140	✓ Provides protection against electric shock. ✓ Common aspects for installation and equipment.	[34]
4	IEC62040-3:2021	✓ Establishes the performance and test requirements applied to movable, stationary, and fixed electronic uninterruptible power systems (UPS).	[35]
5	IEC 60529	✓ Rate and grade the resistance of enclosures of electric and electronic devices against the intrusion of dust and liquids. ✓ Rates how easy it is for individuals to access the potentially hazardous parts within the enclosure.	[36]
6	IEC 60364-7-722:2018	✓ Applies to circuits intended to supply energy to EVs. ✓ Applies to circuits intended for feeding back electricity from EVs. ✓ Applies to circuits covered by this document are terminated at the connecting point.	[37]
7	SAE J1766	✓ Recommended practice for EV and HEV battery systems crash.	[38]
8	ISO/IEC 27000:2018	✓ Provides the overview of information security management systems (ISMS).	[39]
9	AIS 156	✓ Requirements of a vehicle about its electrical safety. ✓ Requirements of a rechargeable electrical energy storage system (REESS) about its safety	[40]

13.6 International test standards for chargers

There are several international test standards for chargers issued by IEC.

S. no.	Category	Details	References
1.	IEC 61851.1: 2017	✓ Applies to EVSE for charging EV, with a rated supply voltage of up to 1,000 V AC or up to 1,500 V DC and a rated output voltage of up to 1,000 V AC or up to 1,500 V DC. ✓ Covers the characteristics and operating conditions of the EVSE.	[41]

(Continues)

(*Continued*)

S. no.	Category	Details	References
		✓ Covers the specification of the connection between the EVSE and the EV. ✓ Covers the requirements for electrical safety for the EVSE.	
2.	IEC 61851-21-1:2017	✓ Gives requirements for conductive connection of an EV to an AC or DC supply. ✓ Applies only to on-board charging units either tested on the complete vehicle or tested on the charging system component level (ESA—electronic sub-assembly). ✓ Covers the electromagnetic compatibility (EMC) requirements for EV in any charging mode while connected to the mains supply.	[42]
3.	IEC 61851-21-2:2018	✓ Defines the EMC requirements for any off-board components or equipment of such systems used to supply or charge EV with electric power by conductive power transfer (CPT), with a rated input voltage, according to IEC 60038:2009, up to 1,000 V AC or 1,500 V DC and an output voltage up to 1,000 V AC or 1,500 V DC.	[43]
4.	IEC 61851-23:2014	✓ Gives the requirements for DC EV charging stations, herein also referred to as "DC charger." ✓ Provides the general requirements for the control communication between a DC EV charging station and an EV.	[44]

13.6.1 *Standards and codes for connectors*

To have generalized charging ports for various EVs, common standards are required. There are several standards and codes for connectors issued by Chinese Standard (GB/T) and IEC.

S. No.	Category	Details	References
1.	GB/ T20234.1	✓ Connectors for Conducting Charging for EVs – Part 1: General Requirements	[45]
2.	IEC 62196-1:2022	✓ provides a general description of the interface between an EV and a charging station as well as general mechanical and electrical requirements and tests for plugs, socket-outlets, vehicle connectors, and vehicle inlets that are intended to be used for EV charging.	[46]
3.	GB/ T20234.2	✓ Connectors for Conducting Charging for EV – Part 2: AC Charging Interfaces	[47]
4.	IEC 62196-2	✓ extends IEC 62196-1 and describes specific designs of plugs, socket-outlets, vehicle connectors, and vehicle	[48]

(*Continues*)

(*Continued*)

S. No.	Category	Details	References
		inlets that are intended to be used for AC charging of EV in modes 1, 2, and 3 as described by IEC 61851-1.	
5.	GB/ T20234.3	✓ Connectors for Conducting Charging for EV – Part 3: DC Charging Interfaces	[49]
6.	IEC 62196-3	✓ describes specific designs of vehicle connectors and vehicle inlets that are intended to be used for DC charging of EV in mode 4 as described by IEC 61851-1 and IEC 61851-23	[50]

13.6.2 Standards and codes for communications

There are several standards and codes for connectors issued by ISO, IEC, SAE, and Chinese standards.

S. no.	Category	Details	References
1.	ISO 15118-20:2022	✓ Road vehicles—V2G communication interface is a proposed international standard defining a V2G communication interface for bi-directional charging/ discharging of EVs.	[51]
2.	SAE J2847	✓ Part 1: Communications between plug-in vehicles and utility grid. ✓ Part 2: Communications between plug-in vehicles and the supply equipment (EVSE). ✓ Part 3: Communications between plug-in vehicles and the utility grid for reverse flow.	[52]
3.	IEC 61851-24:2014	✓ The requirements for digital communication between DC EV charging station and EV for control of DC charging is defined in IEC 61851-24.	[53]
4.	GB/T 27930-2015	✓ Communication Protocol between Off-board Conductive Charger and BMS of EV	[54]
5.	SAE J2931	✓ Establishes the requirements for digital communication between PEV, EVSE, and the utility or service provider, energy services interface (ESI), advanced metering infrastructure (AMI), and home area network (HAN).	[55]
6.	IEC 61850	✓ An international standard defining communication protocols for intelligent electronic devices at electrical substations.	[56]

13.7 Summary

This chapter presents a compilation of available standards across the globe for EVs and associated infrastructure. There are many standards and codes under evolution which may not be included in this chapter. Some standards may have improvements during the course of implementation. This chapter is aimed to provide a generalized overview to the readers regarding the EV standards, worldwide.

Glossary

AEV	all-electric vehicles
AIS	automotive industry standards
AMI	advanced metering infrastructure
BEV	battery electric vehicle
BMS	battery management system
CB	circuit breaker
CPT	conductive power transfer
DER	distributed energy resource
DR	distributed resources
EMC	electromagnetic compatibility
EPS	electric power system
ESA	electronic sub-assembly
ESI	energy service interface
EV	electric vehicle
EVGI	electric vehicle grid integration
EVSE	electric vehicle supply equipment
FCEV	fuel cell electric vehicle
G2V	grid-to-vehicle
HAN	home area network
HEV	hybrid electric vehicle
ISMS	information security management systems
ISO	International Organization for Standardization
IEEE	Institute of Electrical and Electronics Engineers
NEC	national electric code
NFPA	National Fire Protection Association
OBC	on-board charger
OfBC	off-board charger
PHEV	plug-in hybrid electric vehicle
PEV	plug-in electric vehicle
REESS	rechargeable electrical energy storage system
RP	recommended practice
SAE	Society of Automotive Engineers
UL	Underwriters' Laboratories

UN United Nations
USA United States of America
UPS uninterruptible power systems
V2G vehicle-to-grid
WPT wireless power transfer

References

[1] A. Arancibia and K. Strunz, “Modeling of an electric vehicle charging station for fast DC charging,” In: *Electric* Vehicle Conference *(IEVC). 2012 IEEE International*; 2012.

[2] H.S. Das, M.M. Rahman, S. Li, and C.W. Tan, “Electric vehicles standards, charging infrastructure, and impact on grid integration: a technological review,” *Renewable and Sustainable Energy Reviews*, 120, 109618, 2020.

[3] R. Dwivedi, “Investigations on Fast Charging Converter for Electric Vehicle Application,” PhD Thesis, SLIET Longowal, India, May 2023.

[4] A. Foley, I. Winning, and B.O. Gallachoir, “State-of-the-art in electric vehicle charging infrastructure,” In: *IEEE* Vehicle Power *and* Propulsion Conference, 2010.

[5] “AIS-131 Type Approval Procedure for Electric and Hybrid Electric Vehicles introduced in market for Pilot / Demonstration Projects intended for Government Scheme,” https://morth.gov.in/sites/default/files/ASI/19201533750PMAIS-131F.pdf /. [Accessed 31 July 2023]

[6] “Electrically Propelled Road Vehicles – Test Specification for Lithium-Ion Traction Battery Packs and Systems – Part 1: High-Power Applications,” https://www.iso.org/standard/51414.html/. [Accessed 31 July 2023]

[7] “Electrically Propelled Road Vehicles – Test Specification for Lithium-Ion Traction Battery Packs and Systems – Part 2: High-Energy Applications,” https://www.iso.org/standard/55854.html/. [Accessed 31 July 2023]

[8] “Electrically Propelled Road Vehicles — Test Specification for Lithium-Ion Traction Battery Packs and Systems – Part 3: Safety Performance Requirements,” https://www.iso.org/standard/59224.html/. [Accessed 31 July 2023]

[9] “Electrically Propelled Road Vehicles —Test Specification for Lithium-Ion Traction Battery Packs and Systems – Part 4: Performance Testing,” https://www.iso.org/standard/71407.html/. [Accessed 31 July 2023]

[10] “Electric and Hybrid Electric Vehicle Rechargeable Energy Storage System (RESS) Safety and Abuse Testing J2464_202108,” https://www.sae.org/standards/content/j2464_202108/. [Accessed 31 July 2023]

[11] “Safety Standard for Electric and Hybrid Vehicle Propulsion Battery Systems Utilizing Lithium-based Rechargeable Cells J2929,” https://www.sae.org/standards/content/j2929/. [Accessed 31 July 2023]

[12] "UN 38.3 Certification for Lithium Batteries," https://www.intertek.com/batteries/un-38-3-certification/. [Accessed 31 July 2023]

[13] "IEEE Standard Conformance Test Procedures for Equipment Interconnecting Distributed Energy Resources with Electric Power Systems and Associated Interfaces," in *IEEE Std 1547.1-2020*, pp. 1–282, 21 May 2020.

[14] "IEEE Application Guide for IEEE Std 1547(TM), IEEE Standard for Interconnecting Distributed Resources with Electric Power Systems," in *IEEE Std 1547.2-2008*, pp.1–217, 15 April 2009.

[15] "IEEE Guide for Monitoring, Information Exchange, and Control of Distributed Resources Interconnected with Electric Power Systems," in *IEEE Std 1547.3-2007*, pp. 1–160, 16 November 2007.

[16] "IEEE Guide for Design, Operation, and Integration of Distributed Resource Island Systems with Electric Power Systems," in *IEEE Std 1547.4-2011*, pp. 1–54, 20 July 2011.

[17] "IEEE Recommended Practice for Interconnecting Distributed Resources with Electric Power Systems Distribution Secondary Networks," in *IEEE Std 1547.6-2011*, pp. 1–38, 12 September 2011.

[18] "IEEE Guide for Conducting Distribution Impact Studies for Distributed Resource Interconnection," in *IEEE Std 1547.7-2013*, pp. 1–137, 28 February 2014.

[19] "Approved Draft Guide to Using IEEE Standard 1547 for Interconnection of Energy Storage Distributed Energy Resources with Electric Power Systems," in *IEEEP1547.9/D5.6, May 2022 (Approved Draft)*, pp. 1–83, 22 June 2022.

[20] "Inverter Testing and Evaluation for UL 1741," https://www.intertek.com/energy/testing/inverter-and-converter/. [Accessed 31 July 2023]

[21] R. Mahmud, A. Hoke, and D. Narang, "Validating the test procedures described in UL 1741 SA and IEEE P1547.1," in *2018 IEEE 7th World Conference on Photovoltaic Energy Conversion (WCPEC) (A Joint Conference of 45th IEEE PVSC, 28th PVSEC & 34th EU PVSEC)*, Waikoloa, HI, 2018, pp. 1445–1450.

[22] "IEC 61851-1:2017 Electric Vehicle Conductive Charging System – Part 1: General Requirements," https://webstore.iec.ch/publication/33644. [Accessed 31 July 2023]

[23] "IEC 61980-1:2020 Electric Vehicle Wireless Power Transfer (WPT) Systems – Part 1: General Requirements," https://webstore.iec.ch/publication/31657. [Accessed 31 July 2023]

[24] "IEC 62196-1:2022 Plugs, Socket-Outlets, Vehicle Connectors and Vehicle Inlets – Conductive Charging of Electric Vehicles – Part 1: General Requirements," https://webstore.iec.ch/publication/59922/. [Accessed 31 July 2023]

[25] "Energy Transfer System for Electric Vehicles – Part 1: Functional Requirements and System Architectures J2293/1_200807," https:// www.sae.org/standards/content/j2293/1_200807/. [Accessed 31 July 2023]

[26] "SAE Electric Vehicle and Plug in Hybrid Electric Vehicle Conductive Charge Coupler J1772_201710," https://www.sae.org/standards/ content/ j1772_201710/. [Accessed 31 July 2023]
[27] "SAE Electric Vehicle Inductively Coupled Charging (STABILIZED Jun 2014) J1773_201406," https://www.sae.org/standards/content/ j1773_201406/. [Accessed 31 July 2023]
[28] "Communication Between Plug-In Vehicles and Off-Board DC Chargers J2847/2_201504," https://www.sae.org/standards/content/ j2847/2_201504/. [Accessed 31 July 2023]
[29] "Instructions for Using Plug-In Electric Vehicle (PEV) Communications, Interoperability and Security Documents J2836_201807," https://www.sae.org/standards/content/j2836_201807/. [Accessed 31 July 2023]
[30] "Digital Communications for Plug-in Electric Vehicles J2931/1_201412," https://www.sae.org/standards/content/j2931/1_201412/. [Accessed 31 July 2023]
[31] "SAE International Releases J2954 Wireless Charging Standard," https://insideevs.com/news/450548/sae-j2954-wireless-charging-standard/ [Accessed 31 July 2023]
[32] "ISO 6469-1:2019 Electrically Propelled Road Vehicles – Safety Specifications – Part 1: Rechargeable Energy Storage System (RESS)," https://www.iso.org/standard/68665.html/. [Accessed 31 July 2023]
[33] "ISO 17409:2020 Electrically Propelled Road Vehicles – Conductive Power Transfer – Safety Requirements," https://www.iso.org/standard/72880.html. [Accessed 31 July 2023]
[34] "IEC 61140 Protection Against Electric Shock – Common Aspects for Installation and Equipment," https://webstore.iec.ch/preview/info_iec61140 %7Bed3.0%7Den_d.pdf. [Accessed 31 July 2023]
[35] "IEC 62040-3:2021 Uninterruptible Power Systems (UPS) – Part 3: Method of Specifying the Performance and Test Requirements," https://webstore.iec.ch/publication/60140. [Accessed 31 July 2023]
[36] "IEC 60529 Degrees of Protection Provided by Enclosures (IP Code)," https://webstore.iec.ch/preview/info_iec60529%7Bed2.1%7Db.pdf. [Accessed 31 July 2023]
[37] "IEC 60364-7-722:2018 Low-Voltage Electrical Installations – Part 7-722: Requirements for Special Installations or Locations – Supplies for Electric Vehicles," https://webstore.iec.ch/publication/29958. [Accessed 31 July 2023]
[38] "Recommended Practice for Electric, Fuel Cell and Hybrid Electric Vehicle Crash Integrity Testing J1766_201401," https://www.sae.org/ standards/ content/j1766_201401/. [Accessed 31 July 2023]
[39] "ISO/IEC 27000:2018 Information Technology – Security Techniques – Information Security Management Systems – Overview and Vocabulary," https://www.iso.org/standard/73906.html/. [Accessed 31 July 2023]

[40] "AIS 156 Specific Requirements for L Category Electric Power Train Vehicles," https://morth.nic.in/sites/default/files/ASI/AIS-156.pdf. [Accessed 1 August 2023]

[41] "IEC 61851-1:2017 Electric Vehicle Conductive Charging System – General Requirements," https://www.thenbs.com/PublicationIndex/ documents/ details?Pub=IEC&DocID=331802. [Accessed 31 July 2023]

[42] "IEC 61851-21-1:2017 Electric Vehicle Conductive Charging System – Part 21-1: Electric Vehicle On-Board Charger EMC Requirements for Conductive Connection to AC/DC Supply," https://webstore.iec.ch/publication/32045. [Accessed 31 July 2023]

[43] "IEC 61851-21-2:2018 Electric Vehicle Conductive Charging System – Part 21-2: Electric Vehicle Requirements for Conductive Connection to an AC/DC Supply - EMC Requirements for Off Board Electric Vehicle Charging Systems," https://webstore.iec.ch/publication/31282. [Accessed 31 July 2023]

[44] "IEC 61851-23:2014 Electric Vehicle Conductive Charging System – Part 23: DC Electric Vehicle Charging Station," https://webstore.iec.ch/ publication/6032. [Accessed 31 July 2023]

[45] "GB/T 20234.1-2015 Connection Set for Conductive Charging of Electric Vehicles – Part 1: General Requirements," https://www.chinesestandard.net/PDF.aspx/GBT20234.1-2015. [Accessed 31 July 2023]

[46] "IEC 62196-1:2022 Plugs, Socket-Outlets, Vehicle Connectors and Vehicle Inlets – Conductive Charging of Electric Vehicles – Part 1: General Requirements," https://webstore.iec.ch/publication/59922/. [Accessed 31 July 2023]

[47] "GB/T 20234.2-2015 Connection Set for Conductive Charging of Electric Vehicles – Part 2: AC Charging Coupler," https://www.chinesestandard.net/PDF.aspx/GBT20234.2-2015. [Accessed 31 July 2023]

[48] "IEC 62196-2:2022 Plugs, Socket-Outlets, Vehicle Connectors and Vehicle Inlets – Conductive Charging of Electric Vehicles – Part 2: Dimensional Compatibility Requirements for AC Pin and Contact-Tube Accessories," https://webstore.iec.ch/publication/64364. [Accessed 31 July 2023]

[49] "GB/T 20234.3-2015 Connection Set for Conductive Charging of Electric Vehicles – Part 3: DC Charging Coupler," https://www.chinesestandard.net/PDF.aspx/GBT20234.3-2015. [Accessed 31 July 2023]

[50] "IEC 62196-3:2022 Plugs, Socket-Outlets, Vehicle Connectors and Vehicle Inlets – Conductive Charging of Electric Vehicles – Part 3: Dimensional Compatibility Requirements for DC and AC/DC Pin and Contact-Tube Vehicle Couplers," https://webstore.iec.ch/publication/59923. [Accessed 31 July 2023]

[51] "ISO 15118-20:2022 Road Vehicles—Vehicle to Grid Communication Interface—Part 20: 2nd Generation Network Layer and Application Layer Requirements," https://www.iso.org/standard/77845.html. [Accessed 31 July 2023]

[52] "Communication Between Plug-In Vehicles and Off-Board DC Chargers J2847/2_201504," https://www.sae.org/standards/content/ j2847/2_201504/. [Accessed 31 July 2023]

[53] "IEC 61851-24:2014 Electric Vehicle Conductive Charging System – Part 24: Digital Communication Between a d.c. EV Charging Station and an Electric Vehicle for Control of d.c. Charging," https://webstore.iec.ch/ publication/6033. [Accessed 31 July 2023]

[54] "GB/T 27930-2015 Communication Protocols between Off-Board Conductive Charger and Battery Management System for Electric Vehicle," https://www.chinesestandard.net/PDFExample/GBT27930-2015EN.pdf. [Accessed 31 July 2023]

[55] "Digital Communications for Plug-in Electric Vehicles J2931/1_201412," https://www.sae.org/standards/content/j2931/1_201412/. [Accessed 31 July 2023]

[56] "IEC 61850 Communication Networks and Systems in Substations: An Overview for Users," https://www.gegridsolutions.com/multilin/journals/issues/spring09/iec61850.pdf. [Accessed 31 July 2023]

Index

www.ingramcontent.com/pod-product-compliance
Lightning Source LLC
LaVergne TN
LVHW011838200926
848484LV00002B/372

* 9 7 8 1 8 3 9 5 3 6 7 1 7 *